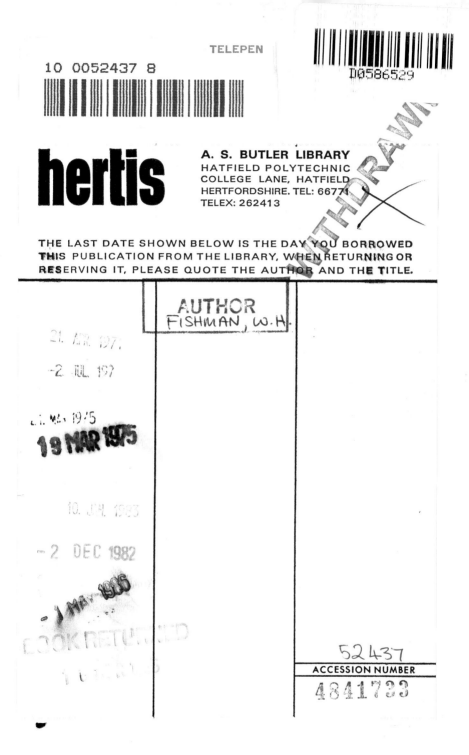

METABOLIC
CONJUGATION
AND
METABOLIC
HYDROLYSIS

VOLUME I

CONTRIBUTORS

EUGENE A. DAVIDSON

K. S. DODGSON

CHARLES C. IRVING

DONALD S. LAYNE

ERKKI LESKINEN

MERLE MASON

TATU A. MIETTINEN

F. A. ROSE

R. L. SMITH

J. L. VAN LANCKER

R. T. WILLIAMS

METABOLIC
CONJUGATION
AND
METABOLIC
HYDROLYSIS

Edited by WILLIAM H. FISHMAN

Tufts University School of Medicine
New England Medical Center Hospitals
Boston, Massachusetts

VOLUME I

ACADEMIC PRESS   New York and London   1970

ACADEMIC PRESS, INC.
111 Fifth Avenue, New York, New York 10003

United Kingdom Edition published by
ACADEMIC PRESS, INC. (LONDON) LTD.
Berkeley Square House, London W1X 6BA

LIBRARY OF CONGRESS CATALOG CARD NUMBER: 79-107556

PRINTED IN THE UNITED STATES OF AMERICA

This Treatise is dedicated to
my dear wife, Lillian, and to
our children, Joel, Nina, and Daniel

CONTENTS

vii

Effects of Conjugated Steroids on Enzymes

Merle Mason

Glucuronic Acid Pathway

Tatu A. Miettinen and Erkki Leskinen

Sulfoconjugation and Sulfohydrolysis

K. S. Dodgson and F. A. Rose

Glycoprotein and Mucopolysaccharide Hydrolysis
(Glycoprotein and Mucopolysaccharide Hydrolysis in the Cell)

Eugene A. Davidson

Hydrolases and Cellular Death

J. L. Van Lancker

LIST OF CONTRIBUTORS

Numbers in parentheses indicate the pages on which the authors' contributions begin.

EUGENE A. DAVIDSON, *Department of Biological Chemistry, The Milton S. Hershey Medical Center of the Pennsylvania State University, Hershey, Pennsylvania* (327)

K. S. DODGSON, *Department of Biochemistry, University College, Cathays Park, Cardiff, England* (239)

CHARLES C. IRVING, *Veterans Administration Hospital and Department of Biochemistry, University of Tennessee, School of Basic Medical Science, Memphis, Tennessee* (53)

DONALD S. LAYNE, *Department of Biochemistry, Faculty of Medicine, University of Ottawa, Ottawa, Canada* (21)

ERKKI LESKINEN, *Department of Clinical Chemistry, Malmi Hospital, Helsinki, Finland* (157)

MERLE MASON, *Department of Biological Chemistry, University of Michigan, Ann Arbor, Michigan* (121)

TATU A. MIETTINEN, *Third Department of Medicine, University of Helsinki, Helsinki, Finland* (157)

F. A. ROSE, *Department of Biochemistry, University College, Cathays Park, Cardiff, England* (239)

R. L. SMITH, *Department of Biochemistry, St. Mary's Hospital Medical School, University of London, London, England* (1)

J. L. VAN LANCKER, *Division of Biology and Medical Science, Brown University, Providence, Rhode Island* (355)

R. T. WILLIAMS, *Department of Biochemistry, St. Mary's Hospital Medical School, University of London, London, England* (1)

xi

PREFACE

Early in the history of biochemistry, it was discovered that organic compounds administered parenterally were excreted in conjugated form in the urine. For many years, this process was referred to as detoxication, since the theory that the body used these conjugation mechanisms to reduce the toxicity of administered compounds was widely accepted.

A lone dissenter to the prevailing view was A. J. Quick (*J. Biol. Chem.* 77, 581, 1927) who stated that "the idea is no longer tenable that these conjugations are more or less unimportant mechanisms concerned solely with the detoxication powers of the organism. If these synthetic processes are looked upon as normal and common chemical reactions made manifest because the body is applying them to a foreign substance, it is possible to perceive how the study of the conjugation of benzoic acid may help to solve various problems of metabolism which at present seem quite unrelated to hippuric acid and glucuronic acid."

This prediction was fulfilled by the isolation of estriol glucuronide and estrone sulfate from urine of the pregnant female thirty-five years ago in Professor G. F. Marrian's laboratory. These metabolites are conjugates of endogenously produced ovarian hormones and not of exogenously administered drugs. It was in this environment that I, as a graduate student, began working on β-glucuronidase and its action on steroid glucuronides, as well as on estrogens and their conjugates as they affect β-glucuronidase in the sex target organs. I found the detoxication concept inadequate for interpreting my experimental results; this experience was reinforced in my subsequent studies in Chicago and in Boston. Finally, other investigators came independently to the same view.

Accordingly, because the "detoxication" label limits our perception of the full significance of conjugating mechanisms and conjugates, I introduced, in 1947, the term "metabolic conjugation" to underline the role of those body reactions which account for the formation of conjugates such as the steroid glucuronides and sulfates.

It follows that "metabolic hydrolysis" describes the enzyme-catalyzed hydrolysis in the body of conjugates produced by metabolic conjugation. In recent years this subject has become relevant to processes in biology and pathology because of the discovery in the cell of the lysosome organelle in which many acid hydrolases are collected. Simultaneous interest appeared in the enterohepatic circulation of metabolites, hormones, and drugs which involve intestinal hydrolysis as an event in the life cycle of the body conjugate.

It is also worthwhile to consider the biochemistry of the conjugation reactions, specifically those which participate in the fabrication of macromolecules. This consideration is particularly relevant to the synthesis of the glycoproteins and mucopolysaccharides which result from a sequence of differing conjugating mechanisms. It is possible that a connection exists between the synthesis of macromolecules and the metabolism and action of smaller molecules such as the steroid hormones.

The relationships to disease of some of these mechanisms are made relevant and clear from the study of genetically produced diseases of the nerve and brain. Here the genetic absence of any one of several glycosidases results in the accumulation of gangliosides in unphysiological amounts within brain, nerve, and other human tissues.

From the biochemical point of view the product of the conjugation reaction becomes a substrate that is hydrolyzed by specific enzymes under other conditions. For many years these two mechanisms of conjugation and hydrolysis were studied independently. In recent years, however, it has become apparent that the phenomena of conjugation and hydrolysis are interrelated mechanisms in physiology and that the balance between these two reactions determines the amount of unconjugated substance that could exert physiological or pharmocological action. Within the cell the locus of the conjugation appears to be in the ribosomes of the endoplasmic reticulum, while the site of the hydrolysis of the reaction product of conjugation is mainly within the lysosomes.

The conjugation of N-hydroxylated compounds is a particularly interesting new type of metabolic conjugation which has relevance to carcinogenesis. An unexpected property, chemical reactivity within tissue nucleophiles, has been found in some conjugates of N-hydroxylated compounds. This reactivity suggests a clear-cut rationale to explain their participation in the carcinogenic mechanism. It may be inferred in this case that the conjugate is the carcinogenic compound and not its hydrolytic product.

Accordingly, it is the intention of this three-volume treatise to deal comprehensively with the main conjugation mechanisms and hydrolytic reactions, so far as knowledge and space permits. The subject matter relates variously to the following: the compound undergoing conjugation, the conjugation itself, the conjugating radical, and enzyme hydrolysis of the conjugate. The state of knowledge and circumstances of scientific interest determined whether emphasis was to be given to individual subjects such as cholic acid conjugation, mercapturic acid synthesis, and β-glucuronidase, or concentrated on a combination of individual topics, as in the case of steroid hormone conjugation and hydrolysis.

For this reason, Volume I begins with the history of the conjugation mechanisms and immediately proceeds to more recent developments of newer metabolic conjugations. The reader is thus enabled to combine a survey of this knowledge with the perspective of his particular field so as to create a definition for himself of the scope of metabolic conjugation and metabolic hydrolysis. It is hoped that the reader will be able to apply these findings to further research aimed at establishing newer knowledge in this general field.

Finally, the order in which manuscripts were received has determined to a large extent the choice of topics assigned to each volume. Accordingly, coverage of the subject may not be balanced in each volume nor nay it be fully completed until all volumes have been published. Nevertheless, I have endeavored to juxtapose those chapters whose subjects now bear or may be expected to bear a closer relationship to each other. In all cases, however, the authors have been given complete freedom to present their particular field of research. I recognize that some overlap of information will undoubtedly occur, but believe that too much editing could destroy the individuality of these presentations.

My sincerest thanks are conveyed to each author for his enthusiastic cooperation which made the publication of this book possible. It is a pleasure to acknowledge the suggestions and views of Dr. Donald S. Layne with whom I have had many pleasant conversations. To the staff of Academic Press, I express my thanks for their efficiency, patience, and consideration. My gratitude goes also to my secretary, Miss Anita Mc-Lellan, and to my editorial assistant, Miss Nina E. Fishman, for their great help.

I am a recipient of Career Research Award 5-K6-CA-18,453-08 of the National Cancer Institute, National Institutes of Health, United States Public Health Service.

HISTORY OF THE DISCOVERY OF THE CONJUGATION MECHANISMS

R. L. SMITH and R. T. WILLIAMS

I. Introduction[1]

The majority of the biochemical reactions called conjugations were discovered during the nineteenth century, largely through the study of the composition of normal and pathological urines. Their discovery was also associated with the rapid development of classical organic chemistry and the possible applications of new organic compounds in medicine. Conjugates are the end products of the metabolism of many natural and foreign chemicals, and it seems logical in view of the methods then used that they should have been found before the oxidation, reduction, and hydrolysis products of foreign compounds which are usually intermediates in the formation of conjugates. The terms "conjugated," "conjugation," and "conjugate" seem to have arisen from the use by Baumann (1876a) of the German word "gepaart" to describe the indigo-forming substances in urine. He showed that they consisted of two parts, one being a carbon-

---

[1] Much historical material concerning the metabolism of foreign compounds is to be found throughout the text of the first and second editions of "Detoxication Mechanisms" by Williams (1947, 1959).

1

containing complex giving rise to indigo and the other sulfuric acid. They were the "gepaarte Schwefelsäure" or paired sulfuric acids, and "gepaart" can be translated as "paired" or "conjugated." Eight of the ten major conjugation reactions were discovered between 1840 and 1900 and three of these were found in the 10 years between 1870 and 1880. They were discovered mainly by German organic and physiological chemists and are associated with the names of Wöhler, Keller, Baumann, von Mering, Jaffé, Schmiedeberg, His, Cohn, Müller, Lang, and Thierfelder. In the present

TABLE I

THE DISCOVERY OF THE MAJOR CONJUGATIONS

Conjugation	Author and date
Glycine	Keller (1842)
Sulfate	Baumann (1876a)
Glucuronic acid	Jaffé (1874)
Ornithine	Jaffé (1877)
Mercapturic acid	Jaffé (1879); Baumann and Preusse (1879)
Methylation	His (1887)
Acetylation	Cohn (1893)
Cyanide detoxication	Lang (1894)
Glutamine	Thierfelder and Sherwin (1914)
Glucoside (1) plants	Miller (1938)
(2) insects	Myers and Smith (1953)

century, two of what we may regard as major conjugation mechanisms were found, namely, the glutamine conjugation in man and the glucoside conjugation in insects. Several apparently rare or infrequent conjugation reactions have been discovered during the past 20 years and no doubt others will be uncovered in the future. The ten major conjugations together with the dates of the publications in which they were described and the authors concerned are listed in Table I.

The discovery of conjugation products in urine naturally raised the questions of where in the animal body they were formed, what was the mechanism of their formation, and of what significance were they to the animal economy. These problems, especially that of formation, were tackled by the early investigators. Thus, Schmiedeberg and Bunge (1877) showed that glycine conjugation was confined to the kidney of the dog and Christiani and Baumann (1878–1879) and Kochs (1880) using perfused organs studied the sulfate conjugation of phenol. Herter and Wakeman (1899) working in the United States toward the end of century found that the ethereal sulfate synthesis took place in several tissues. But satisfactory solutions to the problems of mechanism and tissue location of the conjugation mechanisms were not possible at that time, since these depended upon the development of biochemistry. They had to wait until the techniques of the dealing with isolated tissues and cell fragments and the use of isotopes had been developed. Our knowledge of the mechanisms of conjugation is, therefore, of relatively recent date.

Opinions concerning the significance of the conjugation mechanisms were also expressed during this period (1870–1880). Baumann (1876a) found that the alkali salts of the ethereal sulfate of phenol had a low toxicity and put forward the teleological view that the function of the conjugation mechanisms was to detoxicate deleterious substances that got into the body. It is of interest, however, that Metchnikoff (1910) found that, when injected, potassium phenylsulfate was toxic and was far less innocuous than Baumann had indicated. The concept, therefore, grew that conjugating agents such as glucuronic acid and sulfuric acid combined in the body with foreign compounds or their metabolites which thereby underwent "detoxication" (a translation of the German word "Entgiftung"), a means of removing potentially injurious compounds from the body. At the present time, two variants of this word are used, namely "detoxication" and "detoxification" with possibly a shade of difference in meaning, the former being the process of reducing toxicity and the latter the capacity for reducing toxicity. This concept of detoxication was widely accepted and perhaps reached its zenith in the chemical defense hypothesis of Sherwin (1922). But, as time went on, doubts were expressed about the validity of this hypothesis, especially in connection with the intermediate oxidation, reduction, and hydrolysis products which often preceded the conjugation product during the metabolism of a foreign compound. These intermediates sometimes showed increased toxicity compared with the parent compound. However, Baumann's original idea was probably concerned only with conjugation products. The inclusion of the intermediates in a chemical defense hypothesis was probably due to Sherwin. Nevertheless, it is not entirely clear that all conjugation products are metabolically inert and stable end products of metabolism, for they may function in some circumstances as intermediates in transport processes, possibly in the transport of an active compound in an inactive form from one location in the body to another. However, the more recent discovery that several foreign compounds can stimulate their own metabolism may infuse new life into the defense hypothesis. The general impression one gets from the study of conjugation products is that in the main they are relatively nontoxic, water-soluble excretory products and that their formation usually results in a reduction in toxicity although this process may not be perfect, as might be expected in an imperfect world.

In 1932, Quick (1932a) drew attention to the fact that most conjugates were strongly acidic. It was thus possible that the fundamental change brought about by conjugation was to increase the polarity of the compound and this was causally related to a reduction in toxicity. Many foreign compounds are lipid soluble and therefore tend to be reabsorbed by the kidney tubules and recirculated in the body. However, if they are converted into highly polar compounds, then these tend to be confined to the extracellular fluid and to be readily excreted by the kidney and thereby eliminated from the body. Thus, the lipid-soluble compound, benzene, is first oxidized to phenol which has a pK_a of 10 and is 0.25% ionized at pH 7.4; phenol is then conjugated with glucuronic acid to form phenylglucuronide which has a pK_a of 3.4 and is 99.99% ionized at pH 7.4. It would appear, therefore, that reduction in toxicity is achieved by processes which increase polarity and the consequent ready elimination of the product.

The chemical defense concept as developed by Sherwin in the United States during the first quarter of this century also contained the idea that metabolic conjugations had evolved specifically for detoxicating foreign compounds taken in from the environment. The alternative to this view is that they may represent adaptations of normally existing metabolic processes, and support for this alternative came with the later discovery of the wider physiological roles played by conjugating agents such as glucuronic acid, sulfate, and glycine. Nearly all the enzyme systems involved in conjugation reactions have natural substrates and all the conjugating agents have roles in normal intermediary metabolism apart from those involving foreign compounds.

The major conjugation reactions are relatively few in number and the conjugating agents are derived mainly from the animal's carbohydrate and amino acid sources. The glycoside conjugations are limited mainly to glucose and its derivative, glucuronic acid. Of the 22 amino acids available, relatively few are used by most species in conjugation reactions. These are glycine, cysteine, and glutamic acid in mercapturic acid synthesis, glycine alone in hippuric acid formation, and methionine in methylation. Ornithine, arginine, and glutamine are used only by certain species and serine conjugation appears to be of rare occurrence. Species variations occur more frequently with amino acid conjugations than with the glycoside conjugations and these may reflect evolutionary and ecological pressures.

It is important to point out that the study of the metabolic conjugations which eventually opened up the field of drug metabolism has contributed not only to biochemistry, but also substantially to pharmacology. The study of the acetylation of sulfanilamide led to the discovery of acetyl-CoA and the recognition of the coenzyme's dominant role in intermediary metabolism. The study of the enzymic mechanisms of drug metabolism has brought to the fore in other fields of biochemistry the role of the microsomal particles. Work on glucuronide formation has contributed to the understanding of the biochemistry of bilirubin, and has resulted in the elucidation of a little-known pathway of glucose metabolism involving glucuronic acid and ascorbic acid. In fact, the early study of metabolic conjugations has now developed into the new field of biochemical pharmacology which has a most important bearing on biochemistry, pharmacology, and toxicology today.

The rest of this chapter describes briefly the historical aspects of the glycoside conjugations, the amino acid conjugations, and those reactions which do not belong to either of these.

II. Glycoside Conjugations

The glycoside conjugations utilize a conjugating agent which is carbohydrate in nature, namely, the glucuronic acid used by vertebrates and the glucose used by insects. Ribose and N-acetylglucosamine have also been found to conjugate with certain compounds but these conjugations are very rare. Glucose conjugation can now be regarded as a major mechanism, for it occurs widely in insects and a very large number of species of insects exist. The first of these conjugations to be discovered was glucuronide formation. It is probably the most widespread and well adapted of the various conjugation mechanisms in vertebrates and this may be a reflection of the facility with which glucuronic acid can be produced from carbohydrate sources by several tissues, and of the remarkable variety of chemical groups to which glucuronic acid can be transferred enzymically from the intermediate nucleotide, uridine diphosphate glucuronic acid.

The discovery of glucuronic acid seems to have originated with Schmid (1855) who was investigating euxanthic acid, the main component of the pigment Indian yellow or puree. This pigment was used in certain areas of India for house decoration and was obtained from the urine of cows fed on mango leaves. He found that euxanthic acid gave, on acid hydrolysis, a copper-reducing substance which was later investigated by Baeyer (1870) who described it as a kind of saccharic acid with the formula, $C_6H_{10}O_7$, which is correct for glucuronic acid. At this time (1870–1880) several investigators were administering to animals foreign compounds such as chloral hydrate, nitrobenzene, the nitrotoluenes, and morphine. Some of these were found to give rise to levorotatory and copper-reducing urines. The reducing material was shown not to be glucose because it was not fermented by yeast (Schmiedeberg and Meyer, 1879). In fact at this time the first glucuronic acid conjugates were actually isolated from urine. Jaffé (1874, 1878) fed o-nitrotoluene to dogs and isolated from their urine a substance which he called "Uronitrotoluolsäure." This compound was levorotatory and on acid hydrolysis gave o-nitrobenzyl alcohol and a carbohydrate which was acidic and reducing. Jaffé considered this acid to be an oxidation product of a sugar in which the CH_2OH group had been oxidized into COOH. Mering and Musculus (1875) isolated a conjugate from the urine of humans given chloral hydrate. This was named "Urochloralsäure" (urochloralic acid) which we now know to be trichloroethylglucuronide. Jaffé (1878) pointed out that the "carbohydrate-aldehyde-acid" which he had found as a component of "Uronitrotoluolsäure" was probably also a component of "Urochloralsäure" and this was confirmed by Mering (1882) after glucuronic acid itself had been isolated by Schmiedeberg and Meyer (1879) from the urine of dogs fed with camphor. The latter authors isolated from dog urine material which they called camphoglycuronic acid which on acid hydrolysis yielded "Glycuronsäure," the formula for which was deduced to be, $(CHOH)_4 \begin{cases} CHO \\ COOH \end{cases}$.

The compound actually isolated by Schmiedeberg and Meyer was the lactone of glucuronic acid, that is, glucurone or glucuronolactone. These workers also expressed views regarding the origin of glucuronic acid and thought that it represented an intermediate stage in the oxidation in the body of sugar, presumably glucose, but by "pairing" or conjugating with camphor, the sugar escaped further oxidation.

During the next 30 years many types of compounds such as alcohols, phenols, terpenes, and various drugs were fed to animals and shown to be excreted as glucuronides. All these glucuronides, however, were conjugates of compounds containing hydroxyl groups, i.e., O-ether glucuronides. It was not until 1907 that it was shown unequivocally that carboxylic acids could also form glucuronides when Magnus-Levy (1907) isolated from the urine of sheep given large doses of benzoic acid, the alkaline copper-reducing ester glucuronide, benzoyl glucuronide, which he formulated as an 1-benzoyl glucuronic acid. It is reported (see Ellinger, 1923) that Salkowski in 1877 had observed a copper-reducing nonfermentable substance in mammalian urine after large doses of benzoic acid. However, it appears probable that benzoyl glucuronide was not the first ester glucuronide to be isolated, for 2 years earlier Jaffé (1905) described a reducing glucuronide which he had isolated from the urine of rabbits dosed with p-dimethylaminobenzaldehyde. It melted with decomposition at $205°–206°C$ and broke down on heating in water or on treatment with alkali at room temperature into p-dimethylaminobenzoic acid and glucuronic acid. Silver, calcium, and barium salts of the glucuronide were also described. Many fewer biosynthetic ester glucuronides than O-ether glucuronides have been described to date.

By the end of the last century and the beginning of the present one, the main types of biosynthetic glucuronides had been discovered, namely the O-ether and O-ester glucuronides. In the past 20 years, however, glucuronides of compounds containing amino (NH_2), amide ($CONH_2$), imide ($=NH$), thiol (SH), and carbodithioyl ($CSSH$) groups have been discovered, i.e., N-glucuronides and S-glucuronides (see Smith and Williams, 1966; Williams, 1967).

The question of the mechanism of the formation of glucuronides was not solved until the 1950's. Earlier theories of glucuronic acid conjugation (Sundvik, 1886; Fischer and Piloty, 1891) had suggested that the foreign compound combined with glucose to form a glucoside which was then oxidized at its primary alcohol group to the glucuronide, the more oxidizable aldehyde group being protected since it was combined with the aglycone. A solution to this question was not possible until the modern techniques of the handling of tissue preparations had been developed and the concept of the dominant role of nucleotides in metabolism had been established. The key to the understanding of glucuronide synthesis was the discovery by Dutton and Storey (1953) of uridine diphosphate glucuronic acid (UDPGA) in rabbit liver and of the role of this nucleotide as a cofactor in glucuronide conjugation. Furthermore, it was shown that glucuronic acid was derived by the oxidation of a glucoside but not the glucoside of the foreign compound, as suggested by the earlier workers. This glucoside was uridine diphosphate glucose (UDPG), a cofactor involved in many of the reactions of carbohydrate metabolism. Strominger et al. (1957) were able to show that UDPG was oxidized enzymically to UDPGA, which then transferred its glucuronyl radical to the foreign compound under the influence of glucuronyl transferase to yield a β-glucuronide.

Glucoside conjugation was first found in plants when Miller (1938) showed that ethylene chlorohydrin was converted in gladiolus corms to 2-chloroethyl-β-D-glucoside. Then Myers and Smith (1953), working in this laboratory, showed that locusts converted m-aminophenol into the corresponding glucoside. It was later shown that the glucose conjugation of foreign alcohols, phenols, and aromatic acids is general in insects (see Smith, 1968) and is, in effect, the equivalent of glucuronic acid conjugation in mammals, UDGP, instead of UDPGA, being the intermediate nucleotide.

Ribose and *N*-acetylglucosamine may also be conjugating agents, for Schayer (1956) reported that in rats and mice, imidazole-4-acetic acid was excreted as a compound containing ribose, namely 1-ribosylimidazole-4-acetic acid. Layne *et al.* (1964) described a double conjugate of 17α-estradiol containing glucuronic acid in position 3 and *N*-acetylglucosamine in position 17. These conjugates, if they can be described as such, appear to be unique at the present time.

III. Amino Acid Conjugations

These are mainly reactions of aromatic carboxylic acids. The products are compounds containing an amino acid and the carboxylic acid joined together through a secondary amide link. The most widespread of these reactions is glycine conjugation, which was also the first conjugation to be discovered. The first conjugate to be isolated therefore, was, hippuric acid or benzoylglycine, which was found by Rouelle in 1784 in the urine of cows. However, the first clear proof that hippuric acid could be formed from ingested benzoic acid was given by Keller (1842) more than half a century later. At the suggestion of Wöhler, Keller took benzoic acid himself and isolated hippuric acid from his urine. At that time benzoic acid was tried out as a treatment for gout by the physician, Ure (1841), again at the suggestion of Wöhler. The idea behind this was that on ingesting benzoic acid, the more soluble hippuric acid formed *in vivo* replaced the sparingly soluble uric acid in the urine. The proposed remedy, however, was ineffective and fell into oblivion. The use nearly a century later of hippuric acid formation as a test for liver function was more successful (Delprat and Whipple, 1921; Quick, 1933; Probstein and Londe, 1940).

Schultzen and Gräbe (1867) showed that glycine conjugation was not peculiar to benzoic acid, for they found that chlorobenzoic and anisic acids also formed glycine conjugates in animals. The reaction has subsequently been found to be a general one for most aromatic acids, and it also occurs with some substituted acetic and cinnamic acids and certain steroid acids.

The next amino acid conjugation was discovered by Jaffé (1877) when he found that in hens benzoic acid was not converted into hippuric acid but to ornithuric acid or N, N'-dibenzoylornithine. It was soon found that other aromatic acids also formed ornithine conjugates in hens and the belief arose that ornithine conjugation was common to all birds, until it was found (Efimochkina, 1951; Schachter *et al.*, 1955) that pigeon tissue preparations formed glycine and not ornithine conjugates. The pigeon family of birds (Columbiformes), which includes doves and pigeons, has now been shown to form hippuric acid like mammals (Baldwin *et al.*, 1960).

The third amino acid conjugation to be discovered was that of the glutamine conjugation of arylacetic acids in man. In 1914 Thierfelder and Sherwin described the isolation of phenacetylglutamine from the urine of humans who had ingested phenylacetic acid. At first this reaction appeared to be peculiar to man, since phenylacetic acid formed glycine conjugates in common laboratory animals and, furthermore, it appeared to be limited to phenylacetic acid since Sherwin and his co-workers in the United States about 1920 found that substituted phenylacetic acids were either not conjugated or were conjugated with glycine in man. However, Power (1936) showed that it occurred in the chimpanzee, and in the last decade or so it has been shown to occur in Old World monkeys with other arylacetic acids, namely indolyl-3-acetic acid and 3,4-dihydroxy-5-methoxyphenylacetic acid, a metabolite of mescaline in man. Recently, phenacetylglutamine has been found in traces in cow's milk (Schwartz and Pallansch, 1962) and the glutamine conjugate of diphenylmethoxyacetic acid (which is not an arylacetic acid) has been found as a major metabolite of the antihistaminic, diphenhydramine, in the rhesus monkey (Drach and Howell, 1968).

In recent years, other amino acids have been found as conjugates with certain foreign compounds, but these appear to be rare at the present time. These include the serine conjugate of xanthurenic acid in rats (Rothstein and Greenberg, 1957); the peptide conjugate, containing serine and aspartic acid, of 2,2-bis(*p*-chlorophenyl)acetic acid (DDA) in rats (Pinto *et al.*, 1965); the glycyltaurine and glycylglycine conjugates of quinaldic acid in cats (Kaihara, 1960; Kaihara and Price, 1961), and the arginine conjugates of aromatic acids in arachnids (Hitchcock and Smith, 1964).

The study of the location and mechanism of the amino acid conjugations is largely of recent date and depended upon the discovery of coenzyme A by Lipmann (1945). As already mentioned, Schmiedeberg and Bunge (1877) had concluded that hippuric acid synthesis occurred only in the kidney in the dog. This was confirmed half a century later (Snapper *et al.*, 1923; Quick, 1932b) but it was found that in several other species the synthesis occurred in both the liver and the kidney. With the discovery of CoA, the enzymic mechanism of glycine and glutamine conjugation was rapidly elucidated and shown to have a common basis in that the carboxylic acids involved were activated, forming intermediate nucleotides with CoA from which the acids were transferred to the amino acids under the influence of specific enzymes (Schachter and Taggart, 1953, 1954; Mahler *et al.*, 1953; Moldave and Meister, 1957). The mechanisms of the other amino acid conjugations still have to be elucidated in detail.

The intracellular site of these amino acid conjugations is of interest since, in contrast to many of the other metabolic reactions of foreign compounds, they occur in the mitochondria rather than in the microsomes. This may be related to the ability of the mitochondria to absorb ionic forms, whereas metabolism by microsomes requires, in the main, lipid-soluble nonpolar molecules.

IV. Other Conjugations

Methylation and mercapturic acid synthesis also require amino acids, but these conjugations do not result in the formation of a secondary amide link as in the case of glycine, ornithine, or glutamine conjugation. In methylation, the amino acid methionine takes part in the reaction, but only its methyl group appears in the methylated product. The tripeptide glutathione is involved in mercapturic acid synthesis, but only its cysteine moiety occurs in the final product.

The first instance of biological methylation was discovered by His (1887) who found that pyridine acetate was excreted to a minor extent (about 4%) in dogs as N-methylpyridinium hydroxide. This was confirmed by Cohn (1894) and, to these workers, it appeared that CH_3OH had been added to the pyridine molecule. Thus, the first example of methylation in the animal body was that of the methylation of a tertiary nitrogen atom in an aromatic system. This was followed in 1893–1894 by the discovery of the methylation of selenium and tellurium in the animal body by Hofmeister. Hildebrandt (1900) claimed that the piperidine derivative, 4-piperidinomethylthymol, was methylated at its aliphatic tertiary nitrogen in rabbits, the product excreted in the urine being a glucuronide of 4-(N-methylpiperidinomethyl)thymol. These reactions can be formulated as follows:

Other early examples of N-methylation were the conversion of nicotinic acid to trigonelline (anhydride of N-methylpyridinium hydroxide 3-carboxylic acid) (Ackermann, 1912) and the methylation of quinoline and isoquinoline (Tamura, 1924).

Until the 1950's, methylation appeared to be a relatively rare and minor reaction, mainly of aromatic tertiary nitrogen compounds. Its widespread occurrence was not really appreciated until the metabolic fate of the catecholamines which are O-methylated *in vivo* was studied in detail. The first instance of O-methylation in the body, however, was discovered in another connection by Maclagan and Wilkinson (1954) who were investigating potential antithyroid compounds. They found that butyl 4-hydroxy-3,5-diiodobenzoate was excreted in man, but not in rats or rabbits, partly as 4-methoxy-3,5-diiodobenzoic acid. The methylation of monophenols, however, is a rare reaction and O-methylation is usually characteristic of catechols or polyphenols containing two vicinal hydroxyl groups. This was first shown during the study of the metabolism of the flavones, rutin and quercetin, and of homoprotocatechuic acid in 1955 by Booth *et al.* who found homovanillic acid as a metabolite of these compounds in rats and rabbits. The O-methylation of adrenaline and noradrenaline was discovered 2 years later by Axelrod (1957) and Armstrong and McMillan (1957).

The first case of S-methylation seems to have been recorded by Sarcione and Sokal in 1958 who showed that 2-methylthiouracil was a minor metabolite of thiouracil in the rat.

The mechanism of the methylation of foreign compounds is now largely understood but this depended upon the discovery of the intermediate nucleotide in transmethylations, S-adenosylmethionine, by Cantoni (1953).

It is now clear that in O-, N-, and S-methylations, the methyl group is transferred from S-adenosylmethionine under the influence of methyltransferases, several of which occur and determine the specificity of the various types of methylation. Although the first methylation to be discovered was that of pyridine, it appears that its mechanism has not yet been studied.

The discovery of the mercapturic acids followed closely on that of the ethereal sulfates. That a relationship existed between the ethereal sulfates and mercapturic acids was a view held and discussed for many years during the first quarter of this century. The early studies of Städeler (1851) had shown that human and animal urines contained phenols in a combined form. These studies were followed by attempts to elucidate the nature of the combined phenols and of the indigo-forming substances of urine and by studies on the formation of phenol in animals given benzene. Schultzen and Naunyn (1867) found that phenol was a metabolite of benzene in man and dogs and this was confirmed by Munk (1876). In this latter year Baumann (1876b) was able to show that the phenols and the indigo-forming substances of urine were present as "paired" or "conjugated" sulfuric acids. Baumann eventually isolated phenylsulfate from the urine of a patient who had been treated with carbolic acid (phenol) which had been introduced as an antiseptic into medicine some years earlier by Lister, and found that a number of other substances, including catechol and bromobenzene, caused an increase in the excretion of ethereal sulfates. The term ethereal sulfate arose from the concept that these compounds were ethers of the organic radical with the SO_3H group, i.e., $Ar-O-SO_3H$, and were sometimes erroneously called sulfonates. Baumann's contribution to our early knowledge of the ethereal sulfates was quite impressive. He showed that the urinary ethereal sulfates were associated with putrefactive processes in the intestine since they largely disappeared from the urine of starved dogs fed large doses of calomel or iodoform. Baumann also showed that the salts of phenylsulfate were practically nontoxic and this observation contributed much to the view that conjugation processes were detoxication mechanisms.

Subsequent work has established that sulfate conjugation is a general metabolic reaction of phenols and that it is widely distributed among species. In fact, it appears in almost all species of animals examined (Smith, 1968) and may be the most primitive of the conjugation mechanisms. It has also been shown in recent years to be a minor reaction of primary aliphatic alcohols (Boström and Vestermark, 1960; Vestermark and Boström, 1959) and of some aromatic amines such as aniline, naphthylamines, and sulfonamide drugs which give rise to sulfamates in certain species (Boyland and Manson, 1955; Boyland et al., 1957).

Many early workers studied the site and extent of sulfate conjugation (e.g., Baumann, 1876a,b,c; Baumann and Herter, 1877–1878; De Jonge, 1879; Christiani and Baumann, 1878–1879; Kochs, 1880; Herter and Wakeman, 1899; Embden and Glassner, 1902) and it was deduced that it occurred in several tissues such as the liver, the kidney, and intestine and that quantitatively it was less extensive than glucuronic acid conjugation.

The mechanism of ethereal sulfate synthesis, however, remained obscure until the 1950's. The earlier views were those of Baumann (1876c) and later Hele (1931) who believed that inorganic sulfate reacted directly with phenols and of Sherwin (1922) who thought mercapturic acids might be intermediates. Later studies by De Meio and his co-workers (De Meio and Arnoldt, 1944) proved that inorganic sulfate could be utilized enzymically for sulfate conjugation, not as such, but through an active nucleotide intermediate, 3'-phosphoadenosine-5'-phosphosulfate (PAPS) which was first described by Robbins and Lipmann (1956).

As already mentioned, Baumann found many compounds gave rise to ethereal sulfates and, in 1879, Baumann and Preusse studied their formation from bromobenzene in animals. They found that bromobenzene not only increased the ethereal sulfates in the urine but also produced another sulfur-containing compound. This was isolated and shown to be p-bromophenylmercapturic acid. At about the same time[2] Jaffé (1879) showed that chlorobenzene and iodobenzene also formed mercapturic acids in animals. Little work of importance was done on mercapturic acids, apart from that of Hele and his co-workers in the 1920's, until the late 1930's and early 1940's when Stekol, Young, and Boyland and their co-workers took up their study because of their involvement in the metabolism of the amino acid, cysteine, and of the polycyclic hydrocarbons, some of which were known to be associated with the production of cancer.

The mercapturic acid of naphthalene was isolated in 1934 (Bourne and Young, 1934), of anthracene in 1936 (Boyland and Levi, 1936) and of benzene in 1943 (Zbarsky and Young, 1943). In 1957, it was discovered that some mercapturic acids, i.e., N-acetyl-S-arylcysteines were present in the urine as acid-labile precursors (Young, Boyland and co-workers) which were designated "premercapturic acids." In fact, the work of

[2] Baumann and Preusse's paper begins on p. 806 and Jaffé's paper on p. 1092 of Volume 12 (1879) of Berichte Deutsches Chem. Ges.

Baumann and Jaffé nearly 80 years earlier had indicated that mercapturic acids might be present in urine as labile precursors and these were probably the premercapturic acids or more correctly, the N-acetyl-S-(2-hydroxy-1,2-dihydroaryl)cysteines.

Several aromatic compounds which form mercapturic acids are also converted into phenols which are excreted in part as ethereal sulfates. This observation suggested a relationship between the sulfates and the mercapturic acids as mentioned above. Baumann in 1883 expressed the view that the ethereal sulfate and mercapturic acids formed from such compounds as bromobenzene were formed in separate and independent reactions. However, Sherwin (1922) thought that mercapturic acids might be intermediates in the formation of ethereal sulfates while Rhode (1923) was of the opinion that phenols derived from hydrocarbons were precursors of the mercapturic acids. Present day studies, however, have shown that mercapturic acids arise mainly via glutathione conjugates and that the initial reaction in their formation is between the foreign compound (or some active form of it such as an epoxide) and glutathione, the reaction being catalyzed by one of a group of special enzymes (glutathione S-aryltransferase, S-epoxidetransferase, S-alkyltransferase, etc.), the enzyme involved depending upon the nature of the foreign compound.

The amino acid cysteine (or its oxidized form, cystine) seems to be associated with three conjugation mechanisms. It appears in the acetylated form as a component of the mercapturic acids which are S-aryl- or S-alkyl-N-acetylcysteines. It seems likely that the sulfate used in ethereal sulfate synthesis is derived mainly from cystine (Young and Maw, 1958). The third process in which it may play a role is in the detoxication of cyanide which is converted in the body to thiocyanate (Wood and Cooley, 1956). The conversion of cyanide to thiocyanate was first observed by S. Lang (1894) who found that a fifth to a sixth of small doses of cyanide given orally to dogs over a period of 4 days was excreted in the urine as thiocyanate. It was reported by Treviranus (1814) that human saliva gave a ferric chloride test and that this was due to thiocyanate was shown by Tiedemann and Gmelin (1831). Grober (1901) observed that when human beings were given small doses of cyanide it often, but not always, caused an increase in the thiocyanate content of the saliva. During the 1890's, experiments were also carried out concerning the mechanism of thiocyanate formation, and this led to the discovery that thiosulfate had a slight protective effect against cyanide poisoning. Pascheles (1894) found that the incubation of sodium cyanide with egg albumin or cystine led to the formation of thiocyanate and he concluded that the formation of thiocyanate in the body was a purely chemical reaction between cyanide and the loosely bound sulfur of albumin (or protein). Lang in 1895 studied the effect of sulfur compounds such as sodium sulfide and sodium thiosulfate upon the toxicity of cyanide and discovered that injected thiosulfate had a slight protective effect against cyanide in rabbits. This effect was then investigated by several workers in Europe during the next ten years. Lang also found that a fresh liver brei, but not defibrinated blood, could form thiocyanate from cyanide and thiosulfate and furthermore that the liver brei could destroy the thiocyanate formed. These early observations laid down most of the relevant facts required for the study of the cyanide-thiocyanate detoxication. However, the enzymic conversion of cyanide to thiocyanate was not proved until nearly 40 years after S. Lang's discovery, when K. Lang (1933) working with dried liver powder described many of the properties of the enzyme concerned, which he called rhodanese, or in today's nomenclature, thiosulfate-cyanide sulfurtransferase. The more recent work on this enzyme, including its specificity and mechanism of action, was done by Sörbo in Sweden during the 1950's.

Most of the conjugating agents appear to be derived either from the body's carbohydrate or amino acid sources, but the acetyl group which is used for the acetylation of foreign compounds can be derived from any source, carbohydrate, protein or fat, which can yield acetyl-CoA. It is probable that the first recorded instance of the excretion of acetylated compounds in the animal body is that of the formation of m- and p-acetamidobenzoic acids in rabbits following the administration of the corresponding m- and p-nitrobenzaldehydes observed by Cohn (1893), a pupil of Jaffé. Cohn was following up the earlier discovery of Jaffé and Cohn (1887) that the aldehyde furfural was partly converted in dogs and rabbits to furfuracrylic acid, a reaction considered by Jaffé to be analogous to the Perkin synthesis of cinnamic acid from benzaldehyde and sodium acetate. This reaction seems peculiar to furfural and no other aldehyde, and its mechanism is still obscure.

Acetylation is mainly a reaction of amino groups. From the point of view of acetylation, five types of amino groups can be distinguished, namely, the aliphatic NH_2, the aromatic NH_2, the α-amino acid NH_2, the hydrazino group, and the sulfonamide NH_2 group in sulfanilamide. It is probable that each of these amino groups requires a different transacetylase for its acetylation. The discovery of the acetylation of some natural aliphatic amines such as histamine and serotonin, of hydrazino compounds such as hydrazine itself and isonicotinic hydrazide, and of the sulfonamide group, is of relatively recent date, that is, during the 1950's. The acetylation of unnatural amino acids was found during the 1920's by Knoop who was studying amino acid metabolism. He developed a theory of amino acid synthesis from keto acids, pyruvate, and ammonia, involving the formation of an intermediary N-acetylamino acid. The most important acetylation process, however, is that of aromatic amino compounds, especially the sulfonamide drugs, for the study of this process led to the discovery of the fundamentally important nucleotide coenzyme A by Lipmann (1945). Although the acetylation of several aromatic amino compounds was studied *in vivo* and *in vitro* by Sherwin and his co-workers in the 1920's, the real stimulus to the study of acetylation came when it was discovered in the 1930's that the highly successful drugs, the prontosils, were split in the body to sulfanilamide and that the latter was the active antibacterial agent. Tréfouel *et al.* (1936) suggested that the antibacterial activity of the prontosils was due to sulfanilamide which was formed in the body by reduction of the azo links in the prontosils. This led to the intensive study of the metabolism of sulfanilamide (Marshall *et al.*, 1937) whose main metabolite in most species, except the dog, is N^4-acetylsulfanilamide. The prontosils were thus not only the first really active antibacterial agents, but their study also led to two other fundamental discoveries in biochemistry and pharmacology, namely that of coenzyme A and that of the body itself producing an active antibacterial drug from a relatively inactive precursor. The importance of coenzyme A in intermediary metabolism is well known and the concept of the body producing biological activity in a foreign compound is fundamental to that area where biochemistry, pharmacology, and therapeutics meet.

REFERENCES

Ackermann, D. (1912). *Z. Biol.* **59**, 17.
Armstrong, M. D., and McMillan, A. (1957). *Federation Proc.* **16**, 146.
Axelrod, J. (1957). *Science* **126**, 400.
Baeyer, A. (1870). *Ann. Chem. Liebigs* **155**, 257.
Baldwin, B. C., Robinson, D., and Williams, R. T. (1960). *Biochem. J.* **76**, 595.
Baumann, E. (1876a). *Arch. Ges. Physiol. Pfluegers* **12**, 63, 69.
Baumann, E. (1876b). *Ber. Deut. Chem. Ges.* **9**, 54.
Baumann, E. (1876c). *Arch. Ges. Physiol. Pfluegers* **13**, 285.
Baumann, E., and Herter, E. (1877–1878). *Z. Physiol. Chem. Hoppe-Seylers* **1**, 244.
Baumann, E., and Preusse, C. (1879). *Ber. Deut. Chem. Ges.* **12**, 806.
Booth, A. N., Murray, C. W., Jones, F. T., and De Eds, F. (1955). *Federation Proc.* **14**, 321.
Boström, H., and Vestermark, A. (1960). *Acta Physiol. Scand.* **48**, 88.
Bourne, M. C., and Young, L. (1934). *Biochem. J.* **28**, 803.
Boyland, E., and Levi, A. A. (1936). *Biochem. J.* **30**, 728.
Boyland, E., and Manson, D. (1955). *Biochem. J.* **60**, ii.
Boyland, E., Manson, D., and Orr, S. F. D. (1957). *Biochem. J.* **65**, 417.

Cantoni, G. L. (1953). *J. Biol. Chem.* **204**, 403.
Christiani, A., and Baumann, E. (1878–1879). *Z. Physiol. Chem. Hoppe-Seylers* **2**, 350.
Cohn, R. (1893). *Z. Physiol. Chem. Hoppe-Seylers* **17**, 274.
Cohn, R. (1894). *Z. Physiol. Chem. Hoppe-Seylers* **18**, 112.
De Jong, D. (1879). *Z. Physiol. Chem. Hoppe-Seylers* **3**, 177.
Delprat, G. D., and Whipple, G. H. (1921). *J. Biol. Chem.* **49**, 229.
De Meio, R. H., and Arnoldt, R. I. (1944). *J. Biol. Chem.* **156**, 577.
Drach, J. C., and Howell, J. P. (1968). *Biochem. Pharmacol.* **17**, 2125.
Dutton, G. J., and Storey, I. D. E. (1953). *Biochem. J.* **53**, 37.
Efimochkina, E. F. (1951). *Dokl. Akad. Nauk SSSR* **80**, 793.
Ellinger, A. (1923). *In* "Handbuch der Experimentellen Pharmakologie" (A. Heffter, ed.), vol. 1, p. 979. Springer, Berlin.
Embden, G., and Glassner, K. (1902). *Beitr. Chem. Physiol. Pathol.* **1**, 310.
Fischer, E., and Piloty, O. (1891). *Ber. Deut. Chem. Ges.* **24**, 521.
Grober, J. A. (1901). *Deut. Arch. Klin. Med.* **69**, 243.
Hele, T. S. (1931). *Biochem. J.* **25**, 1736.
Herter, C. A., and Wakeman, A. J. (1899). *J. Exptl. Med.* **4**, 307.
Hildebrandt, H. (1900). *Arch. Exptl. Pathol. Pharmakol. Naunyn-Schmiedebergs* **44**, 278.
His, W. (1887). *Arch. Exptl. Pathol. Pharmakol. Naunyn-Schmiedebergs* **22**, 253.
Hitchcock, M., and Smith, J. N. (1964). *Biochem. J.* **93**, 392.
Hofmeister, F. (1893–1894). *Arch. Exptl. Pathol. Pharmakol. Naunyn-Schmiedebergs* **33**, 198.
Jaffé, M. (1874). *Ber. Deut. Chem. Ges.* **7**, 1673.
Jaffé, M. (1877). *Ber. Deut. Chem. Ges.* **10**, 1925.
Jaffé, M. (1878). *Z. Physiol. Chem. Hoppe-Seylers* **2**, 47.
Jaffé, M. (1879). *Ber. Deut. Chem. Ges.* **12**, 1092.
Jaffé, M. (1905). *Z. Physiol. Chem. Hoppe-Seylers* **43**, 374.
Jaffé, M., and Cohn, R. (1887). *Ber. Deut. Chem. Ges.* **20**, 2311.
Kaihara, M. (1960). *J. Biol. Chem.* **235**, 136.
Kaihara, M., and Price, J. M. (1961). *J. Biol. Chem.* **236**, 508.
Keller, W. (1842). *Ann. Chem. Liebigs* **43**, 108.
Kochs, W. (1880). *Arch. Ges. Physiol. Pfluegers* **23**, 161.
Lang, K. (1933). *Biochem. Z.* **259**, 243.
Lang, S. (1894). *Arch. Exptl. Pathol. Pharmakol. Naunyn-Schmiedebergs* **34**, 247.
Lang, S. (1895). *Arch. Exptl. Pathol. Pharmakol. Naunyn-Schmiedebergs* **36**, 75.
Layne, D. S., Sheth, N. A., and Kirdoni, R. Y. (1964). *J. Biol. Chem.* **239**, 3221.
Lipmann, F. (1945). *J. Biol. Chem.* **160**, 173.
Maclagan, N. F., and Wilkinson, J. H. (1954). *Biochem. J.* **56**, 211.
Magnus-Levy, A. (1907). *Biochem. Z.* **6**, 502.
Mahler, H. R., Wakil, S. J., and Bock, R. M. (1953). *J. Biol. Chem.* **204**, 453.
Marshall, E. K., Emerson, K., and Cutting, W. C. (1937). *Science* **85**, 202.
Mering, E. (1882). *Z. Physiol. Chem. Hoppe-Seylers* **6**, 489.
Mering, E., and Musculus, E. (1875). *Ber. Deut. Chem. Ges.* **8**, 662.
Metchnikoff, E. (1910). *Ann. Inst. Pasteur* **24**, 755.
Miller, L. P. (1938). *Contrib. Boyce Thompson Inst.* **9**, 425.
Moldave, K., and Meister, A. (1957). *Biochim. Biophys. Acta* **24**, 654.
Munk, I. (1876). *Arch. Ges. Physiol. Pfluegers* **12**, 146.

Myers, C. M., and Smith, J. N. (1953). Biochem. J. 54, 276.
Pascheles, W. (1894). Arch. Exptl. Pathol. Pharmakol. Naunyn-Schmiedebergs 34, 281.
Pinto, J. D., Camien, M. N., and Dunn, M. S. (1965). J. Biol. Chem. 240, 2148.
Power, F. W. (1936). Proc. Soc. Exptl. Biol. Med. 33, 598.
Probstein, J. G., and Londe, S. (1940). Ann. Surg. 111, 230.
Quick, A. J. (1932a). J. Biol. Chem. 97, 403.
Quick, A. J. (1932b). J. Biol. Chem. 96, 73.
Quick, A. J. (1933). Am. J. Med. Sci. 185, 630.
Rhode, H. (1923). Z. Physiol. Chem. Hoppe-Seylers 124, 15.
Robbins, P. W., and Lipmann, F. (1956). J. Am. Chem. Soc. 78, 2652.
Rothstein, M., and Greenberg, D. M. (1957). Arch. Biochem. Biophys. 68, 206.
Sarcione, E. J., and Sokal, J. E. (1958). J. Biol. Chem. 231, 605.
Schachter, D., and Taggart, J. V. (1953). J. Biol. Chem. 203, 925.
Schachter, D., and Taggart, J. V. (1954). J. Biol. Chem. 208, 263.
Schachter, D., Manis, J. G., and Taggart, J. V. (1955). Am. J. Physiol. 182, 537.
Schayer, R. W. (1956). Brit. J. Pharmacol. 11, 472.
Schmid, W. (1855). Ann. Chem. Liebigs 93, 83.
Schmiedeberg, O., and Bunge, G. (1877). Arch. Exptl. Pathol. Pharmakol. Naunyn-Schmiedebergs 6, 233.
Schmiedeberg, O., and Meyer, H. (1879). Z. Physiol. Chem. Hoppe-Seylers 3, 422.
Schultzen, O., and Gräbe, C. (1867). Arch. Anat. Physiol. p. 166.
Schultzen, O., and Naunyn, B. (1867). Arch. Anat. Physiol. p. 349.
Schwartz, D. P., and Pallansch, M. J. (1962). Nature 194, 186.
Sherwin, C. P. (1922). Physiol. Rev. 2, 264.
Smith, J. N. (1968). Advan. Comp. Physiol. Biochem. 3, 173.
Smith, R. L., and Williams, R. T. (1966). In "Glucuronic Acid: Free and Combined. Chemistry, Biochemistry, Pharmacology, and Medicine" (G. J. Dutton, ed.), pp. 457–491. Academic Press, New York.
Snapper, I., Grünbaum, A., and Neuberg, J. (1923). Ned. Tijdschr. Geneesk. 1, 426.
Städeler, G. (1851). Ann. Chem. Liebigs 77, 17.
Strominger, J. L., Kalckar, H. M., Axelrod, J., and Maxwell, E. S. (1954). J. Am. Chem. Soc. 76, 6411.
Sundvik, E. (1886). Jahresber. Fortschr. Tierchem. 16, 76.
Tamura, S. (1924). Acta Schol. Med. Univ. Kioto. 6, 449, 459.
Thierfelder, H., and Sherwin, C. P. (1914). Ber. Deut. Chem. Ges. 47, 2630.
Tiedemann, F., and Gmelin, L. (1831). Die Verdauung nach Versuchen Heidelberg v. Leipzig 1, 8.
Tréfouel, J., Tréfouel, M., Nitti, F., and Bovet, D. (1936). Compt. Rend. Soc. Biol. 120, 756.
Treviranus, G. R. (1814). Biologie 4, 330.
Ure, A. (1841). London Med. Gaz., N. S. 27 (I), 735.
Vestermark, A., and Boström, H. (1959). Exptl. Cell Res. 18, 174.
Williams, R. T. (1947). "Detoxication Mechanisms. The Metabolism of Drugs and Allied Organic Compounds" (1st ed.). Chapman & Hall, London.
Williams, R. T. (1959). "Detoxication Mechanisms. The Metabolism and Detoxication of Drugs, Toxic Substances and Other Organic Compounds" (2nd ed.). Chapman & Hall, London.

Williams, R. T. (1967). *In* "Biogenesis of Natural Compounds" (P. Bernfield, ed.), pp. 589–639. Macmillan (Pergamon), New York.

Wood, J. L., and Cooley, S. L. (1956). *J. Biol. Chem.* **218**, 449.

Young, L., and Maw, G. A. (1958). "The Metabolism of Sulphur Compounds," p. 108. Methuen, London.

Zbarsky, S. H., and Young, L. (1943). *J. Biol. Chem.* **151**, 487.

NEW METABOLIC CONJUGATES
OF STEROIDS

DONALD S. LAYNE

I. Introduction

In a recent review Williams (1967) points out that our knowledge of the processes of conjugation and detoxication is derived mainly from investigations of the fate of foreign organic compounds in the body. The sterols, bile acids, and steroid hormones, however, together with their metabolites, constitute a group of naturally occurring compounds whose study has provided a great deal of fundamental information on metabolic conjugation. Bernstein and his co-workers (1966, 1968) have compiled a valuable literature survey of steroid conjugates, and a catalog of their physical properties.

For almost two decades, glucosiduronates and sulfates were the only steroid conjugates known to occur naturally in animals, with the exception of the conjugates formed by the bile acids, by peptide-type linkages, with glycine and taurine. The biological methylation of phenolic steroids was established with the isolation of 2-methoxyestrone by Kraychy and Gallagher in 1957, while the results of Weichselbaum and Margraf (1960) indicated the occurrence *in vivo* of steroid acetylation. Since then, ring A-phenolic steroids have been shown to undergo conjugation with glutathione (see Section IV, D).

The foregoing conjugation reactions are well known in connection with compounds other than steroids. In contrast, the observation that steroids are excreted in the urine in combination with N-acetylglucosamine (Section II) represents a completely novel metabolic conjugation reaction, and the recent characterization of a steroid glucoside in rabbit urine (Section III) indicates that glucosyl transfer to molecules other than sugars, a process previously observed only in plants, bacteria, insects, and mollusks, occurs in at least one mammalian species. In addition, the steroid double conjugates, such as the sulfoglycosides, are unique insofar as they display the employment of two separate conjugating mechanisms in the metabolism of a single molecule of natural origin.

The present chapter deals in detail with the isolation, recognition, enzymic formation and hydrolysis, and the possible significance of steroid N-acetylglucosaminides and glucosides. The other newer steroid conjugates will also be described, and an attempt will be made to integrate the data into the total field of metabolic conjugation and hydrolysis.

Finally, work on steroids has made a prime contribution to the study of the *in vivo* metabolism of conjugates. In fact, it is now recognized that some steroid conjugates, in particular the sulfates, are not necessarily mere end products of metabolism, but are secreted by endocrine tissue and are active metabolic intermediates. The significance of steroid conjugate metabolism is the subject of another contribution to this treatise.

II. Conjugation with N-Acetylglucosamine

A. GENERAL

Within the past few years steroid N-acetylglucosaminides have been found in the urine of rabbits (Layne et al., 1964; Layne, 1965) and of the human (Arcos and Lieberman, 1967), as well as in human bile (Jirku and Levitz, 1969). The isolation of these compounds establishes the presence in animals of a hitherto unknown form of metabolic conjugation, for although N-acetylglucosamine transfer takes place in animal tissues during the formation of mucopolysaccharides, the attachment of this compound to small molecules other than in the formation of saccharides and nucleotides has not previously been recorded.

All the steroid N-acetylglucosaminides so far isolated from animal sources are double conjugates, in which N-acetylglucosamine is attached at one of the positions in the D-ring or on the 17 side chain, while either glucuronic or sulfuric acid is attached to the hydroxyl on carbon 3 of the steroid. The isolation, characterization, and some aspects of the biosynthesis of these conjugates are detailed below.

B. N-ACETYLGLUCOSAMINE CONJUGATION IN THE RABBIT

1. Nature of the Conjugates

In 1964, Layne et al. injected pregnant and nonpregnant rabbits with estrone-16-^{14}C and detected in the urine a conjugate which was only partially cleaved by β-glucuronidase. This conjugate was isolated in crystalline form from the urine of rabbits which had received large doses of estrone benzoate, and was rigorously identified as the 17α-N-acetylglucosaminide of 17α-estradiol. Since this conjugate could be extracted with ethyl acetate from the urine only after the latter had been treated with β-glucuronidase, it seemed probable that 17α-estradiol was excreted by the rabbit as a double conjugate, with glucuronic acid at position 3 in addition to the N-acetylglucosamine at position 17. This indication was verified (Layne, 1965) and the structure of the double conjugate was established as shown in Fig. 1, with the reservation that the β-orientation of the glycosidic linkages was assumed from the hydrolysis of the conjugate by β-glucuronidase and by the β-N-acetylglucosaminidase present in almond emulsin. Subsequent work (Collins *et al.*, 1967b) established by optical rotation experiments that both the glycosidic linkages were indeed of the β-configuration, provided that the sugars were the usual D-isomers.

In several experiments it has been shown that the double conjugate of 17α-estradiol is by far the major excretory product in rabbit urine after the administration of either estrone, 17β-estradiol, or 17α-estradiol (Layne

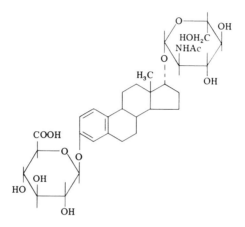

Fig. 1.   Estradiol-3-β-glucuronide-17α-β-N-acetylglucosaminide.

et al., 1964; Layne, 1965; Williams *et al.*, 1968). Neither the 3-mono-glucosiduronate nor the 17α-mono-N-acetylglucosaminide have been detected in the urine, nor has there been any indication that glucuronic acid might be attached to position 17 or N-acetylglucosamine to position 3.

A series of studies was carried out to determine whether the administration of other steroids to rabbits would lead to the excretion of metabolites conjugated with N-acetylglucosamine. In all cases the administered steroids were labeled with tritium, and the nature of the radioactive products in the urine was examined. Neither progesterone, testosterone, nor epitestosterone gave rise to N-acetylglucosaminides in urine (Mellin, Collins, Williams, and Layne, unpublished results). However, estriol and 17α-ethynylestradiol were both metabolized to compounds which were excreted as double conjugates with glucuronic acid and N-acetylglucosamine. The metabolite of 17α-ethynylestradiol which forms the aglycone of the conjugate has been identified as D-homoestradiol-17-aα (Abdel-Aziz and Williams, 1969), and this work represents the first demonstration of the *in vivo* formation of a D-homosteroid. Small amounts of 17α-estradiol are formed in rabbits treated orally with 17α-ethynyl-estradiol. This presumably results from removal of the ethynyl group and metabolism of the steroid molecule to 17α-estradiol, which is excreted as the double conjugate shown in Fig. 1. When estriol was administered to rabbits (Collins and Layne, 1968) there was considerable metabolism of the molecule to the two isomers containing an α-oriented hydroxyl at position 17, namely 17-epiestriol and 16,17-epiestriol. Only these isomers were excreted as double conjugates. Estriol itself, in which the 17-hydroxyl is β-oriented, was excreted mainly as a monoglucosiduronate, with some evidence for the presence of a diglucosiduronate (see Section V, C). These results indicate that, in the rabbit, the 17α-hydroxyl group on phenolic steroids is a preferred site for conjugation with N-acetylglucosamine. A summary of the steroid N-acetylglucosaminides so far isolated from rabbit urine is given in Table I.

2. *Formation of Steroid N-Acetylglucosaminides by Rabbit Tissues in Vitro*

The isolation from rabbit urine of the double conjugate shown in Fig. 1 indicated that, in addition to the well-known enzyme system which transfers glucuronic acid from uridine diphosphate (UDP)-glucuronic acid, rabbit tissues must contain a mechanism for the transfer of N-acetylglucosamine to steroids. Jirku and Layne (1965) found that rabbit liver homogenates, in the presence of UDP-N-acetylglucosamine, would effect this transfer to the 17α-, but not to the 17β-hydroxyl group of estradiol. Furthermore, these authors found that in order for the transfer of N-acetylglucosamine to take place, UDP-glucuronic acid as well as UDP-N-acetylglucosamine had to be added to the homogenates, or else preformed estradiol-3-glucosiduronate had to be used as substrate. The product obtained from the incubations was the double conjugate shown in Fig. 1, and N-acetylglucosamine was not transferred to the steroid in the absence of glucuronic acid at position 3.

These earlier experiments were carried out with tritiated 17β-estradiol, and the yields of the double conjugate were very poor because the 17β-estradiol had to be converted by the liver homogenate, via estrone, to 17α-estradiol before the transfer of N-acetylglucosamine could take place. With the availability (Layne *et al.*, 1965) of 17α-estradiol-6,7-³H of high specific activity, Collins *et al.* (1968) confirmed and extended these observations and were able to obtain the double conjugate in high yields. They showed that the steroid N-acetylglucosaminyl transferase was, like glucuronyl transferase, located in the microsomes, and they also examined

TABLE I

Steroid Conjugates with N-Acetylglucosamine

Steroid aglycone + other conjugating group	Compound administered	Steroid hydroxyl to which N-acetyl-glucosamine attached	Source of conjugate	Reference
17α-Estradiol-3-glucosiduronate	Estrone or 17β-estradiol	17α	Rabbit urine	Layne (1965)
		17α	Rabbit bile	Collins et al. (unpublished)
17-Epiestriol-3-glucosiduronate	Estriol	17α	Rabbit urine	Collins and Layne (1968)
16,17-Epiestriol-3-glucosiduronate	Estriol	17α	Rabbit urine	Collins and Layne (1968)
17α-Estradiol-3-sulfate	Estrone-3-sulfate	17α	Rabbit urine	Collins and Layne (1969)
5-Pregnene-3β-20α-diol-3-sulfate	Pregnenolone	20α	Human urine	Arcos and Lieberman (1967)
15α-Hydroxyestrone-3-sulfate	Estrone-3-sulfate	15α	Human bile	Jirku and Levitz (1969)
15α-Hydroxyestradiol-3-sulfate	Estrone-3-sulfate	15α	Human bile	Jirku and Levitz (1969)
D-Homoestradiol-17aα-3-glucosiduronate	17α-Ethynyl-estradiol-17β	17aα	Rabbit urine	Abdel-Aziz and Williams (1969)

its distribution in various rabbit tissues (Table II). While liver was by far the most abundant source of both glucuronyl transferase and N-acetylglucosaminyl transferase, the latter was the preponderant of the two enzymes in kidney tissue.

TABLE II

ABILITY OF HOMOGENIZED TISSUES TO TRANSFER GLUCOSIDURONIC ACID TO
17α-ESTRADIOL COMPARED TO THEIR ABILITY TO TRANSFER N-ACETYL-
GLUCOSAMINE TO 17α-ESTRADIOL-3-GLUCURONOSIDE[a]

Tissue[b]	17α-Estradiol conjugated with glucuronic acid ($10^6 \times$ μmoles/hour)	17α-Estradiol-3-glucosiduronate conjugated with N-acetylglucosamine ($10^6 \times$ μmoles/hour)	Ratio of N-acetylglucosaminyl transferase activity to glucuronyl transferase activity
Uterus	10.9	0.0	—
Large intestine	14.6	11.7	0.8
Small intestine	16.9	2.9	0.2
Kidney	2.3	22.6	9.8
Liver	1112.0	588.0	0.5

[a] Figures are for amounts of homogenate equivalent to 0.5 mg of wet tissue.

[b] Ovary, adrenal, and spleen homogenates had no detectable ability to transfer either sugar, nor did whole blood, plasma, or erythrocytes.

3. Characteristics of the Steroid N-Acetylglucosaminyl Transferase of Rabbit Liver

Some of the characteristics of this enzyme in rabbit liver microsomes have been studied and compared with those of the phenolic steroid glucuronyl transferase from the same source (Collins et al., 1968). The results confirmed the original observations of Jirku and Layne (1965) that glucuronic acid and N-acetylglucosamine are added to 17α-estradiol in an irreversible sequence, and that 17α-estradiol-3-glucosiduronate is the substrate for the N-acetylglucosaminyl transferase, with UDP-N-acetylglucosamine as the source of the sugar. In addition, Table III shows that, of a large series of steroids tried as substrates, both before and after prior formation of a glucosiduronate, only the 3-glucosiduronates of 17α-estradiol, the 17α-epimers of estriol, and 17β-methyl-17α-estradiol formed N-acetylglucosaminides. These findings are in accord with those obtained by examination of rabbit urine after the administration of various steroids (see Section II, B,1). Thus, the N-acetylglucosaminyl transferase of rabbit

TABLE III

COMPOUNDS TESTED AS SUBSTRATES FOR TRANSFERASES OF RABBIT LIVER MICROSOMES

Compound	Trivial name	Formation of glucosiduronate	Formation of N-acetyl-glucosaminide
3-Hydroxyestra-1,3,5(10)-trien-17-one-6,7-³H	Estrone	+	—
Estra-1,3,5(10)-trien-3,17β-diol-6,7-³H	17β-Estradiol	+	—
Estra-1,3,5(10)-trien-3,17α-diol-6,7-³H	17α-Estradiol	+	+[a]
17β-Methylestra-1,3,5(10)-trien-3,17α-diol	17β-Methyl-17α-estradiol	+	+[a]
3,15α-Dihydroxyestra-1,3,5(10)-trien-17-one-4-¹⁴C	15α-Hydroxyestrone	+	—
Estra-1,3,5(10)-trien-3,16α,17β-triol-6,7-³H	Estriol	+	—
Estra-1,3,5(10)-trien-3,16β,17β-triol-16-¹⁴C	16-Epiestriol	+	—
Estra-1,3,5(10)-trien-3,16α,17α-triol-6,7-³H	17-Epiestriol	+	+[a]
Estra-1,3,5(10)-trien-3,16β,17α-triol-6,7-³H	16,17-Epiestriol	+	+[a]
3α,17α,21-Trihydroxy-5β-pregnan-11,20-dione-1,2-³H	Tetrahydrocortisone	+	—
3α,11β,17α,21-Tetrahydroxy-5β-pregnan-20-one-1,2-³H	Tetrahydrocortisol	+	—
3β,17α-Dihydroxy-5β-pregnan-20-one-7α-³H	17α-Hydroxypregnenolone	+	—
17α-Hydroxyandrost-4-en-3-one-1,2-³H	Epitestosterone	—	—
Androst-4-en-3α,17α-diol-1,2-³H	3α,17α-Androstenediol	+	—
Androst-4-en-3β,17α-diol-1,2-³H	3β,17α-Androstenediol	+	—
3,4-Bis(p-hydroxyphenyl)-3-hexene-(ethyl-1-¹⁴C)	Diethylstilbestrol	+	—
17α-Hydroxyestra-1,3,5(10)-trien-6,7-³H-3-yl-β-D-glucopyranosiduronic acid	17α-Estradiol-3-glucosiduronate	—	+
3-Hydroxyestra-1,3,5(10)-trien-6,7-³H-17β-yl-β-D-glucopyranosiduronic acid	Estradiol-17β-glucosiduronate	—	—
3-Hydroxyestra-1,3,5(10)-trien-6,7-³H-17α-yl-2'-acetamido-2'-deoxy-β-D-glucopyranoside	Estradiol-17α-N-acetylglucosaminide	—	—

[a] These compounds formed N-acetylglucosaminids only after prior formation of a monoglucosiduronate.

liver microsomes appears to have a high specificity for the 3-glucosiduronates of steroids containing a phenolic ring A and an α-oriented hydroxyl group at position 17.

One exception to the above specificity is that if sulfate is substituted for glucosiduronate at position 3 of 17α-estradiol, transfer of N-acetylglucosamine to the 17α-hydroxyl can still take place. This possibility was investigated because, while there had been no evidence in the above

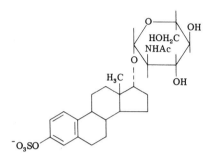

FIG. 2.   Estradiol-3-sulfate-17α-β-N-acetylglucosaminide.

work for the excretion in rabbit urine of phenolic steroids as sulfate conjugates, double conjugates with sulfate and N-acetylglucosamine have been found in the human, as discussed later in Section II, C. Collins and Layne (1969) were able to demonstrate the transfer of N-acetylglucosamine to 17α-estradiol-3-sulfate *in vitro* by rabbit liver microsomes, and when estrone sulfate doubly labeled with ^{35}S and tritium was injected into rabbits, substantial amounts of a double conjugate were excreted. This conjugate was identified as estradiol-3-sulfate-17α-N-acetylglucosaminide (Fig. 2). The results of the *in vitro* experiments indicated that the 3-sulfate of 17α-estradiol was less efficiently conjugated with N-acetylglucosamine than was the 3-glucosiduronate.

The steroid glucuronyl transferase and the N-acetylglucosaminyl transferase of rabbit liver microsomes have different tissue distributions (Table II), specificity (Table III), and pH-activity relationships (Collins *et al.*, 1968). It seems likely, therefore, that they may be separate proteins, although definitive evidence on this point requires their solubilization and separation.

4. *Effect of Steroid Treatment on Steroid Transferase Levels in Rabbit Tissues*

The N-acetylglucosaminyl transferase level in rabbit kidney is increased by treatment with either 17α-estradiol, 17β-estradiol, or diethyl-

Donald S. Layne

stilbestrol at levels of 1.5–5.0 mg per kilogram live weight (Himaya *et al.*, 1969). This effect is illustrated by Fig. 3. Similar doses of cortisol, testosterone, or progesterone did not affect the enzyme level, neither did treatment with any of the steroids, including the estrogens, significantly alter the levels of N-acetylglucosaminyl transferase in liver, or of phenolic

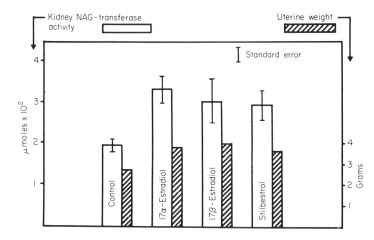

FIG. 3.   Comparison of the kidney N-acetylglucosaminyl transferase activity and uterine weight of rabbits treated with estrogens (1.5 mg/kg live weight). Values for enzyme activities are in terms of substrate conjugated in the assay described by Collins *et al.* (1968). Values for treated animals are significantly different from those of controls ($p = 0.05$). (Data from Himaya *et al.*, 1969).

steroid glucuronyl transferase in either liver or kidney. The estrogen-induced increase in kidney N-acetylglucosaminyl transferase seems to be due, at least in part, to a stimulation of synthesis of the enzyme, since the transferase activity is significantly increased in relation to total protein in the kidney homogenates of the estrogen-treated animals.

No explanation is immediately apparent for the sensitivity of kidney N-acetylglucosaminyl transferase to estrogen treatment, and the lack of this effect on the liver enzyme. The observation is of interest in view of the preponderance of N-acetylglucosaminyl transferase over glucuronyl transferase in kidney tissue (Table II). It seems possible that the rabbit kidney is unable to excrete the 3-monoglucosiduronate of 17α-estradiol without the addition of N-acetylglucosamine at position 17, since only the double conjugate is found in urine (Collins *et al.*, 1967b). Thus, the kidney may actively transfer N-acetylglucosamine to any 3-glucosiduronate presented to it by the circulation. 17β-Estradiol is converted by rabbit tissues *in vivo* to the 17α-isomer, which is excreted in conjugated form in the urine. Thus, the response of the kidney N-acetylglucosaminyl transferase to both these compounds could be due to the fact that they both give rise to the specific substrate for the enzyme, namely 17α-estradiol-3-glucosiduronate. However, the fact that treatment with diethylstilbestrol also increases the enzyme level (Fig. 3) cannot be thus explained, since this compound is excreted by the rabbit largely as the monoglucosiduronate, and neither diethylstilbestrol nor its monoglucosiduronate serve as substrates for the steroid N-acetylglucosaminyl transferase *in vitro* (Table III). It is possible, therefore, that the effects of all three compounds on the kidney enzyme are due to their common action as estrogens. At the high levels employed in the experiment shown in Fig. 3, 17α-estradiol appeared to be as potent as either 17β-estradiol or stilbestrol in increasing the weight of the rabbit uteri, although in more nearly physiological doses 17β-estradiol is much more potent than the 17α-epimer in promoting uterine growth in this animal (Saldarini and Hilliard, 1969).

C. N-ACETYLGLUCOSAMINE CONJUGATES IN THE HUMAN

After the administration of pregnenolone-³H to normal human subjects, Arcos and Lieberman (1967) found that about 8% of the radioactivity excreted in the urine was in the form of a double conjugate, which they crystallized and identified as the 3β-sulfate-20α-N-acetylglucosaminide of 5-pregnene-3β, 20α-diol. On the basis of optical rotatory dispersion measurements, they concluded that the N-acetylglucosamine was linked to the steroid by an α- rather than a β-oriented glycosidic bond. In the absence of a large series of steroid α- and β-N-acetylglucosaminides for comparison, this assignation cannot be taken as unequivocally proven; but should it be correct, then the conjugate isolated by Arcos and Lieberman, and shown in Fig. 4, would be a glycoside unique in the animal kingdom. In fact, while the above sentence was in press, Matsui and Fukushima (1969) provided evidence from total synthesis which indi-

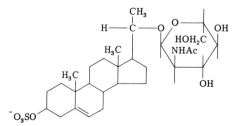

FIG. 4.   5-Pregnene-3β-sulfate-20α-α-N-acetylglucosaminide.

cates that the configuration at C-1′ in this conjugate is really β rather than α.

Jirku and Levitz (1969), in the course of a study of the metabolites of estrone-³H-sulfate-³⁵S in a woman with a bile fistula, found that the bile contained a significant proportion of the excreted radioactivity in the form of sulfo-N-acetylglucosaminides of 15α-hydroxyestrone and 15α-hydroxyestradiol. It is virtually certain that the N-acetylglucosamine is attached to the 15α-hydroxyl group of these steroids. Of the several other steroid metabolites isolated, none was attached to N-acetylglucosamine. Accordingly, under the conditions of this experiment, considerable specificity of this sugar for the 15α-group on the steroid must be assumed. The configuration of the glycosidic linkage in these conjugates has not yet been determined. Table I summarizes the steroid N-acetylglucosaminides so far identified in animals.

D. POSSIBLE SIGNIFICANCE OF STEROID N-ACETYLGLUCOSAMINIDES

Conjugation with N-acetylglucosamine results in an appreciable increase in the water solubility of steroids, but this increase is not comparable in degree with that effected by conjugation with glucuronic or sulfuric acids to form strongly acidic, ionizable compounds. For this reason, it is not surprising that the steroids so far found in bile or urine as N-acetylglucosaminides are also coupled at another position on the molecule to either glucuronic or sulfuric acid. An insufficient number of these conjugates is presently known for it to be apparent whether any significance should be attached to the fact that the acid is at position 3 while the N-acetylglucosamine is attached either directly to a ring D hydroxyl or to a hydroxyl on a side chain on this ring. It appears possible that the addition of N-acetylglucosamine as well as glucuronic acid to 17α-estradiol may be a prerequisite for kidney clearance of this compound by the rabbit, although the double glycoside is also present in large amounts in rabbit bile (Collins, Williams, and Layne, unpublished), so that its formation is not exclusively for purposes of urinary excretion. Further, the sulfo-N-acetylglucosaminides of 15α-hydroxylated estrogens in humans are apparently excreted, at least initially, largely in the bile (Jirku and Levitz, 1969). While nothing is known as yet about the enterohepatic circulation of these conjugates and their possible final excretion in the urine, kidney clearance does not appear to be a primary reason for their formation in the human, whatever may be the situation in the rabbit.

Whittemore and Layne (1965) have shown that both estradiol-3-

glucosiduronate and estradiol-17α-N-acetylglucosaminide are effectively hydrolyzed by the glycosidases of rabbit tissues during *in vitro* incubations. Although β-N-acetylglucosaminidase is widely distributed in animal tissues (Conchie *et al.*, 1959), it has previously been studied with synthetic substrates, and the 17α-N-acetylglucosaminide of estradiol is the first such glycoside of a nonsugar to be found in animals. Whether or not the steroid might be liberated from the glycoside *in vivo* by β-N-acetylglucosaminidase remains to be investigated. In this connection it is pertinent that the presence of a specific α-N-acetylglucosaminidase in mammalian tissues is well established (Roseman and Dorfman, 1951; Weissman *et al.*, 1967). Steroid α-N-acetylglucosaminides might also serve as substrates for such an enzyme.

The structure of the conjugate shown in Fig. 1, and those of the steroid double conjugates containing sulfuric acid and N-acetylglucosamine (Figs. 2 and 4), present a certain analogy with the basic structural units of mucopolysaccharides. Hyalobiuronic acid, the repeating unit of hyaluronic acid, consists of glucuronic acid and N-acetylglucosamine joined by β-1→4 linkages. It seems possible that some of the enzymes involved in mucopolysaccharide synthesis may also function in the synthesis of steroid conjugates, and it may be that involvement of the steroid molecule in a mucopolysaccharide matrix might take place *in vivo*. This possibility bears exploration in view of the known effects of steroids on mucopolysaccharide metabolism.

III. Conjugation with Glucose

As described in Section II, Collins *et al.* (1968) showed that microsomal preparations from rabbit tissues could effect the transfer of N-acetylglucosamine from UDP-N-acetylglucosamine to the 17α-hydroxyl group of 17α-estradiol-3-glucosiduronate. In further experiments it was found that substitution of UDP-glucose for UDP-N-acetylglucosamine in incubations with liver microsomes also led to the formation of a double conjugate of 17α-estradiol. Treatment of this double conjugate with β-glucuronidase (Ketodase) to remove the glucuronic acid at position 3 yielded a monoconjugate which was easily distinguishable from estradiol-17α-N-acetylglucosaminide on thin-layer chromatograms. These results suggested that liver microsomes could effect the transfer of glucose to the 17α-hydroxyl group of 17α-estradiol-3-glucosiduronate. Since only small amounts of the conjugate could be obtained *in vitro*, a search was made for it in the urine of a rabbit which had been injected with 17β-estradiol-

6,7-³H. After treatment of a crude extract of the urinary conjugates with
Ketodase, a tritiated compound extracted from the aqueous extract with
ethyl acetate was found to be chromatographically similar to the suspected
estradiol-17α-glucoside formed *in vitro.* This material was eventually
obtained in crystalline form from the urine of rabbits treated with
large doses of estrone benzoate, and its structure was definitively established, by comparison with authentic material, as estra-1,3,5(10)-trien-3-
ol-17α-yl-β-D-glucopyranoside (Williamson *et al.,* 1969). The structure is
shown in Fig. 5.

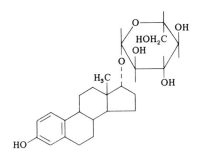

FIG. 5.   Estradiol-17α-β-glucoside.

The release of the monoconjugate by Ketodase from the urinary extract
was inhibited by the presence of saccharo-1,4-lactone, which indicates
strongly that, in the urine, the 3-hydroxyl of the steroid is also conjugated,
probably with glucuronic acid. The glucoside was found to be remarkably
stable to β-glucuronidase, since it was cleaved to the extent of only 4%
when incubated with Ketodase for 24 hours. By comparison, about 25%
of a similar sample of estradiol-17α-N-acetylglucosaminide is cleaved by
Ketodase under these conditions (Layne *et al.,* 1964).

Many species of plants and insects form β-glucosides (Smith, 1964),
and this type of conjugation occurs also in bacteria (Tabone and Tabone,
1956) and in mollusks (Dutton, 1966). The only nucleotide so far implicated as the source of the glucose for these reactions is UDP-glucose.
In recent unpublished experiments in the author's laboratory, attempts
were made, using *in vitro* incubations with liver microsomes, to effect the
transfer of glucose to 17α-estradiol-3-glucosiduronate from ADP-glucose,
CDP-glucose, GDP-glucose, and TDP-glucose. No transfer from any of
these nucleotides could be demonstrated under conditions which gave
effective transfer from UDP-glucose. The UDP-nucleotide thus appears
to be a specific substrate for the steroid glucosyl transferase of rabbit

tissues, and in this respect the enzyme is similar, therefore, to those of plants and lower animals. Further exploration of the characteristics and specificity of the transferase has not yet been carried out, and its possible relationship to those involved in other metabolic transfers of glucose, such as glycogen synthesis, is an intriguing possibility.

IV. Conjugations Other than Glycoside or Sulfate

A. ACETYLATION

Steroid acetates have been used in synthetic work and in therapy for many years, and their hydrolysis by several enzymes, particularly cholinesterase, has been reported (Billiar and Eik-Nes, 1965). The first indication that metabolic acetylation of steroids might occur in vivo was the relatively recent evidence (Weichselbaum and Margraf, 1960; Margraf et al., 1963) that corticosteroid acetates are present in normal human blood. Evidence for steroid acetylation by animal tissues in vitro has been provided by King et al. (1964), who reported that testosterone was acetylated in small yield by incubation with tissue from the mammary gland of pregnant rats, and from mammary tumors induced in rats by dimethylbenzanthracene. Similar results were obtained with tissue from the mammary gland of pregnant mice, and from spontaneous mammary tumors in mice. The testosterone acetate formed was rigorously characterized by comparisons with authentic material, so that the site of acetylation must be the 17β-hydroxyl group. No attempt was made to identify the acetylating agent in the reaction. Schubert and Wehrberger (1965), Young Lai and Solomon (1967), and Nambara and Numazawa (1969) have all reported the isolation from urine of steroid acetates which they do not consider to have been formed artifactually.

Significant quantities of the C-21 acetates of cortisol and cortisone are formed when cortisol is incubated with minced brain tissue from 3-day-old rats (Grosser and Axelrod, 1967), and from fetal and newborn baboons (Grosser and Axelrod, 1968). Considerably less acetylation was observed in adult baboon brain, and the authors speculate that this might be due, not to a decrease in acetylating capacity, but to increase of esterase activity as the brain matures. The soluble corticosteroid-21-O-acetyltransferase from the young baboon brain has been partially purified by Purdy and Axelrod (1968a). A radiochemical assay of the enzyme has been developed which employs incubation in the presence of acetyl-1-^{14}C CoA, cortisol, and physostigmine, with the addition of cortisol acetate-

1-³H at the end of a 1-hour incubation to measure the recovery of the cortisol acetate-1-¹⁴C product in a benzene extract (Purdy and Axelrod, 1968b). The dialyzed enzyme, which is widely distributed in separate anatomical areas of the brain (Purdy *et al.*, 1968) except for the pituitary gland (Purdy, personal communication) is only active in the presence of acyl-thiol ester. The enzyme is stable at $-20°C$, and appears, on the basis of its intercellular distribution and its behavior in the presence of thio-glycollate and cyanide, to be distinct from choline acetyltransferase.

It is not possible on the basis of present evidence to ascribe a specific physiological role to corticosteroid acetates in the blood and in brain. The relative ability of corticosteroids and corticosteroid acetates to pass the blood–brain barrier would undoubtedly be dependent upon their relative binding by plasma proteins and their binding interactions with macromolecular components in the brain. The acetylation of free corticosteroids reaching the brain may provide a means of both altering their effects upon the central nervous system and changing the dynamics of total corticosteroid clearance rates in nervous tissue. In this respect it is interesting to note that aldosterone and the potent synthetic corticoid, dexamethasone, are both very poor substrates of brain corticosteroid-21-*O*-acetyltransferase (Purdy, personal communication).

The acetylation of steroid hydroxyl groups is of mechanistic interest, since, with the notable exception of choline and serine, *in vivo* acetylation in animals usually involves amino groups. Purdy and Axelrod (1968b) have provided information which is consistent with the initial formation of an acyl-thiol enzyme intermediate in the corticosteroid-21-*O*-acetyltransferase reaction. At pH 7.4 where the reaction is not appreciably reversible, and in the absence of added acetyl CoA, the incubation of unlabeled cortisol acetate and cortisol-4-¹⁴C with an ammonium sulfate-precipitated and dialyzed enzyme preparation results in the formation of ¹⁴C-labeled cortisol acetate. This product formation is markedly inhibited by Ellman's thiol reagent, 5,5′-dithio-bis(2-nitrobenzoic acid), and does not occur unless a steroid acetate is used. The steroid substrate specificity for this group transfer reaction requires the presence of an α-ketol side chain (Purdy, personal communication).

B. PHOSPHORYLATION

No endogenously produced steroid phosphate ester has been isolated from, or unequivocally characterized in animal tissues or excreta. However, Oertel and Eik-Nes (1958) found that dehydroepiandrosterone was liberated from human blood plasma following treatment with commercial acid and alkaline phosphatases. In further experiments, these workers (Oertel and Eik-Nes, 1959) obtained from human blood plasma complexes which contained dehydroepiandrosterone and phosphate in lipid containing conjugates of 17-ketosteroids. Synthetic steroid phosphates are used therapeutically and appear to be readily hydrolyzed *in vivo* to the parent steroids. Furthermore, it has been shown (Scardi *et al.*, 1962) that estrogen phosphates are more effective than the sulfates as inhibitors of the reconstitution of kynurenine-aminotransferase and aspartate aminotransferase obtained from rat kidney and pig heart, respectively.

Because of the above, there has been persistent speculation that steroid phosphates may be formed in animals *in vivo* and may have a specific physiological role. This speculation has received support from recent work (Botte and Koide, 1968a,b) which indicates that steroid phosphates are hydrolyzed with marked specificity by human plasma acid and alkaline phosphatases. Of particular interest is the observation (Di Pietro, 1968) that human placental acid phosphatase III shows marked selectivity for estradiol-3-phosphate over several nonsteroid and steroid phosphates, including estradiol-17β-phosphate. It should, however, be noted that Purdy *et al.* (1961), during the characterization of estrone sulfate from human plasma, searched for estrone phosphate but did not detect it. On balance, it appears possible that a steroid phosphokinase, or more than one such enzyme, may be found to function in the human and other animals, but concrete evidence for the presence of such an enzyme or its products is lacking.

C. METHYLATION

The isolation and characterization of 2-methoxyestrone (Kraychy and Gallagher, 1957) were among the first indications that methylation of oxygen might have significance in the animal organism. Methoxy derivatives of 17β-estradiol (Frandsen, 1959; Axelrod and Goldzieher, 1961) and estriol (Fishman and Gallagher, 1958) are formed in man, while administered 17α-estradiol is partially converted to the 2-methoxy derivative (Williams and Layne, 1967). The formation of 2-methoxyestrogens proceeds by way of the corresponding catechols, or 2-hydroxyestrogens and, like the methylation of catecholamines, the reaction is mediated by an oxygen-methyltransferase and requires S-adenosylmethionine as the methyl group donor (Breuer *et al.*, 1961; King, 1961b). An enzyme preparation from rat liver, purified as described by Axelrod and Tomchick (1958), methylated 2-hydroxyestradiol-17β at a rate somewhat less than that at which it methylated adrenaline (Breuer *et al.*, 1962). The steroid-

O-methyltransferase must, therefore, be very similar to, if not identical with, catechol-*O*-methyltransferase. In further work, Knuppen and Breuer (1966) found that methylation of 2-hydroxyestrogens can also be effected at the 3 position by rat liver preparations to yield 2-hydroxy-3-methoxy-derivatives. Small amounts of the 2,3-dimethoxy derivative of estrone

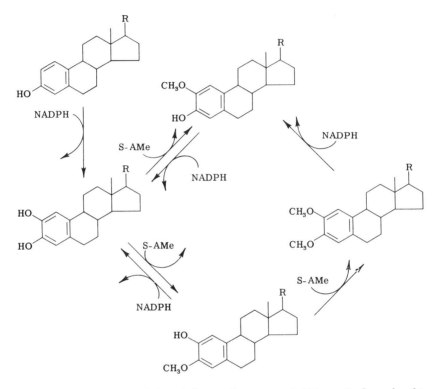

Fɪɢ. 6.   Methylation and demethylation of estrogens. S-AMe = S-adenosylmethionine. (After Knuppen and Breuer, 1966.)

were also detected. The biological orthomethylation of 2-hydroxyestrogens has also been demonstrated in tissues of the human female genital tract (Lucis, 1965).

Breuer *et al.* (1964) have shown that an NADPH-dependent enzyme in rat liver microsomes can demethylate methoxyestrogens *in vitro*, and have obtained evidence that this reaction also takes place in the human *in vivo*. On the basis of these results, they suggest that a cyclic interconversion of 2-hydroxy- and 2-methoxyestrogens takes place under physiological conditions as shown in Fig. 6, and that this cycle may have a regulatory function in the intermediary metabolism of estrogens. In this connection it should be noted, however, that both Knuppen and Breuer (1966) and Fishman *et al.* (1967) have found that, *in vivo*, no detectable 3-methylated estrogens appear in the urine after administration of estradiol. This might be due to effective demethylation of the 3-position (Fig. 6), or, as suggested by Fishman *et al.* (1967), to competition by other groups, such as sulfate, for this hydroxyl *in vivo*.

From the comparative viewpoint, two interesting observations should be mentioned. Collins *et al.* (1967a) found that when estrone or 17β-estradiol was administered to hamsters a large percentage of the dose was excreted in the urine as 2-hydroxyestrone, while comparatively little 2-methoxyestrone was excreted. It is interesting to speculate as to whether these results might be due to a low level of steroid-O-methyltransferase in this animal, or to a very effective demethylation of the methoxy-estrogens *in vivo* (see Knuppen and Breuer, 1966). In the rabbit, neither radioactive 2-hydroxy nor 2-methoxy derivatives of radioactive estrone, 17β-estradiol, or estriol have been detected in the urine in many experiments (Layne *et al.*, 1964; Collins *et al.*, 1967b; Collins and Layne, 1968) involving the administration of these compounds. This is a curious finding in view of the fact that King (1961a) used rabbit liver to obtain 2-hydroxyestriol from estriol *in vitro*. If metabolism of these estrogens at position 2 does take place *in vivo* in the rabbit, the metabolites must be entirely excreted by routes other than the urine, but it is pertinent that 17β-methyl-17α-estradiol, in which compound metabolism and conjugation in ring D may be hindered, is metabolized *in vivo* by the rabbit to the 2-hydroxy derivative (K. I. H. Williams, personal communication).

D. GLUTATHIONE CONJUGATION

Evidence has been found repeatedly for the formation by the rat of metabolites of the steroidal estrogens which are water-soluble and which are neither glucosiduronates nor sulfates (Ryan and Engel, 1953; Valcourt *et al.*, 1955; Wotiz *et al.*, 1958). Jellinck and his collaborators (1967) have investigated the products of the incubation of estradiol-^{14}C with sub-cellular fractions from rat liver, and have obtained evidence for the formation of an estrogen–glutathione conjugate. Kuss (1967) confirmed this finding, and in further work (Kuss, 1968, 1969) identified the compounds formed by incubation of estradiol-4-^{14}C with rat liver homogenates as 2-hydroxyestrone and 2-hydroxyestradiol, each joined by a thioether linkage at either the C-4 or C-1 of the steroid to the cysteine moiety of glutathione (Fig. 7). This confirms previous indications (Lazier and Jellinck, 1965; Marks and Hecker, 1968) that formation of the water-soluble conjugates took place after 2-hydroxylation of the molecule.

Mechanisms of the metabolism of estrone in rat liver are discussed by Marks and Hecker (1969).

Liver microsomes from male rats show a greater ability to form the water-soluble conjugates than do those of females. This difference was apparent at maturity, and could be influenced by castration or by treatment with gonadal hormones (Jellinck and Lucieer, 1965).

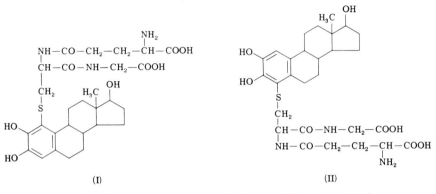

(I)   (II)

FIG. 7.   Glutathione conjugates of 17β-estradiol. (Kuss, 1969.)

The formation of glutathione conjugates of foreign compounds in the liver is followed by removal of the glutamic acid and glycine moieties and subsequent acetylation of the cysteine residue of the glutathione (see Parke, 1968). The resulting mercapturic acids are excreted in the urine. The formation of glutathione conjugates of the estrogens in rat liver makes it seem probable that the unidentified water-soluble urinary products of the steroidal estrogens in this species may be mercapturic acids, and the results indicate the employment of this mechanism in the metabolism of a naturally occurring, as distinct from an administered, foreign compound. It remains to be seen whether the reaction will be found to occur in animals other than the rat. It is of interest that, when radioactive estrone or estradiol is injected into the Mongolian gerbil, the urinary excretion products consist of highly polar materials which do not appear to be glucosiduronates or sulfates (Collins, Williams, and Layne, unpublished).

V. Conjugation with Glucuronic Acid

Among the very large number of alcoholic and phenolic glucosiduronates which have been found in animals, those involving a steroid aglycone constitute a significant and fairly representative group. Glucuronic acid may be attached to any of a number of positions on a variety of steroids. These have been documented recently (Jayle and Pasqualini, 1966). Only those steroid glucosiduronates which have special interest as novel metabolic conjugates will be considered here.

A. GLUCURONIDATION AT SPECIFIC SITES ON THE STEROID MOLECULE

Considerable evidence has been obtained on the specificity of steroid substrates in the transfer of glucuronic acid from UDP-glucuronic acid. Recently it has been shown (Dahm *et al.*, 1966) that the microsomal fraction of human intestine contains a UDP-glucuronyltransferase which catalyzes the formation of testosterone 17β-glucosiduronate. The fraction conjugates phenolic steroids and ring A saturated steroids to a lesser degree, but has no activity toward Δ⁵-3β-hydroxysteroids. Dahm and Breuer (1966a,b) have found that there is a different intracellular distribution in human intestine of the enzymic activities which effect the glucuronidation of estriol at the 16α- or at the 17β-positions. This indicates strongly that different enzymes may be responsible for the glucuronidation of estriol at these two positions, and that yet another enzyme may effect the glucuronidation of the phenolic hydroxyl at position 3 (Slaunwhite *et al.*, 1964; Dahm and Breuer, 1966b).

A glucosiduronate of aldosterone has been characterized (Underwood and Tait, 1964; Pasqualini, 1965) and shown to be of a novel type, both in regard to its chemical properties and to the site of attachment of the sugar to the steroid. This attachment is to the hydroxyl present on the angular 18 carbon atom of aldosterone, and the linkage appears to be glycosidic. The conjugate is stable to both bacterial and hepatic β-glucuronidases, but is cleaved by an enzyme from the digestive tract of *Helix pomatia*, and this cleavage is inhibited by saccharodilactone (Pasqualini, 1965). This glucosiduronate has the unique property of being readily hydrolyzed at room temperature at pH 1.0, and long before characterization of the conjugate, this property became the basis for most of the methods presently in use for the estimation of aldosterone in urine and of aldosterone secretion rates in various clinical states (Tait and Tait, 1962).

B. ENOL-GLUCOSIDURONATES

Schubert (1958) observed that androstene-3,17-dione is a product of β-glucuronidase hydrolysis of material isolated from the urine of individuals injected with testosterone. In order for this steroid to form a glucosiduronate, it must have been conjugated through an enolized ketone at position 3. Wotiz and Fishman (1963) obtained good evidence for the presence of 11β-hydroxy-17-oxoandrosta-3,5-dien-3-yl-β-D-glucopyranosiduronic acid in the urine of rats after the administration of adrenosterone. These results established the existence of enolglucosiduronates as *in vivo* metabolites of steroids. Wotiz (1962) and Wotiz and Fishman (1963) incubated synthetic steroid enol-glucosiduronates with rat liver tissues, and showed that these conjugates may undergo further metabolic change without obligatory cleavage. Recently, Harkness *et al.* (1969) have obtained evidence for the presence of small amounts of progesterone in the form of an unstable enol-glucosiduronate in the urine of women to whom progesterone had been administered.

C. Double Conjugates Involving Glucuronic Acid

Small amounts of diglucosiduronates of 17β-estradiol (Breuer and Wessendorf, 1966) and of estriol and 17-epiestriol (Collins *et al.*, 1968) have been reported to be formed by rabbit liver *in vitro*, while Felger and Katzman (1961) tentatively identified a diglucosiduronate of estriol in commercial preparations of estriol glucosiduronate from pregnancy urine. Kirdani *et al.* (1968) incubated estriol and UDP-glucuronic acid with guinea pig liver homogenate, and then incubated the product with mouse liver homogenate, again in the presence of UDP-glucuronic acid. In this way they obtained estriol-3,16-diglucosiduronate.

Several steroid double conjugates have been identified in which glucuronic acid is attached to one position on the steroid, with a different conjugating group at another position. Of these, the double glycosides of 17α-estradiol and of estriol epimers, in which the other group is N-acetylglucosamine, have already been discussed in Section II. In addition Straw *et al.* (1955) found evidence from sequential enzyme hydrolyses that about 5% of the estriol excreted in the urine in late pregnancy was in the form of a double conjugate, probably the 3-sulfate, 16(17)-glucosiduronate. Other investigators, cited by Levitz *et al.* (1965), have found similar evidence for the presence of estriol-sulfoglucosiduronate in human body fluids, among them the urine of the newborn, and maternal blood, as well as in bile following the administration of labeled estrone sulfate. The fact that a large proportion of the circulating estriol in late pregnancy is in the form of this sulfoglucosiduronate (Touchstone, 1965) suggests that the double conjugation may have metabolic significance.

In all the above cases, the formation of these double conjugates appears, on the basis of *in vitro* evidence, to take place in a sequence which is not reversible. Levitz *et al.* (1965) found that estriol-sulfoglucosiduronate could be biosynthesized by incubating estriol with human liver homogenate in the presence of UDP-glucosiduronic acid, and then incubating the resultant estriol-16-glucosiduronate with guinea pig liver homogenate in the presence of ATP. If estriol-3-sulfate was incubated with liver preparations in the presence of UDP-glucosiduronic acid, *no* formation of the double conjugate was observed. This is analogous to the finding (Jirku and Layne, 1965; Collins *et al.*, 1968) that rabbit liver preparations can transfer N-acetylglucosamine to the 17α-group of estradiol only after the prior transfer of glucosiduronic acid to position 3. These results indicate that, in the formation of these double conjugates, one of the enzymes may be specific for a preformed monoconjugate, rather than for the steroid molecule itself. In the case of the diglucosiduronate biosynthesized by Kirdani *et al.* (1968), preformed estriol-3-glucosiduronate was glucuronidated at position 16 by a mouse liver preparation which also glucuronidates estriol at this position. However, no estriol diglucosiduronate was formed when the 16-glucosiduronate was incubated with a guinea pig liver homogenate which readily glucuronidates estriol at position 3.

VI. Conjugation with Sulfuric Acid

A. BIOSYNTHESIS AND METABOLISM OF STEROID SULFATES

Sulfate conjugation is a widely distributed biochemical process in mammalian and other species (Williams, 1967). Until recently, it was assumed that the liver was the major site of formation of steroid sulfates, as it is for most other conjugation processes. In 1962, Baulieu found a relatively high content of dehydroepiandrosterone sulfate in adrenal vein blood from a patient with an adrenal tumor, and the work of several groups then established that human adrenal tissue *in vitro* could effect the sulfation of dehydroepiandrosterone, and that the sulfate was secreted by the adrenal. This work has been reviewed by Vande Wiele *et al.* (1963). The synthesis *in vitro* by bovine adrenal tissue of estrone sulfate was shown by Sneddon and Marrian (1963) and the work of Wallace and Silberman (1964) established that another endocrine tissue, namely the ovary, could synthesize steroid 3-monosulfates. Subsequently, several other endocrine tissues have been shown to possess this ability, notably the testis, placenta, and corpus luteum (Payne and Mason, 1965). A tabulation of steroids sulfated by various animal tissues has been provided by Wang and Bulbrook (1968).

The fact that steroid sulfates are synthesized in active endocrine tissue is in contrast to the situation, as far as it is presently known, with other steroid conjugates, and strongly supports the hypothesis that the steroid sulfates have a specific metabolic role other than the provision of an ionizable form of the steroid suitable for clearance by the kidney. The presence of steroid sulfatases has been demonstrated in reproductive tissues as well as the liver in the human, and the placenta is particularly rich in these enzymes (Warren and French, 1965). The possibility exists, therefore, that a dynamic relationship between free steroid and steroid sulfate in peripheral circulation may be maintained by means of the steroid sulfatases.

In the past 5 years it has been firmly established that cholesterol sulfate occurs *in vivo* and can serve as an intermediate in the biosynthesis of steroid sulfates. These compounds, whether formed by this route or by direct sulfation, undergo extensive metabolism without prior removal of the sulfate group. These findings have been reviewed (Döllefeld and Breuer, 1966; Wang and Bulbrook, 1968) and are discussed in another contribution to this treatise (Baulieu, 1969).

B. Double Conjugates Involving Sulfate

A number of steroid disulfates have been prepared synthetically and these are cataloged by Bernstein *et al.* (1968), but only a few have been isolated from natural sources. Among these are 5-pregnene-3β, 20α-diol-3, 20-disulfate (Arcos and Lieberman, 1967), which was isolated from human urine and fully characterized. The presence of other steroid disulfates has been established both in urine (Pasqualini and Jayle, 1962; Baulieu and Corpechot, 1965) and in bile (Laatikainen *et al.*, 1968). Wengle and Böstrom (1963) found evidence that disulfation of androst-5-en-3β,17β-diol is accomplished by human liver *in vitro* in two steps, the 3β-sulfate being formed first, followed by sulfurylation of the 17β-hydroxy group.

The detection of estriol-3-sulfate-16-glucosiduronate in body fluids and the sequential synthesis of this compound by liver preparations *in vitro* has been discussed in Section V, C. Four other steroid sulfoglycosides which have recently been identified involve the novel conjugation mechanism by which N-acetylglucosamine is added to steroids. These are dealt with in Section II and appear in Table I.

C. Comparative Aspects of Sulfate Conjugation

Some recent data on the formation of sulfates of the phenolic steroids by the hen are of considerable comparative significance. Raud and

Hobkirk (1968) found that cell-free preparations of the liver, oviduct, and vagina of laying hens, in the presence of ATP, readily formed the 3-monosulfates of estrone and 17β-estradiol. In a study of the type of urinary conjugates excreted by the laying hen after the administration of 17β-estradiol-4-^{14}C, Mathur et al. (1969) observed that most, if not all, of the excreted radioactivity was in the form of sulfate conjugates. Monosulfates of estrone, 17α-estradiol, 17β-estradiol, estriol, and 16-epiestriol were detected, and all these steroids except estrone were present in the disulfate fraction as well. These results indicate that sulfation is the predominant, and perhaps the only, conjugation mechanism employed by the hen in the excretion of phenolic steroids, despite the fact that hen tissues in vitro are known to form glucosiduronates, for instance that of o-aminophenol (Dutton and Ko, 1966). The hen is unique, therefore, among the species in which estrogen conjugation has been studied in any detail, and provides an interesting contrast to the rabbit, in which animal the in vitro synthesis of estrogen sulfate takes place in liver (Nose and Lipmann, 1958), although no radioactive sulfate conjugates have been detected in the urine in repeated experiments involving the injection of radioactive estrogens (Layne et al., 1964; Collins et al., 1967b; Collins and Layne, 1968).

Williams (1967) has reviewed the zoological distribution of sulfate conjugation, and has pointed out that its distribution throughout large segments of the phylogenetic tree makes it probable that it may be the most primitive of the detoxication mechanisms. An interesting addition to knowledge in this regard is the finding by Creange and Szego (1967) that sulfation of estradiol takes place in the gut of the echinoderm Strongylocentrotus franciscanus. It should also be noted that the chief bile salt in hagfishes, which are rather primitive chordates, is the disulfate of the bile alcohol myxinol (Haslewood, 1965), while Hutchins and Kaplanis (1969) have recently isolated sterol sulfates from an insect, namely the tobacco hornworm Manduca sexta.

VII. Miscellaneous Steroid Conjugates or Complexes

A. CONJUGATION WITH AMINO ACIDS

In addition to glycocholic and taurocholic acids, which are the well-known conjugates of cholic acid with glycine and taurine, respectively, other common bile acids form conjugates with these amino acids by means of a peptide link (Haslewood and Wooton, 1950). Since this form of conjugation requires the presence of a carboxylic acid group on the steroid, it is unlikely that it would be of frequent occurrence among steroids other than the bile acids. However, preliminary reports which claim the possible existence of amino acid complexes with adrenal steroids have appeared. Thus, Eades *et al.* (1954) found substances in normal male urine, which, on the basis of tentative colorimetric and chromatographic evidence, they suggested were such complexes. Similar substances were reported by Hudson and Lombardo (1955) to be present on paper chromatograms of extracts of adrenal glands and of adrenal vein blood, while Voigt and Schroeder (1955) recorded evidence for nitrogen-containing, ninhydrin-positive steroid complexes in aqueous adrenal cortical extracts. However, none of these reports has been followed by definitive work, and they must be discounted as strong evidence for steroid amino acid conjugates.

A negative experiment has recently been reported by Mathur (1969) who injected estrone-^{3}H and ornithine-^{14}C simultaneously into a domestic hen in an attempt to see whether the bird might excrete phenolic steroids in conjugation with ornithine. In the urine, the tritium and ^{14}C could be completely separated from each other chromatographically. This indicates that, under these conditions, no labeled ornithine conjugates of metabolites of estrone were excreted in the urine. However, the author points out that this experiment does not obviate the presence of steroid–ornithine conjugates if these were formed from an active form of ornithine which did not derive from the injected ^{14}C-labeled amino acid. However, since Mathur *et al.* (1969) have shown that phenolic steroids in the hen are excreted in the urine largely as sulfates, the possibility of significant amounts of other types of conjugates seems fairly remote, unless these were excreted exclusively in the bile, which was not examined in these experiments.

B. PRELIMINARY REPORTS OF NOVEL CONJUGATES

In addition to the suggested complexes of steroids with amino acids, reports of other nitrogen-containing substances which possess steroidlike properties, or which can be split to yield steroids, are most persistent. However, with the exception of the N-acetylglucosamine conjugates discussed earlier (Section II), and the urea complex described below (Section VII, C), only preliminary evidence, and this in some cases extremely tentative, has been presented to substantiate the identifications. One of the earliest such reports is that of Andrew and Fenger (1936), who described a compound of probable empirical formula $C_{20}H_{42}O_2N$ which they obtained from ovarian concentrates, presumably bovine, and which they claimed possessed high estrogenic properties. More recently,

Nishizawa and Eik-Nes (1963) observed what appeared to be a complex of cortisone with caffeine in the urine of four out of eleven subjects infused with cortisone-^{14}C. Kornel and Fujiki (1966) have suggested, in a nonarchival publication, that corticosteroid nucleosides occur in human urine. These results, however, await confirmation, and speculation on their significance would be premature.

A conjugate of testosterone in the adrenal venous blood of the brush-tailed possum (*Trichosurus vulpecula* Kerr) has been described by Chester Jones *et al.* (1964). The conjugating group is removed by boiling with acid, but its nature is unknown, except that it appears to be phenolic.

C. UREASTERONE

During an investigation of the metabolism of 11β-hydroxy-Δ^4-andro-stene-3,17-dione in man, Noguchi and Fukushima (1966) isolated a highly polar metabolite from the urine, which was shown to be a compound of the administered steroid with urea, and was called ureasterone. Complete chemical and spectroscopic identification established the identity of the isolated material as 3α-ureido-11β-hydroxy-Δ^4-androsten-17-one, in which urea is attached by means of one of its amino groups to the 3 position of the steroid. Fukushima *et al.* (1966) showed that ureasterone could be formed nonenzymically by incubation of urea with the allylic alcohol 3α,11β-dihydroxy-Δ^4-androsten-17-one and concluded that the compound was formed as an artifact when the urine was incubated at pH 5 to cleave the glucosiduronate of this alcohol, which is a normal metabolite of 11β-hydroxy-Δ^4-androstene-3,17-dione. This investigation, while it indicated the artifactual nature of a material which at first seemed to be of great physiological interest, is an excellent example of the careful characterization of an isolated material and proper investigation of its origin.

D. "SULFATIDYL" CONJUGATES

In addition to the possible phosphate-containing steroid conjugates described in Section IV, B, Oertel and his co-workers have described fractions located in blood plasma by solvent partition and chromatography which are lipophilic, and which they suggest represent compounds of ketosteroids with glycerides, or "sulfatidyl" conjugates. They have further suggested that these conjugates could be transport forms for steroid sulfates. This work has been reviewed in part by Oertel (1964); Bernstein *et al.* (1966) provides a further source of reference. No such conjugate has been rigorously characterized, however, so that the type of association which might exist between steroids or steroid sulfates and glyceride remains speculative.

VIII. Conclusion

In 1962 evidence began to appear (see Section VI) that steroid sulfates were secreted by endocrine tissue and metabolized without removal of the sulfate group. This caused the precipitate abandonment of the feeling commonly held, and often curtly stated, by steroid biochemists that all steroid conjugates were metabolic end products formed solely for purposes of excretion. The interest awakened by the work on sulfates has been strengthened by the isolation of some of the newer conjugates described above.

It seems appropriate to point out that, although some of these conjugates may be metabolic intermediates, a large number probably *are* formed for excretion by way of the urine or bile. It is also pertinent to emphasize that the reactions which enable the body to excrete the steroids, far from lacking biological interest, are of critical importance, not only with reference to the elimination of the metabolites of endogenously produced steroids, but also in the context of the use of exogenous steroids for various forms of therapy, including their increasing application as contraceptives.

Wang and Bulbrook (1968) have remarked that the important recent work on steroid sulfates has not so far resulted in the delineation of any unified concept of their physiological role. With regard to the other new conjugates, it is likely that the interest of some, as for instance the glutathione conjugates of estrogens in the rat, rests in the demonstration that steroids may serve as substrates for conjugations hitherto well known with other compounds. In the case of the steroid N-acetylglucosaminides, the peculiar specificity shown for this sugar by certain groups on the steroid molecule, and the sequential addition of the conjugating sugars to the steroid to form the double conjugate shown in Fig. 1, make it seem possible that the molecule has a specific role *in vivo*. The possible similarity between the enzymic processes used in forming the steroid N-acetylglucosaminides and those involved in mucopolysaccharide metabolism has been pointed out in Section II. In the case of the steroid glucoside described in Section III, the possible similarity of the enzyme effecting this glucosylation with those involved in other metabolic glucosyl transfers has been suggested. In the absence of experimental evidence, these possibilities cannot be discussed further without raising the problem of teleological argument in a peculiarly acute form. For this reason, the present chapter has not attempted to be very much other than descriptive. Since, however, the present author's interests have been in the newer steroid glycosides, a few brief concluding remarks on future work in this area may, perhaps, be permissible.

First, the separation and purification of the microsomal steroid glycosyl transferases is essential for work on their specificity. In the rabbit, for which animal most data are presently available, there are at least three such enzyme activities, involved, respectively, in the transfer of glucuronic acid, N-acetylglucosamine, and glucose to the steroid molecule. Their solubilization and separation could be a problem of some magnitude.

Second, further work on the possible *in vivo* significance of the steroid glycosides at the molecular level requires the use of doubly labeled material, as for instance glycosides in which the steroid molecule is labeled with tritium and the sugar with ^{14}C. Such doubly labeled compounds should facilitate work on the possible role of the glycosides in transport of steroids and the dynamics of their metabolism. The *in vivo* relationship between the steroid glycosides and the glycosidases in tissues might be effectively studied with doubly labeled compounds.

ACKNOWLEDGMENTS

The author's research has been supported initially by grant AM-10216 from the U.S. Public Health Service, and presently by grant MA-3287 from the Medical Research Council of Canada. Dr. H. Breuer, Dr. R. H. Common, Dr. P. H. Jellinck, and Dr. R. H. Purdy read and made helpful criticisms of parts of the draft. The author owes much to a continued exchange of ideas with Dr. K. I. H. Williams and Dr. D. C. Collins.

REFERENCES

Abdel-Aziz, M. T., and Williams, K. I. H. (1969). *Steroids* 13, 809.
Andrew, R. H., and Fenger, F. (1936). *Endocrinology* 20, 563.
Arcos, M., and Lieberman, S. (1967). *Biochemistry* 6, 2032.
Axelrod, J., and Tomchick, R. (1958). *J. Biol. Chem.* 233, 702.
Axelrod, L. R., and Goldzieher, J. W. (1961). *J. Clin. Endocrinol. Metab.* 21, 211.
Baulieu, E. E. (1962). *J. Clin. Endocrinol. Metab.* 22, 501.
Baulieu, E. E., and Corpechot, C. (1965). *Bull. Soc. Chim. Biol.* 47, 443.
Bernstein, S., Cantrall, E. W., Dusza, J. P., and Joseph, J. P. (1966). "Steroid Conjugates," Chem. Abstr. Serv., Am. Chem. Soc., Washington, D. C.
Bernstein, S., Dusza, J. P., and Joseph, J. P. (1968). "Physical Properties of Steroid Conjugates," Springer Verlag, New York.
Billiar, R. B., and Eik-Nes, K. B. (1965). *Anal. Biochem.* 13, 11.
Botte, V., and Koide, S. S. (1968a) *Biochim. Biophys. Acta* 152, 396.

Donald S. Layne

Botte, V., and Koide, S. S. (1968b). *Endocrinology* **82**, 1062.

Breuer, H., and Wessendorf, D. (1966). *Z. Physiol. Chem.* **345**, 1.

Breuer, H., Pangels, G., and Knuppen, R. (1961). *J. Clin. Endocrinol. Metab.* **21**, 1331.

Breuer, H., Vogel, W., and Knuppen, R. (1962). *Z. Physiol. Chem.* **327**, 217.

Breuer, H., Knuppen, M., Gross, D., and Mittermayer, C. (1964). *Acta Endocrinol.* **46**, 361.

Chester Jones, I., Vinson, G. P., Jarrett, I. G., and Sharman, G. B. (1964). *J. Endocrinol.* **30**, 149.

Collins, D. C., and Layne, D. S. (1968). *Can. J. Biochem.* **46**, 1089.

Collins, D. C., and Layne, D. S. (1969). *Steroids* **13**, 783.

Collins, D. C., Williams, K. I. H., and Layne, D. S. (1967a). *Endocrinology* **80**, 893.

Collins, D. C., Williams, K. I. H., and Layne, D. S. (1967b). *Arch. Biochem. Biophys.* **121**, 609.

Collins, D. C., Williams, K. I. H., and Layne, D. S. (1966). Unpublished.

Collins, D. C., Jirku, H., and Layne, D. S. (1968). *J. Biol. Chem.* **243**, 2928.

Conchie, J., Findlay, J., and Levvy, G. A. (1959). *Biochem. J.* **71**, 318.

Creange, J. E., and Szego, C. M. (1967). *Biochem. J.* **102**, 898.

Dahm, K., and Breuer, H. (1966a). *Acta Endocrinol.* **52**, 43.

Dahm, K., and Breuer, H. (1966b). *Z. Klin. Chem.* **4**, 153.

Dahm, K., Breuer, H., and Lindlau, M. (1966). *Z. Physiol. Chem.* **345**, 139.

Di Pietro, D. L. (1968). *J. Biol. Chem.* **243**, 1303.

Döllefeld, E., and Breuer, H. (1966). *Z. Vitamin-, Hormon- Fermentforsch.* **14**, 193.

Dutton, G. J. (1966). *Arch. Biochem. Biophys.* **116**, 399.

Dutton, G. J., and Ko, V. (1966). *Biochem. J.* **99**, 550.

Eades, H. C., Jr., Pollack, R. L., and King, S. J., Jr. (1954). *Federation Proc.* **13**, 201.

Felger, C. B., and Katzman, P. A. (1961). *Federation Proc.* **20**, 199.

Fishman, J., and Gallagher, T. F. (1958). *Arch. Biochem. Biophys.* **77**, 511.

Fishman, J., Miyazaki, M., and Yoshizawa, I. (1967). *J. Am. Chem. Soc.* **89**, 7147.

Frandsen, V. A. (1959). *Acta Endocrinol.* **31**, 603.

Fukushima, D. K., Noguchi, S., Bradlow, H. L., Zumoff, B., Nozuma, K., Hellman, L., and Gallagher, T. F. (1966). *J. Biol. Chem.* **241**, 5336.

Grosser, B. I., and Axelrod, L. R. (1967). *Steroids* **9**, 229.

Grosser, B. I., and Axelrod, L. R. (1968). *Steroids* **11**, 827.

Harkness, R. A., Davidson, D. W., and Strong, J. A. (1969). *Acta Endocrinol.* **60**, 221.

Haslewood, G. A. D. (1965). *Gastroenterology* **49**, 6.

Haslewood, G. A. D., and Wooton, V. (1950). *Biochem. J.* **47**, 584.

Himaya, A., Collins, D. C., Williamson, D. G., and Layne, D. S. (1969). *Biochem. J.* **113**, 445.

Hudson, P. B., and Lombardo, M. E. (1955). *J. Clin. Endocrinol.* **15**, 324.

Hutchins, R. F. N., and Kaplanis, J. N. (1969). *Steroids* **13**, 605.

Jayle, M. F., and Pasqualini, J. R. (1966). *In* "Glucuronic Acid: Free and Combined. Chemistry, Biochemistry, Pharmacology, and Medicine" (G. J. Dutton, ed.), pt. III, pp. 507–543. Academic Press, New York.

Jellinck, P. H., and Lucieer, I. (1965). *J. Endocrinol.* **32**, 91.

Jellinck, P. H., Lewis, J., and Boston, F. (1967). *Steroids* **10**, 329.

Jirku, H., and Layne, D. S. (1965). *Biochemistry* **4**, 2126.

Jirku, H., and Levitz, M. (1969). *J. Clin. Endocrinol. Metab.* **29**, 615.

King, R. J. B. (1961a). *Biochem. J.* **79**, 355.

King, R. J. B. (1961b). *Biochem. J.* **79**, 361.

King, R. J. B., Gordon, J., and Smith, J. A. (1964). *J. Endocrinol.* **28**, 345.

Kirdani, R. Y., Slaunwhite, W. R., Jr., and Sandberg, A. A. (1968). *Steroids* **12**, 171.

Knuppen, R., and Breuer, H. (1966). *Z. Physiol. Chem.* **346**, 114.

Kornel, L., and Fujiki, T. (1966). *Program 48th Meeting Endocrine Soc., Chicago* p. 46.

Kraychy, S., and Gallagher, T. F. (1957). *J. Biol. Chem.* **229**, 519.

Kuss, E. (1967). *Z. Physiol. Chem.* **348**, 1707.

Kuss, E. (1968). *Z. Physiol. Chem.* **349**, 1234.

Kuss, E. (1969). *Z. Physiol. Chem.* **350**, 95.

Laatikainen, T., Peltokallio, P., and Vihko, R. (1968). *Steroids* **12**, 407.

Layne, D. S. (1965). *Endocrinology* **76**, 600.

Layne, D. S., Sheth, N. A., and Kirdani, R. Y. (1964). *J. Biol. Chem.* **239**, 3221.

Layne, D. S., Roberts, J. B., Gibree, N., and Williams, K. I. H. (1965). *Steroids* **6**, 855.

Lazier, C., and Jellinck, P. H. (1965). *Can. J. Biochem.* **43**, 281.

Levitz, M., Katz, J., and Twombly, G. H. (1965). *Steroids* **6**, 553.

Lucis, O. J. (1965). *Steroids* **5**, 163.

Margraf, H. W., Margraf, C. O., and Weichselbaum, T. E. (1963). *Steroids* **2**, 155.

Marks, F., and Hecker, E. (1968). *Z. Physiol. Chem.* **349**, 523.

Marks, F., and Hecker, E. (1969). *Z. Physiol. Chem.* **350**, 69.

Matsui, M., and Fukushima, D. K. (1969). *Biochemistry* **8**, 2997.

Mathur, R. S. (1969). *Can. J. Biochem.* **47**, 535.

Mathur, R. S., Common, R. H., Collins, D. C., and Layne, D. S. (1969). *Biochim. Biophys. Acta* **176**, 394.

Mellin, T. N., Collins, D. C., Williams, K. I. H., and Layne, D. S. (1966). Unpublished results.

Nambara, T., and Numazawa, M. (1969). *Chem. Pharm. Bull.* **17** (6), 1200.

Nishizawa, E. E., and Eik-Nes, K. B. (1963). *J. Chromatog.* **10**, 493.

Noguchi, S., and Fukushima, D. K. (1966). *J. Biol. Chem.* **241**, 761.

Nose, Y., and Lipmann, F. (1958). *J. Biol. Chem.* **233**, 1348.

Oertel, G. W. (1964). *In* "Structure and Metabolism of Corticosteroids" (J. R. Pasqualini and M. F. Jayle, eds.), pp. 65–76. Academic Press, New York.

Oertel, G. W., and Eik-Nes, K. B. (1958). *Acta Endocrinol.* **28**, 293.

Oertel, G. W., and Eik-Nes, K. B. (1959). *Acta Endocrinol.* **30**, 93.

Parke, D. V. (1968). "The Biochemistry of Foreign Compounds," 1st ed., pp. 92–96. Macmillan (Pergamon), New York.

Pasqualini, J. R. (1965). *Biochim. Biophys. Acta* **104**, 515.

Pasqualini, J. R., and Jayle, M. F. (1962). *J. Clin. Invest.* **41**, 981.

Payne, A. H., and Mason, M. (1965). *Steroids* **5**, 21.

Purdy, R. H., and Axelrod, L. R. (1968a). *Steroids* **11**, 851.

Purdy, R. H., and Axelrod, L. R. (1968b). *Proc. 3rd Intern. Congr. Endocrinol., Mexico, D. F.* p. 170. Excerpta Med. Found., New York.

Purdy, R. H., Engel, L. L., and Oncley, J. L. (1961). *J. Biol. Chem.* **236**, 1043.

Purdy, R. H., Grosser, B. I., and Axelrod, L. R. (1968). *Steroids* **11**, 837.

Raud, H. R., and Hobkirk, R. (1968). *Can. J. Biochem.* **46**, 749.

Roseman, S., and Dorfman, A. (1951). *J. Biol. Chem.* **191**, 607.

Ryan, K. J., and Engel, L. L. (1953). *Endocrinology* **52**, 277.

Saldarini, R. J., and Hilliard, J. (1969). *Federation Proc.* **28**, 704.

Scardi, V., Iaccarino, M., and Scarano, E. (1962). *Biochem. J.* **83**, 413.

Schubert, K. (1958). *Acta Endocrinol.* **27**, 36.

Schubert, K., and Wehrberger, K. (1965). *Endocrinologie* **47**, 290.

Slaunwhite, W. R., Jr., Lichtman, M. A., and Sandberg, A. A. (1964). *J. Clin. Endocrinol. Metab.* **24**, 638.

Smith, J. N. (1964). *In* "Comparative Biochemistry" (M. Florkin and H. S. Mason, eds.), Vol. 6, p. 415. Academic Press, New York.

Sneddon, A., and Marrian, G. F. (1963). *Biochem. J.* **86**, 385.

Straw, R. F., Katzman, P. A., and Doisy, E. A. (1955). *Endocrinology* **57**, 87.

Tabone, D., and Tabone, J. (1956). *Compt. Rend.* **242**, 302.

Tait, S. A. S., and Tait, J. F. (1962). *In* "Methods in Hormone Research" (R. I. Dorfman, ed.), Vol. 1, pp. 265–336. Academic Press, New York.

Touchstone, J. C. (1965). *In* "Estrogen Assays in Clinical Medicine" (C. A. Paulsen, ed.), pp. 164–172. Univ. of Washington Press, Seattle, Washington.

Underwood, R. H., and Tait, J. F. (1964). *J. Clin. Endocrinol. Metab.* **24**, 1110.

Valcourt, A. J., Thayer, S. A., Doisy, E. A., Jr., Elliot, W. H., and Doisy, E. A. (1955). *Endocrinology* **57**, 692.

Vande Wiele, R. L., MacDonald, P. C., Gurpide, E., and Lieberman, S. (1963). *Recent Progr. Hormone Res.* **19**, 275.

Voigt, K. D., and Schroeder, W. (1955). *Nature* **176**, 599.

Wallace, E., and Silberman, N. (1964). *J. Biol. Chem.* **239**, 2809.

Wang, D. Y., and Bulbrook, R. D. (1968). *In* "Advances in Reproductive Physiology" (A. McLaren, ed.), Vol. 3, pp. 113–146. Logos Press, London.

Warren, J. C., and French, A. P. (1965). *J. Clin. Endocrinol. Metab.* **25**, 278.

Weichselbaum, T. E., and Margraf, H. W. (1960). *J. Clin. Endocrinol. Metab.* **20**, 1341.

Weissman, B., Rowin, G., Marshall, J., and Friederici, D. (1967). *Biochemistry* **6**, 207.

Wengle, B., and Böstrom, H. (1963). *Acta Chem. Scand.* **17**, 1203.

Whittemore, K., and Layne, D. S. (1965). *Nature* **208**, 288.

Williams, K. I. H., and Layne, D. S. (1967). *J. Clin. Endocrinol. Metab.* **27**, 159.

Williams, K. I. H., Henry, D. H., Collins, D. C., and Layne, D. S. (1968). *Endocrinology* **83**, 113.

Williams, R. T. (1967). *In* "Biogenesis of Natural Compounds" (P. Bernfeld, ed.), 2nd ed., pp. 589–639. Macmillan (Pergamon), New York.

Williamson, D. G., Collins, D. C., Layne, D. S., Conrow, R. B., and Bernstein, S. (1969). *Biochemistry* **8**, 4299.

Wotiz, H. H. (1962). *Biochim. Biophys. Acta* **60**, 28.

Wotiz, H. H., and Fishman, W. H. (1963). *Steroids* **1**, 211.

Wotiz, H. H., Ziskind, B. S., and Ringler, I. (1958). *J. Biol. Chem.* **231**, 593.

Young Lai, E., and Solomon, S. (1967). *Biochemistry* **6**, 2040.

CONJUGATES OF *N*-HYDROXY COMPOUNDS

CHARLES C. IRVING

I. Introduction

Before 1960 (Cramer *et al.*, 1960), metabolic conjugates of *N*-hydroxy compounds were unknown. This class of conjugates, involving an N-O-X linkage (e.g., X = glucuronyl or sulfonate), has become of unique interest and importance in studies of the metabolism and mechanism of action of carcinogenic aromatic amines. Conjugates of N-hydroxy compounds are the only type of metabolically formed conjugates with chemical reactivity (nonenzymic) leading to the formation of covalent bonds under physiological conditions.

Metabolism of a drug or other foreign compound usually results in the termination of any biological or pharmacological action which the parent compound might have possessed. This is particularly true in the case of metabolic reactions involving conjugation (Williams, 1959; Parke, 1968). The metabolism of aromatic amines, however, represents an exception. Aromatic amines are among the most frequent causes of ferrihemoglobin (methemoglobin) formation *in vivo*, but in general, these compounds do not oxidize hemoglobin unless they undergo certain biochemical changes. The formation of ferrihemoglobin by aromatic amines is due to their conversion *in vivo* to the corresponding arylhydroxylamine by N-hydroxylation (reviewed by Kiese, 1966). In some instances, free arylhydroxylamine derivatives of carcinogenic aromatic amines may also be directly involved in the mechanism of carcinogenesis by these compounds. In many cases, N-hydroxy metabolites of N-acyl derivatives of aromatic amines are further metabolized by conjugation reactions. Although these conjugates do not appear to be acutely toxic in moderate doses, there are indications that they may not be completely innocuous. Recent studies have shown that some conjugates of N-hydroxy compounds possess unexpected chemical reactivity with tissue nucleophiles, such as methionine and tryptophan residues in proteins and with guanine residues in nucleic acids (see Miller and Miller, 1969a,b).

The objectives of this article are the following: (1) to give a brief review of mechanisms involved in the enzymic formation of N-hydroxy compounds; (2) to summarize and discuss pathways, both known and hypothetical, by which N-hydroxy compounds may be further metabolized *in vivo*; and (3) to describe our present knowledge of the chemical properties of conjugates of N-hydroxy compounds, particularly as these properties are related to the generation of reactive cations or free radicals *in vivo*. Mechanisms of carcinogenesis per se do not fall within the scope of this review. However, because of the reasons indicated above, the chapter will be concluded with a brief discussion of the possible role of conjugates of N-hydroxy compounds in the mechanism of carcinogenesis by aromatic amines and their N-acyl derivatives.

II. Biochemical Formation of N-Hydroxy Compounds

N-Hydroxy compounds are synthesized in biological systems by three general pathways. The direct N-hydroxylation of an amine or the N-acyl derivative results in the formation of the corresponding hydroxylamine or N-acylhydroxylamine,[1] respectively. Arylhydroxylamines are also formed by the reduction of the corresponding nitro compounds. A third mechanism, demonstrated in only one instance in microorganisms, involves the direct addition of hydroxylamine to the double bond of fumaric acid, yielding N-hydroxyaspartic acid.

A. N-HYDROXYLATION

1. Introduction

Cramer *et al.* (1960) provided the first unequivocal demonstration of N-hydroxylation *in vivo*. N-Acetyl-N-2-fluorenylhydroxylamine (I), which was found in urine almost entirely as the glucuronide conjugate, was excreted as a major metabolite of the carcinogen 2-acetylamino-fluorene (II).

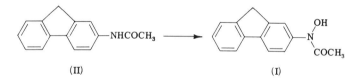

(II)   (I)

[1] In order to avoid confusion, particularly in the case of naming conjugates of N-hydroxy compounds, the following nomenclature is used throughout this article. Structures of the type

are named as derivatives of hydroxylamine (Y = X = H), except in the case of N-hydroxyurethane (Y = CH₃CH₂-O-C(O)-; X = H) or related carbamate derivatives which are the only examples to be described where Y is not an aryl group. When Y = aryl and X = H or alkyl (such as —CH₃), the compound is referred to as an arylhydroxylamine. When Y = aryl and X = acyl (such as —COCH₃), the compound is referred to as an N-acyl-N-arylhydroxylamine rather than being named as a substituted hydroxamic acid. Conjugates are then named as derivatives of the N-arylhydroxylamine or the N-acyl-N-arylhydroxylamine.

Irving (1962a) detected the formation of the N-hydroxy derivative I from the amide II by rabbit liver microsomes *in vitro*. This was the first instance in which the initial product of an N-hydroxylation reaction *in vitro* had been demonstrated.

A year earlier, Uehleke had reported that 2-aminofluorene (III) was N-hydroxylated by rat liver microsomes *in vitro* (Uehleke, 1961). However, the initial product of the N-hydroxylation of III, i.e., N-2-fluorenyl-hydroxylamine (IV), was not directly identified in Uehleke's experiments. Instead, 2-nitrosofluorene (V), formed by oxidation of IV, was detected in the system.

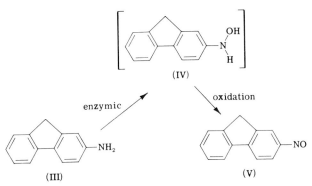

(IV)

enzymic   oxidation

(III)   (V)

Uehleke's report on the N-hydroxylation of III *in vitro* has been criticized because he did not give any data which substantiated his claims to have synthesized V or to have detected V in suspensions of rat liver microsomes incubated with III (Gutmann, 1964; Lotlikar *et al.*, 1965; Kiese *et al.*, 1966). Uehleke (1968) later described in more detail the method which he had used for the synthesis of V.

Because of the instability of arylhydroxylamines, indirect methods (i.e., measurement of the nitroso derivative) have been used in many studies on the N-hydroxylation of aromatic amines, both *in vivo* and *in vitro* (Kiese, 1966). In a relatively few instances products of the N-hydroxylation of aromatic amines have been identified directly. For example, Thauer *et al.* (1965), showed that rat liver microsomes catalyzed the N^4-hydroxylation of sulfanilamide (VI). In these studies, p-hydroxyl-aminobenzenesulfonamide (VII) was isolated from the reaction mixture and identified by thin-layer chromatography.

(VI)   (VII)

2. Requirements of the Enzyme System

The N-hydroxylation of aromatic amines and their amide derivatives is carried out by enzymes located in the endoplasmic reticulum (Kiese and Uehleke, 1961; Irving, 1962a, 1964; Booth and Boyland, 1964). The enzyme system, which is NADPH-dependent and requires molecular oxygen, appears to belong to the group of mixed function oxidases or hydroxylases (Mason, 1957) of the endoplasmic reticulum. These hydroxylases catalyze the incorporation of one atom of oxygen into the substrate, the other oxygen atom being reduced by hydrogen. It has not yet been demonstrated that the N-hydroxylase enzyme system involves

$$ArN \begin{matrix} H \\ \diagup \\ \diagdown \\ X \end{matrix} + O_2 + NADPH + H+ \longrightarrow ArN \begin{matrix} OH \\ \diagup \\ \diagdown \\ X \end{matrix} + H_2O + NADP+$$

the participation of cytochrome P-450. Kampffmeyer and Kiese (1965) obtained no inhibition by carbon monoxide of the N-hydroxylation of aniline or N-ethylaniline by rabbit liver microsomes. On the other hand, N-hydroxylation of aniline and N-ethylaniline by microsomes from the livers of the guinea pig, dog, and rat is at least partially inhibited by carbon monoxide (Appel et al., 1965). The role of cytochrome P-450 in the enzymic N-hydroxylation of aromatic amines and their N-acyl derivatives needs additional investigation.

The mechanism of the postulated pathway of the activation of oxygen in hydroxylase reactions, and electron transport via cytochrome P-450 are discussed by Omura et al. (1965), Kato (1966), and Remmer et al. (1968).

Types of substituted amines which have been shown to be N-hydroxylated in vitro are listed in Table I. The information given in Table I is not intended to be exhaustive, but rather indicative of the kinds of structures reported to be N-hydroxylated. A more detailed account and comprehensive list of compounds which have been reported to be N-hydroxylated is available (Kiese, 1966).

Many investigators feel that there are separate enzymes involved in N-hydroxylation and C-hydroxylation reactions. It is difficult to prove that the enzyme system which is responsible for the N-hydroxylation of aromatic amines and their N-acyl derivatives is a distinct and separate enzyme system from that which C-hydroxylates aromatic compounds. This problem is related to the controversial question of whether separate enzyme systems are involved in the o- and p-hydroxylation of aromatic amines:

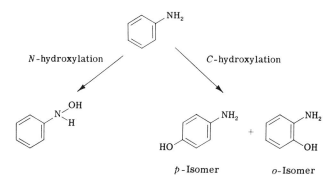

p-Isomer   o-Isomer

TABLE I

TYPES OF SUBSTRATES N-HYDROXYLATED *in Vitro*

General structure		Specific examples	Reference
Substrate	Product		
Ar-NH₂ →	→OH Ar-N⟨ H	p-Chloraniline	Kiese and Uehleke (1961)
		2-Aminofluorene	Uehleke (1961); Kiese *et al.* (1966)
		4-Aminobiphenyl	Booth and Boyland (1964)
		Sulfanilamide	Thauer *et al.* (1965)
H Ar-N⟨ COCH₃	OH Ar-N⟨ COCH₃	2-Acetylaminofluorene 2-Acetylaminobiphenyl	Irving (1962a, 1964) Booth and Boyland (1964)
		4-Acetylaminostilbene	Baldwin and Smith (1965)
H Ar-N⟨ R	OH Ar-N⟨ H	N-Methylaniline N-Ethylaniline[a]	Uehleke (1962) Kiese and Rauscher (1963)
O ‖ R-O-C⟨ NH₂	O ‖ R-O-C⟨OH N⟨ H	Urethane (R = C₂H₅⁻)	Boyland and Nery (1965)

[a] The N-dealkylation of N-alkyl arylamines, such as N-ethylaniline, may involve the intermediate formation of the N-hydroxy-N-alkyl derivative, followed by loss of the alkyl group and formation of the unsubstituted arylhydroxylamine (Kiese and Rauscher, 1963; Kiese, 1966).

The evidence that there are multiple hydroxylase enzymes in the liver endoplasmic reticulum is based primarily on differences in positional hydroxylation of aromatic ring systems and aromatic amines in various species and upon the influence of various agents affecting hydroxylating reactions within a single species. Kiese (1966) feels that induction experiments with a number of compounds, such as phenobarbital, 3, 4-benzopyrene, 3-methylcholanthrene or DDT, strongly suggest that the N-hydroxylation of aniline as well as the o- and p-hydroxylation of aniline are catalyzed by different microsomal factors. Some evidence suggesting that the enzymes hydroxylating aniline at the C- and N-positions

TABLE II

COMPARISON OF N-HYDROXYLATION AND C-HYDROXYLATION OF ANILINE BY RAT LIVER MICROSOMES in Vitro[a]

	N-Hydroxylation	C-Hydroxylation
Aniline concentration	Strongly dependent	Less dependent
Activity in liver microsomes	Strong	Strong
Activity in lung microsomes	Weak	Strong
Effect of snake venom phospholipase	Strong inhibition	No inhibition
2,4-Dichlorophenol, 1 mM	No inhibition	Inhibition
p-Chloromercuribenzoate, 1 mM	No inhibition	Inhibition
Semicarbazide, 10 mM	No inhibition	No inhibition
Induction by 3,4-benzpyrene	Slight	Slight

[a] Data from Kampffmeyer and Kiese (1963).

are not the same is given in Table II. Other studies have shown that the N- and p-hydroxylation of aniline and N-ethylaniline were stimulated to different degrees by liver microsomes from rabbits treated with phenobarbital or DDT (Lange, 1967). The N-hydroxylation of N-ethylaniline was not stimulated by phenobarbital or DDT whereas the N-hydroxylation of aniline and the p-hydroxylation of N-ethylaniline were increased manyfold after administration of phenobarbital or DDT to rabbits. Phenobarbital or DDT did not have any effect on any of these reactions when added to rabbit liver in vitro. Lange's results support the assumption that different enzymes in the endoplasmic reticulum catalyze the N- and p-hydroxylation of aromatic amines.

Lotlikar et al. (1967b) have examined the N- and C- hydroxylation of 2-acetylaminofluorene (II) in various species and the effect of pretreatment with 3-methylcholanthrene on these reactions. Their data also support the hypothesis that separate enzymes are involved in the N- and C-hydroxylation of aromatic amines and amides (Table III). The

Charles C. Irving

relatively specific increase in N-hydroxylation of II, as a consequence of the administration of methylcholanthrene to the hamster, appears to differentiate the N-hydroxylation system from the ring hydroxylation systems (Table III). Furthermore, comparison of the ratios of 3-, 5-, and 7-hydroxylation of II by liver microsomes from various species showed

TABLE III

EFFECTS OF PRETREATMENT WITH 3-METHYLCHOLANTHRENE (MC) ON THE
N- AND C-HYDROXYLATION OF 2-ACETYLAMINOFLUORENE
(AAF; II) BY LIVER MICROSOMES[a]

Species	MC	mμmoles formed/gm liver/20 minutes			
		N-Hydroxy-AAF	3-Hydroxy-AAF	5-Hydroxy-AAF	7-Hydroxy-AAF
Rat	—	n.d.[b]	24 ± 5	20 ± 4	57 ± 12
	+	37 ± 8	480 ± 100	395 ± 90	580 ± 190
Hamster	—	69 ± 18	47 ± 7	140 ± 19	640 ± 60
	+	950 ± 145	90 ± 21	186 ± 27	590 ± 130
Mouse	—	37 ± 6	31 ± 17	37 ± 5	120 ± 11
	+	236 ± 53	175 ± 33	220 ± 35	550 ± 210
Rabbit	—	16 ± 7	n.d.	n.d.	160 ± 85
	+	77 ± 15	n.d.	n.d.	200 ± 45
Guinea pig	—	n.d.	n.d.	43 ± 15	1150 ± 320
	+	n.d.	n.d.	73 ± 20	1800 ± 300

[a] Data from Lotlikar *et al.* (1967b). Weanling male animals were injected with methylcholanthrene (2–10 mg/100 gm weight) 24 hours before assay for hydroxylase activity.
[b] n.d. = not detected.

that the 7-hydroxylation activity varied independently of 3- and 5-hydroxylation.

There might be alternative explanations to account for the species differences in the N- and C-hydroxylation of aromatic amines and amides and for effects of various inducing agents on these enzymic reactions (Remmer *et al.*, 1968). For example, the lipid layer of the endoplasmic reticulum, which is in close contact with the hydroxylating enzyme, may concentrate the compound to be hydroxylated, permitting access to the enzyme. Consequently, differences in the lipid composition in various species or differences in the lipid composition caused by the administration of inducing agents might influence metabolic rates (Remmer *et al.*, 1968). However, the consensus appears to be that there *are* multiple enzymes affecting N- and C-hydroxylation of aromatic amines, although the systems dealing with the activation of oxygen might be identical.

3. Species and Tissue Distribution of the Enzyme System

Kiese (1966) has compiled a list of aromatic amines and some *N*-acyl derivatives which are *N*-hydroxylated *in vitro* by NADPH-dependent enzymes in liver microsomes of various species. Most of these species, which include the rat, rabbit, hamster, dog, cat, chicken, and guinea pig, also excrete *N*-hydroxylated metabolites in the urine, either in free or conjugated form. More recently it has been shown that microsomes from human liver are also capable of *N*-hydroxylating 2-acetylaminofluorene (II) (Enomoto and Sato, 1967). Weisburger *et al.* (1964b) had reported earlier that the human excreted *N*-acetyl-*N*-2-fluorenylhydroxylamine (I) in conjugated form after administration of tracer doses of II. The steppe lemming *Lagarus lagarus* (Weisburger *et al.*, 1965) and the mastomys *Rattus natalensis* (Yamamoto *et al.*, 1968) excrete I in conjugated form after administration of II.

A major item of controversy in this area involves the question of whether the guinea pig is capable of *N*-hydroxylation of aromatic amines. Guinea pigs are refractory to the carcinogenic action of 2-acetylamino-fluorene (II) and it has been postulated that the reason for this is the apparent inability of the guinea pig to *N*-hydroxylate this amide (Miller *et al.*, 1964b). The *N*-hydroxy metabolite I was not detected in the urine of guinea pigs fed a diet containing II or injected with this compound (Miller *et al.*, 1960; Miller *et al.*, 1964b). Furthermore, Irving (1964) could not find any evidence that II was *N*-hydroxylated by guinea pig liver microsomes *in vitro*. These results were confirmed by Lotlikar *et al.* (1967b) who were unable to detect the *N*-hydroxylation of II by liver microsomes obtained from normal guinea pigs or from guinea pigs pretreated with methylcholanthrene. On the other hand, Appel *et al.* (1965) reported that guinea pig liver microsomes *N*-hydroxylate aniline and *N*-ethylaniline at rates comparable to those found with rabbit liver microsomes. Kiese *et al.* (1966) showed that guinea pig liver microsomes *N*-hydroxylated 2-aminofluorene (III), the reaction proceeding at least as rapidly as the *N*-hydroxylation of aniline. Furthermore, a small fraction of the amine III injected intraperitoneally into guinea pigs was excreted in the urine as *N*-2-fluorenylhydroxylamine (IV). The arylhydroxylamine IV was identified by indirect methods involving oxidation to the nitroso analog (V); V was identified by its uv absorption, by thin-layer chromatography, and by the formation of a diazo compound on reaction with nitrous acid. Kiese and Wiedemann (1968) suggest that the rapid disposal of *N*-hydroxy derivatives produced by microsomal *N*-hydroxy-lation of arylamines plays an important role in the refractoriness of guinea pigs to the carcinogenic and ferrihemoglobin-forming activity of III. The apparent discrepancy between the positive data on the N-hydroxylation of the amine III as opposed to negative results with the N-acyl derivative II might be due to a lower rate of N-hydroxylation of II and/or the very rapid rate of deacetylation of the N-hydroxy metabolite I by guinea pig liver microsomes (Irving, 1966; see Section V,D,3).

Only one tissue besides the liver has been reported to N-hydroxylate aromatic amines *in vitro*. Bladder mucosa of hogs and guinea pigs catalyzed the N-hydroxylation of p-chloraniline or 4-aminobiphenyl (Uehleke, 1966). The products were identified as the nitroso derivatives. However, the extent of N-hydroxylation by bladder mucosa was very low. Using 3 gm of tissue, only 3.6 μg of p-chloronitrosobenzene or 2.4 μg of 4-nitrosobiphenyl were found after 60 minutes' incubation with the parent aromatic amine.

4. Factors Influencing the Activity of the Enzyme System

The activities of a number of enzymes in the endoplasmic reticulum of the liver can be altered by dietary and nutritional factors, hormonal changes, and by the administration of a variety of foreign chemicals such as phenobarbital or methylcholanthrene (Conney, 1967). Remmer *et al.* (1968) and Orrenius and Ernster (1967) feel that there is a common hydroxylating system for various drugs. They suggest that the penetration of these drugs into the membranes of the endoplasmic reticulum might be the rate-limiting step in the binding of drugs to the endoplasmic reticulum. The binding site for the foreign compound might be identical with the terminal oxidase of the hydroxylating system, i.e., cytochrome P-450 (Orrenius and Ernster, 1967). The binding of the foreign chemical is postulated to be irreversible unless hydroxylation of the drug takes place or the cytochrome P-450 is modified by interaction with carbon monoxide. The actual inducer when animals are pretreated with a variety of "inducing agents" may not be the administered compound but a substance (e.g., a steroid) which is a natural substrate of the hydroxylating system and which is released from or is prevented from binding to the endoplasmic reticulum when the foreign compound is bound (Orrenius and Ernster, 1967).

The enzyme system in the liver endoplasmic reticulum responsible for the N-hydroxylation of aromatic amines or their N-acyl derivatives can be stimulated or induced by treatment of animals with phenobarbital, methylcholanthrene, DDT, or 3,4-benzypyrene. The N-hydroxylation of N-ethylaniline by rat liver microsomes was stimulated markedly by pretreatment of the animals with benzpyrene (Kampffmeyer and Kiese, 1963), but was not stimulated in liver microsomes obtained from rabbits given either phenobarbital or DDT (Lange, 1967). On the other hand, the N-hydroxylation of aniline by rabbit liver microsomes was markedly stimulated by pretreatment with phenobarbital or DDT (Lange, 1967), while the N-hydroxylation of aniline by rat liver microsomes was only weakly stimulated by benzpyrene (Kampffmeyer and Kiese, 1963). Uehleke (1967) reported that the N-hydroxylation of N-ethylaniline by rat liver microsomes was not appreciably affected by giving either phenobarbital, methylcholanthrene, or DDT, but was decreased about 50% after giving all three of these agents in combination. The N-hydroxylation of p-chloraniline by rat liver microsomes was not significantly affected after giving the rats phenobarbital or DDT, but was increased about 50% after administration of methylcholanthrene (Uehleke, 1967).

The N-hydroxylation of 2-acetylaminofluorene (II) by liver microsomes of the hamster, mouse, and rabbit was stimulated 5–15-fold by prior injection of methylcholanthrene (Lotlikar et al., 1967b). The N-hydroxylation of II could not be detected in liver microsomes from normal rats, whereas this reaction was detectable in liver microsomes obtained from rats pretreated with methylcholanthrene. Enomoto et al. (1968a) have also observed that the N-hydroxylation of II by hamster liver microsomes was increased by a single intraperitoneal dose of methylcholanthrene.

A variety of factors, such as administration of ACTH and growth hormone (Shirasu et al., 1967a), partial hepatectomy (Margreth et al., 1964), or prolonged feeding of 2-acetylaminofluorene (II) (Miller et al., 1960) increased the urinary excretion of the glucuronide of N-acetyl-N-2-fluorenylhydroxylamine in the rat. This was attributed to an increase in the formation of the N-hydroxylated metabolite (I) in the animal. However, alterations in the amounts of urinary metabolites do not necessarily reflect differences in the activities of enzymes metabolizing these compounds. Measurement of urinary levels of metabolites is not a valid criterion for determining the extent of N-hydroxylation in the rat, and probably in other species (Irving et al., 1967b; see Section III, E).

B. REDUCTION OF NITRO COMPOUNDS

Aromatic nitro compounds such as chloramphenicol, p-nitrophenol, and p-nitrobenzoate, are reduced to the corresponding primary aromatic amines in mammalian tissues. This reduction has been attributed to the action of one or more enzymes present mainly in the endoplasmic reticulum of the liver (Fouts and Brodie, 1957). Reduction of *p*-nitrobenzoate to *p*-aminobenzoate was catalyzed by an enzyme localized in liver microsomes, with the transfer of electrons to the aromatic compound being mediated by cytochrome P-450 (Gillette *et al.*, 1968). However, boiled preparations of various fractions of liver homogenate caused an increase in nitroreductase activity, which was blocked by carbon monoxide, suggesting the possibility that some intermediate was required to mediate the transfer of electrons between cytochrome P-450 and the substrate (Gillette *et al.*, 1968). Earlier studies by Otsuka (1961) showed that the liver nitro reducing system could be separated into two enzyme fractions: a nitroreductase which catalyzed the reduction of *p*-nitrophenol (VIII) to *p*-nitrosophenol (IX) and a second fraction, containing a nitrosoreductase, which was responsible for the reduction of the nitroso derivative to *p*-hydroxylaminophenol (X). The conversion of X to the

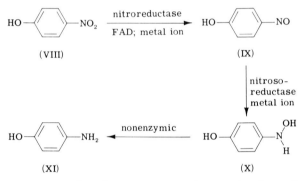

amine XI was proposed to be a nonenzymic reaction brought about by reduced pyridine nucleotides or by ferrous ion. The nitroreductase and nitrosoreductase activities could be distinguished by the finding that the nitrosoreductase did not require FAD. Otsuka purified these enzymes from a 10,000 × g supernatant of a homogenate from swine liver.

4-Nitroquinoline-1-oxide (XII) was reduced to 4-hydroxylaminoquinoline-1-oxide (XIII) by rat liver (Sugimura *et al.*, 1966) and by subcutaneous tissues of the rat (Matsushima *et al.*, 1968):

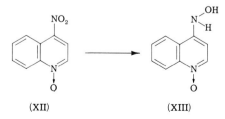

The conversion of XII to XIII, occurring in the supernatant fraction (105,000 × g) of homogenates of rat liver and lung, required either NADH or NADPH as a hydrogen donor. XIII was not further reduced to 4-aminoquinoline-1-oxide by this system with either NADH or NADPH. The enzyme system also catalyzed the reduction of dichloro-phenolindophenol by NADH or NADPH and appeared to be identical with DT diaphorase. Dicoumarol inhibited the partially purified enzyme.

Uehleke (1963) found that a supernatant fraction (78,000 × g) of a homogenate of rat liver reduced nitrobenzene to nitrosobenzene and phenylhydroxylamine. The reduction of nitrobenzene required NADPH and FMN. Uehleke referred to this system as a "nitroreductase." Uehleke and Nestel (1967) demonstrated the formation of 4-aminobiphenyl during anaerobic incubation of 4-nitrobiphenyl with rat liver-soluble proteins, NADPH, and FMN. 4-Nitrosobiphenyl was formed during the reaction. The rate of reduction of 4-nitrosobiphenyl was four times higher than that of 4-nitrobiphenyl. 2-Nitrofluorene was reduced to 2-aminofluorene to a small extent in this system.

Although the detailed mechanism and the various enzymes involved have not been clearly defined and differentiated, it is clear that liver and other tissues of the rat, such as lung and subcutaneous tissue, reduce aromatic nitro compounds to arylhydroxylamines. The reduction involves the intermediate formation of the nitroso derivative, and in many cases the hydroxylamine is reduced further to the amine derivative.

C. OTHER MECHANISMS

Cylic hydroxamic acids and N-acyl derivatives of some N-hydroxyamino acids have been isolated from microbial fermentations. Such compounds include N-formyl-N-hydroxyglycine (Hadacidin) from *Penicillium fre-quentans* (Kaczka *et al.*, 1962), mycelianamide obtained from the mycelium of strains of *Penicillium griseofulvin* (Birch *et al.*, 1956), and aspergillic acid, a cyclic hydroxamic acid produced by *Aspergillus flavus* (Dutcher, 1947). δ-N-Hydroxyornithine is a component of the ferrichrome compounds described by Emery and Neilands (1961) and ε-N-hydroxy-lysine has been found as a degradation product of myobactin, a growth factor for *Mycobacterium johnei* (Snow, 1954). Pulcherrimin, a red pig-ment isolated from *Cardide pulcherrima*, also contains a cyclic hy-droxamic acid (Kluyver *et al.*, 1953).

Little is known about the mechanisms involved in the biosynthesis of these N-hydroxylated compounds. Free N-hydroxyamino acids are un-stable at neutral pH and have not been demonstrated in living cells (Emery, 1963). Aspartase, partially purified from *Bacillus cadaveris*, was found to catalyze the condensation of hydroxylamine with fumaric acid to yield N-hydroxy-L-aspartic acid (XIV) (Emery, 1963). There is no evidence for the occurrence or formation of N-hydroxylamino acids in mammalian tissues.

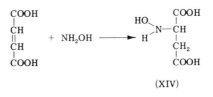

(XIV)

III. Glucuronide Conjugates of N-Hydroxy Compounds

Glucuronide conjugates are quantitatively the most important type involved in the excretion of N-hydroxy compounds. N-Acetylated-N-arylhydroxylamines are found in the urine and bile, after administration of the aglycone or the amine or amide precursor, almost entirely as conjugates yielding the corresponding aglycone upon treatment of the urine or bile with β-glucuronidase. Only a few glucuronide conjugates of N-hydroxy compounds have been isolated or synthesized and their structures unequivocally established by conventional chemical procedures. Many others have been identified indirectly.

A. CHEMISTRY OF KNOWN GLUCURONIDE CONJUGATES

1. *General Structural Features*

For comparative purposes, brief consideration will be given to some general features of the structures of O-glucuronides. O-Glucuronides can be formed from a variety of phenolic compounds, from primary, secondary, or tertiary alcohols, or from carboxylic acids (Williams, 1959; Marsh, 1966; Dutton, 1966; Parke, 1968). Ether-type glucuronides are formed from alcohols and phenolic compounds and have a —C_1—O—C— glycosidic linkage (—C_1 represents the anomeric carbon of the glucuronyl moiety). In general, these compounds are stable in alkali and do not reduce alkaline copper reagents. Ester-type glucuronides (—C_1—O—CO—linkage) are labile in dilute alkali and reduce alkaline copper solutions. A third type of O-glucuronide, the enol-type, is formed from pseudo-acids such as 4-hydroxycoumarin; these are also labile in alkaline solutions and reduce alkaline copper reagents.

Glucuronides of N-hydroxy compounds, which have a —C_1—O—N— linkage, represent a new class of O-glucuronides. Several types of N-substituted O-glucuronides of N-hydroxy compounds are possible. These are illustrated by structures XV, XVI, and XVII.

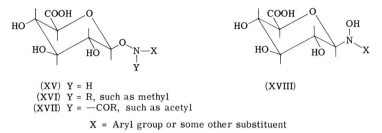

(XV) Y = H   (XVIII)
(XVI) Y = R, such as methyl
(XVII) Y = —COR, such as acetyl

X = Aryl group or some other substituent

Since N-glucuronides are formed chemically or enzymically (Williams, 1959; Marsh, 1966; Dutton, 1966; Parke, 1968), an N-glucuronide of an N-hydroxy compound (XVIII) should be considered if there is a replaceable hydrogen atom on the nitrogen of the N-hydroxy derivative (compare XV with XVIII).

Of the possible glucuronides represented (XV, XVI, XVII, XVIII) only XVII (X = aryl, Y = acetyl) has been identified unequivocally as a metabolite of an N-hydroxy compound. XV (X = 2-fluorenyl) has been synthesized but the product has not been shown to be a metabolite of N-2-fluorenylhydroxylamine (Section III, A, 2).

2. Synthesis, Isolation, and Characterization of Specific Conjugates

The first synthesis of an N-hydroxy compound with a glycosidic linkage involved the condensation of 1-bromo-1-deoxy-tetra-O-acetyl-D-glucose with phenylhydroxylamine. A product identified as tetraacetyl-D-glucosidophenylhydroxylamine was obtained (Utzinger, 1944). Utzinger did not indicate whether an N-glucoside or an O-glucoside of phenylhydroxylamine was formed. On the basis of known methods for synthesis of N-glucosides (Marsh, 1966), Utzinger probably synthesized the N-glucoside (XIX). More recently, several carbohydrates, including glucuronolactone, glucose, galactose, ribose, and rhamnose were reacted with phenylhydroxylamine. These products likewise appeared to be secondary hydroxylamines (Boyland and Nery, 1963). The product of the

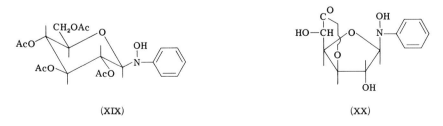

(XIX)   (XX)

reaction of phenylhydroxylamine with glucuronolactone was characterized as 1-deoxy-1-(*N*-hydroxy-*N*-phenylamino)ᴅ-glucuronolactone (XX).

The synthesis of an *O*-glucuronide of an *N*-hydroxy compound was first attempted by this author (Irving, 1965). A number of conventional procedures (Bollenback *et al.*, 1955; Conchie *et al.*, 1957; Whistler and Wolfrom, 1963) were tried in efforts to prepare the *O*-glucuronide of *N*-

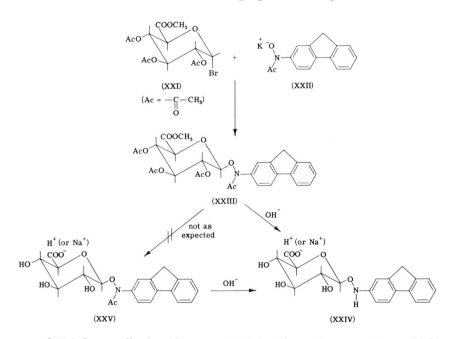

acetyl-*N*-2-fluorenylhydroxylamine (XXV). The only procedure which yielded the required intermediate, methyl (*N*-acetyl-*N*-2-fluorenylhydroxylamine 2,3,4-tri-*O*-acetyl-β-ᴅ-glucosid)uronate (XXIII), in sufficient amounts involved reaction of the potassium salt of *N*-acetyl-*N*-2-fluorenylhydroxylamine (XXII) with methyl (tri-*O*-acetyl-α-ᴅ-glucopyranosyl bromide)uronate (XXI). However, attempts to obtain the desired glucuronide conjugate (XXV) from XXIII by alkaline hydrolysis or by catalytic deesterification were not successful. It has not been possible to prepare *O*-glucuronide conjugates of *N*-acetyl-*N*-arylhydroxylamines by conventional synthetic procedures. Instead, biosynthetic procedures have been used. Compound XXV was isolated from the urine of rabbits fed either 2-acetylaminofluorene (II) or *N*-acetyl-*N*-2-fluorenylhydroxylamine (I) and was characterized as crystalline sodium (*N*-acetyl-*N*-2-fluorenyl-hydroxylamine-β-ᴅ-glucosid)uronate (Hill and Irving, 1966, 1967), which was later converted to the free acid (Hill, 1968). Subsequent investigations revealed that, under carefully controlled conditions, XXIV could be isolated as a product of the alkaline hydrolysis of XXIII or of XXV. Compound XXIV has been characterized as the crystalline sodium salt.

Sodium (*N*-acetyl-*N*-phenylhydroxylamine-β-D-glucosid)uronate (XXVI) has been isolated in crystalline form from the urine of rabbits given

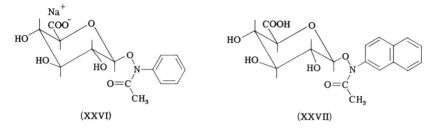

(XXVI)   (XXVII)

N-acetyl-*N*-phenylhydroxylamine (Kato *et al.*, 1967). The structure of XXVI was further confirmed by comparison of the properties of its tri-*O*-acetyl methyl ester derivative with an authentic sample of methyl (*N*-acetyl-*N*-phenylhydroxylamine 2,3,4-tri-*O*-acetyl-β-D-glucosid)uronate which had been synthesized by reaction of the bromo ester XXI with *N*-acetyl-*N*-phenylhydroxylamine in benzene in the presence of silver carbonate (Kato *et al.*, 1967).

The glucuronide conjugate of *N*-acetyl-*N*-2-naphthylhydroxylamine (XXVII) was obtained in a crude gum from rabbit urine after administration of *N*-acetyl-*N*-2-naphthylhydroxylamine. The glucuronide XXVII was not isolated; instead, this glucuronide was characterized as the crystalline triacetyl methyl ester derivative (Boyland and Manson, 1966).

A summary of the physical and chemical properties of known glucuronides of *N*-hydroxy compounds is given in Table IV.

Two glucoside derivatives of *N*-acetyl-*N*-2-fluorenylhydroxylamine have been synthesized (Hill and Irving, 1967; Hill, 1968): *N*-acetyl-*N*-2-fluorenylhydroxylamine 2, 3, 4, 6-tetra-*O*-acetyl-β-D-glucopyranoside (XXVIII), m.p. 166°–167°C and *N*-acetyl-*N*-2-fluorenylhydroxylamine 2,3,4,6-tetra-*O*-methyl-β-D-glucopyranoside (XXIX), m.p. 148°–149°C.

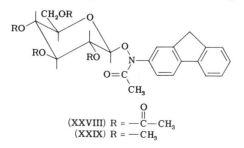

(XXVIII) R = —C̈—CH₃
(XXIX) R = —CH₃

TABLE IV

O-Glucuronides of N-Hydroxy Compounds Which Have Been Isolated or Synthesized and Adequately Characterized

Aglycone	Structure of conjugate	Species (source)	Characterized as	Reference
N-Acetyl-N-phenylhydroxylamine	XXVI	Rabbit (urine)	Sodium salt, m.p. 158°C	Kato et al. (1967)
			TAME[a] derivative, m.p. 157°–158°C	Kato et al. (1967)
N-Acetyl-N-2-naphthylhydroxylamine	XXVII	Rabbit (urine)	TAME derivative m.p. 124°–125°C	Boyland and Manson (1966)
N-Acetyl-N-2-fluorenylhydroxylamine	XXV	Rabbit (urine) or rat (bile)	Sodium salt, m.p. 195°–196°C	Hill and Irving (1967)
			TAME derivative, m.p. 166°–167°C	Irving (1965)
			Free acid, m.p. 118°–119°C	Hill (1968)
N-2-Fluorenylhydroxylamine	XXIV	b	Sodium salt	Irving and Russell (1969)

[a] TAME = triacetyl methyl ester.
[b] Synthesized from XXV or its TAME derivative (see Section III,A,3).

3. Alkaline Sensitivity

In contrast to the known stability of ether-type-O-glucuronides (Section III, A, 1), O-glucuronides of N-acetyl-N-hydroxy compounds are labile in alkaline solution. This property was first observed in the attempted synthesis of the O-glucuronide of N-acetyl-N-2-fluorenylhy-

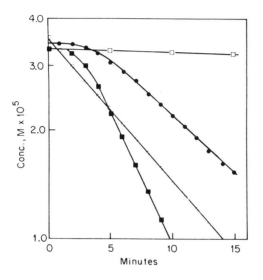

FIG. 1. Comparison of the alkaline sensitivity of some —C—O—N glycosides. Solutions of methyl (N-acetyl-N-2-fluorenylhydroxylamine 2,3,4-tri-O-acetyl-β-D-glucosid)uronate (XXIII, ●———●), sodium (N-acetyl-N-2-fluorenylhydroxylamine-β-D-glucosid)uronate (XXV, ○———○), N-acetyl-N-2-fluorenylhydroxylamine 2,3,4,6-tetra-O-acetyl-β-D-glucoside (XXVIII, ■———■), and N-acetyl-N-2-fluorenylhydroxylamine 2,3,4,6-tetra-O-methyl-β-D-glucoside (XXIX, □———□) in 50% ethanol were made 0.01 M in respect to NaOH at zero time and the decrease in the concentration of starting material was recorded.

droxylamine, XXV (Irving, 1965). The alkaline lability of XXV was confirmed after the compound had been prepared biosynthetically (Hill and Irving, 1966, 1967). Alkaline sensitivity appears to be a general property of glucuronides of this type (see XVII; X = aryl, Y = —COCH₃), since both XXVI (Kato et al., 1967) and XXVII (Boyland and Manson, 1966) are also unstable in alkali. Since these compounds are glucuronides of substituted hydroxamic acids and may be considered esterlike, it is not surprising that they are labile in alkaline media. However, alkaline hydrolysis of ester-type glucuronides ordinarily yields the aglycone as one of the products. The product of the action of dilute alkali on XXV was not the aglycone (I); hence, the alkaline lability was not due to hydrolysis of XXV.

Studies by Hill and Irving (1966, 1967) and by Hill (1968) on the mechanism of the sensitivity of XXV to alkali have revealed that the alkaline instability is due to migration of the N-acetyl group. Results of

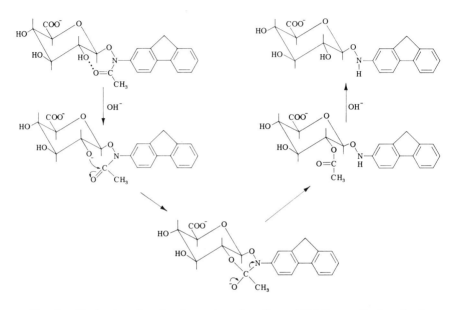

FIG. 2. Mechanism for the formation of sodium (N-2-fluorenylhydroxylamine-β-D-glucosid)uronate (XXIV) by the action of alkali on sodium(N-acetyl-N-2-fluorenylhydroxylamine-β-D-glucosid)uronate (XXV). Glucuronides of the type exemplified by XXV (see structure XVII) may be stabilized by hydrogen bonding as shown.

kinetic studies on the alkaline degradation of XXIII, XXV, XXVIII, and XXIX are shown in Fig. 1. Upon addition of alkali to solutions of the O-acetylated glycoside derivatives (XXIII and XXVIII), there was a lag phase of several minutes before a decrease in the concentration of the starting glycoside could be observed (Fig. 1). There was an immediate decrease in the concentration of XXV upon addition of alkali. On the other hand, the tetra-O-methyl glycoside derivative (XXIX) was stable in dilute alkali. The hydrolysis of XXV appears to proceed by the mechanism outlined in Fig. 2. With the O-acetylated glycoside derivatives, XXIII and XXVIII, the lag in the decrease in concentration upon addition of alkali (Fig. 1) was probably due to the time required for deesterification of the O-acetyl groups which were involved in the migration of the N-acetyl group. The stability of XXIX in dilute alkali is due to the stability of the O-methyl group in alkali; migration of the N-acetyl group could not occur in the O-methylated glycoside XXIX.

The facile removal of the N-acetyl group from XXV in dilute alkali led to the synthesis of sodium (N-2-fluorenylhydroxylamine β-D-glucosid)-uronate (XXIV). Compound XXIV, which is very unstable in aqueous solution, has also been synthesized by the action of alkali on XXIII (Irving and Russell, 1969).

4. Reaction with Nucleophilic Compounds

Generally, O-glucuronide conjugates are considered to be rather inert chemically and are regarded as end products in the metabolism of foreign compounds capable of forming such conjugates. However, XXV was discovered to have unusual and unexpected reactivity with certain nucleophilic compounds, particularly those contained in proteins and nucleic acids (Lotlikar et al., 1967a; Irving et al., 1967a, 1969a,b; Miller et al., 1968). Whether or not this type of reactivity is characteristic of other compounds of this class remains to be determined. Compound XXV reacted in vitro at pH 7 with methionine and tryptophan (Miller et al., 1968); the reactions were similar in some respects to those reported previously for synthetic esters, such as N,O-diacetyl-N-2-fluorenylhydroxylamine (LXIX) (see Section V, D,1), but differed considerably in rate and in the nature of the products formed (Table V). At pH 7, the major end products of the reaction of either XXV or N,O-diacetyl-N-2-fluorenylhydroxylamine with methionine were 1- and 3-methylmercapto-2-acetylaminofluorene (Fig. 3). At pH greater than 7, XXV, but not N,O-diacetyl-N-2-fluorenylhydroxylamine, reacted with methionine to yield considerable amounts of the deacetylated products, 1- and 3-methylmercapto-2-aminofluorene. In the reaction of XXV with methionine the proportionate yield of the deacetylated products increased with increasing pH (Table V). Neither the aglycone of XXV nor the triacetyl methyl ester derivative of XXV reacted with methionine at pH 5–9. With N-2-fluorenylhydroxylamine (IV), the reaction with methionine to yield 1- and 3-methylmercapto-2-aminofluorene increased below pH 5.5 but did not increase above pH 7. The probable mechanism involved in the formation of the o-methylmercapto-2-acetylaminofluorene and -aminofluorene derivatives from XXV or N,O-diacetyl-N-2-fluorenylhydroxylamine (LXIX) and methionine is shown in Fig. 3. The reaction of methionine with LXIX is discussed further by Lotlikar et al. (1966) (see Section V,

D,1). The products of the reaction of XXV with tryptophan have not been characterized.

Compound XXV also reacted with guanosine (Lotlikar *et al.*, 1967a; Miller *et al.*, 1968), guanosine 5′-monophosphate (GMP) (Irving *et al.*, 1969b), and with nucleic acids (Irving *et al.*, 1967a, 1969a,b,c). The re-

TABLE V

REACTIONS OF SODIUM (*N*-ACETYL-*N*-2-FLUORENYLHYDROXYLAMINE-β-D-GLUCOSID)-
URONATE (*N*-GlO-AAF; XXV) AND *N,O*-DIACETYL-*N*-2-FLUORENYLHYDROXYLAMINE
(*N*-ACETOXY-AAF; LXIX) WITH METHIONINE[a]

		Percent fluorene derivative converted to					
		o-Methylmercapto-2-acetylaminofluorene			*o*-Methylmercapto-2-aminofluorene		
Fluorene derivative	pH	Water	13% Ethanol	30% Acetone	Water	13% Ethanol	30% Acetone
N-GlO-AAF	5	—	0.6	0.1	—	0.01	0.01
	7	1.3	0.7	0.1	0.05	0.01	0
	8	1.3	0.7	—	2.5	2.5	—
	9	1.1	0.7	0.1	5.6	5.3	1.2
N-Acetoxy-AAF	5	—	64	66	—	0	0
	7	—	56	72	—	0	0
	9	—	30	24	—	0.3	0.1

[a] Data from Miller *et al.* (1968). Eight μmoles of *N*-GlO-AAF or *N*-acetoxy-AAF, 200 μmoles of L-methionine, and 200 μmoles of buffer were incubated under nitrogen in 3.0 ml of water, 13% ethanol, or 30% acetone for 22 hours at 37°C. *o*-Methylmercapto-2-acetylaminofluorene and -2-aminofluorene were determined by gas–liquid chromatography after extraction into benzene–hexane.

action of XXV with guanosine yielded a mixture of *N*-(guanosin-8-yl)-2-acetylaminofluorene (XXX) and *N*-(guanosin-8-yl)-2-aminofluorene (XXXI); reaction of XXV with GMP yielded a mixture of the corresponding GMP derivatives, XXXII and XXXIII (Table VI). Compound XXV reacted with rat liver rRNA, yeast sRNA, or calf thymus DNA. Although the rate of reaction of XXV with nucleic acids was slow, the reaction was linear with time (Fig. 4). *N*-Acetyl-*N*-2-fluorenylhydroxylamine (I), the aglycone of XXV, reacted more slowly with nucleic acids (Fig. 4). The reaction of XXV with RNA or DNA was pH-dependent with a higher rate of reaction occurring as the pH increased (Figs. 5 and 6). With increased reactivity of XXV with nucleic acids at higher pH values, there was an increase in the extent of loss of *N*-acetyl groups from the fluorenylamino residue which was bound (Table VII). The loss of the

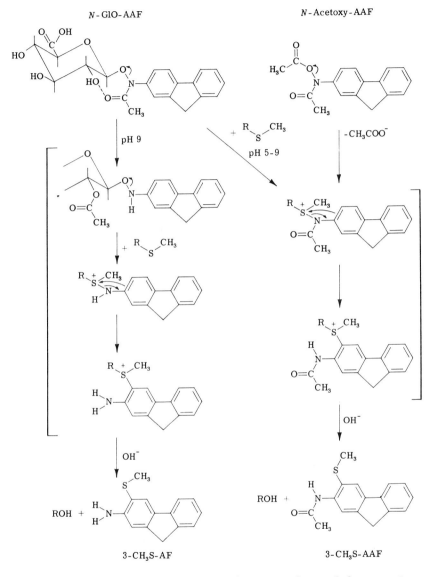

FIG. 3. Suggested mechanisms for the formation of 3-methylmercapto-2-ace-tylaminofluorene (3-CH$_3$S-AAF), 3-methylmercapto-2-aminofluorene (3-CH$_3$S-AF), and homoserine (ROH) by reaction of sodium (N-acetyl-N-2-fluorenylhydroxylamine-β-D-glucosid)uronate (N-GlO-AAF; XXV) with methionine (R-S-CH$_3$). (From Miller *et al.*, 1968.) The reactions of methionine and other nucleophiles wth N,O-diacetyl-N-2-fluorenylhydroxylamine (N-acetoxy-AAF; LXIX) are discussed further in Section V,D,1.

TABLE VI

REACTION OF SODIUM (*N*-ACETYL-*N*-2-FLUORENYLHYDROXYLAMINE-β-D-
GLUCOSID)URONATE(*N*-GlO-AAF; XXV) WITH GUANOSINE (GuO)
OR GUANOSINE 5′-MONOPHOSPHATE (GMP)[a]

Product	Percent reaction with *N*-GlO-AAF	
(structure)	Guanosine	GMP
GuO-AAF (XXX)	0.5	—
GuO-AF (XXXI)	1.4	—
GMP-AAF (XXXII)	—	2.8
GMF-AF (XXXIII)	—	5.8

[a] The data for the reaction of *N*-GlO-AAF(XXV) with guanosine are from Miller *et al.* (1968). Compound XXV (1 μmole) was incubated with 5 μmoles of guanosine at pH 7.0 for 20 hours at 37°C.

The data for the reaction of *N*-GlO-AAF (XXV) with GMP are from Irving *et al.* (1969b) and Irving and Russell (1969). Compound XXV (1 μmole) was incubated with GMP-[14]C (25 μmoles) at pH 7.4 for 20 hours at 37°C. The amounts of GMP-AAF (XXXII) and GMP-AF (XXXIII) were determined as described in Table VIII.

N-acetyl group from the residue bound and the increased reactivity of XXV at higher pH have at least two interpretations. Since XXV is alkaline labile (Section III,A,3), a more reactive species, specifically the glucuronide of *N*-2-fluorenylhydroxylamine (XXIV), might have been generated

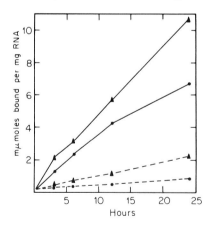

FIG. 4. Reaction of sodium (*N*-acetyl-*N*-fluorenylhydroxylamine β-D-glucosid)-uronate (XXV: ————) and *N*-acetyl-*N*-2-fluorenylhydroxylamine (I: – – – – –) with yeast tRNA (▲) and rat liver rRNA (●). (From Irving *et al.*, 1969a.) The reaction mixture contained 2 mg of RNA and 0.56 μmole of either XXV (————) or I (– – – – –) in 1 ml of 0.10 *M* NaCl–0.001 *M* tris-HCl; pH 7.6. The samples were incubated in air at 37°C.

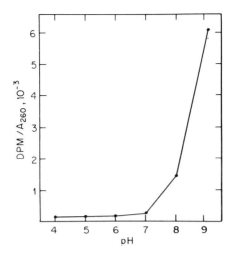

FIG. 5. Effect of pH on the reaction of sodium (N-acetyl-N-2-fluoren[9-^{14}C]ylhydroxylamine β-D-glucosid)uronate (XXV) with yeast tRNA. (From Irving *et al.*, 1969a.)

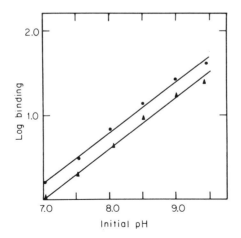

FIG. 6. Kinetics of the pH-dependence of the reaction of sodium (N-acetyl-N-2-fluoren[9-^{14}C]ylhydroxylamine-β-D-glucosid)uronate (XXV) with calf thymus DNA (●) and yeast tRNA (▲). (From Irving *et al.*, 1969a.) Binding is expressed as mμmoles fluorene residue bound per milligram of nucleic acid.

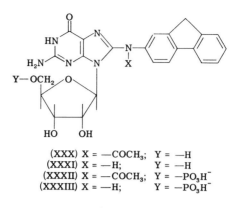

(XXX) X = —COCH₃; Y = —H
(XXXI) X = —H; Y = —H
(XXXII) X = —COCH₃; Y = —PO₃H⁻
(XXXIII) X = —H; Y = —PO₃H⁻

at higher pH or even at neutral pH on prolonged incubation. Indeed, XXIV has been synthesized from XXV (Irving and Russell, 1969) and does react at a faster rate with GMP than does XXV (Table VIII). Alter-

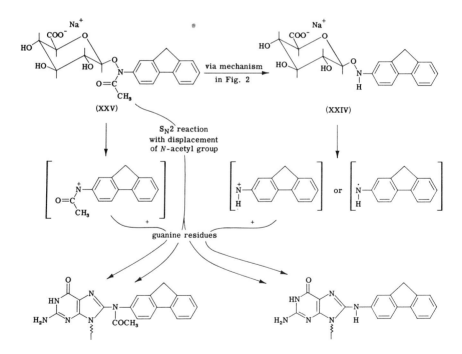

Fig. 7. Suggested mechanisms involved in the reactions of sodium (*N*-acetyl-*N*-2-fluorenylhydroxylamine β-ᴅ-glucosid)uronate (XXV) and sodium (*N*-2-fluorenylhydroxylamine-β-ᴅ-glucosid)uronate (XXIV) with nucleic acids (guanine residues) or with guanosine or GMP.

TABLE VII

INFLUENCE OF pH ON THE DIFFERENCE IN BINDING OF ^{3}H AND ^{14}C TO YEAST tRNA UPON REACTION *in Vitro* WITH SODIUM $(N\text{-}[2'\text{-}^3\text{H}]\text{-ACETYL-}N\text{-}2\text{-}$FLUOREN$[9\text{-}^{14}\text{C}]$YLHYDROXYLAMINE-β-D-GLUCOSID)URONATE (XXV)[a]

pH of reaction mixture	mμmoles Fluorene compound bound per mg to RNA		Percent of bound fluorene residues with N-acetyl group
	^{3}H	^{14}C	
7.0	0.39	0.89	43.8
7.5	0.41	1.47	27.9
8.0	0.34	3.64	9.3
8.5	0.34	8.96	3.8
9.0	0.34	16.8	2.0
9.5	0.24	21.1	1.1

[a] Data from Irving *et al.* (1969b,c). The reaction was carried out as described in Fig. 3, except the incubation was for 22 hours at 25°C.

TABLE VIII

REACTION OF SODIUM $(N\text{-}2\text{-FLUORENYLHYDROXYLAMINE-}β\text{-D-GLUCOSID})$URONATE $(N\text{-GlO-AF}; XXIV)$ WITH GUANOSINE 5'-MONOPHOSPHATE (GMP)[a]

Time incubated (minutes)	$\dfrac{[GMP]}{[N\text{-GlO-AF}]}$	GMP-AF (XXXIII) formed (expressed as % reaction of GMP with N-GlO-AF)
5	25	31.5
30	25	33.2
120	25	32.4
180	25	33.8
15	1	6.8
15	2.5	12.7
15	5	16.7
15	10	22.6
15	25	34.1
15	50	34.1

[a] Data from Irving and Russell (1969). A solution containing 1.3–65 μmoles of GMP-^{14}C(uniformly labeled) in 0.2 ml of 0.1 M tris-HCl, pH 7.4, was added to 0.5 mg (1.3 μmoles) of solid N-GlO-AF. About 3–4 minutes were required for the N-GlO-AF to dissolve completely. At the time indicated, the solution was chromatographed on a column (0.9 × 60 cm) of Sephadex G-25 equilibrated with 0.1 M ammonium bicarbonate–2 M urea, pH 9.0. The column was washed with the same buffer (flow rate, 20 ml/hour) and fractions of 1 ml each were collected. Unreacted GMP was eluted in fractions 15–35, carrier GMP-AAF (XXXII) in fractions 40–60, and carrier GMP-AF (XXXIII) in fractions 70–100. The amount of product, GMP-AF, was calculated from the radioactivity in fractions 70–100.

TABLE IX

O-GLUCURONIDES OF N-HYDROXY COMPOUNDS WHICH HAVE NOT BEEN ISOLATED OR SYNTHESIZED BUT FOR WHICH
THERE IS GOOD EVIDENCE FOR THEIR FORMATION *in Vivo*

Aglycone (Ar-N⟨ OH / COCH$_3$)

Ar =

Ar =	Compound administered	Species	Evidence[a]	Reference
	7-Fluoro-2-acetylaminofluorene	Rat	A,B	Miller *et al.* (1960); Miller *et al.* (1966b)
	Aglycone 4-Aminobiphenyl 4-Acetylaminobiphenyl	Rat, dog	A,B,C	Miller *et al.* (1961)
	Aglycone 4-Aminostilbene 4-Acetylaminostilbene	Rat	A,B,C	Andersen *et al.* (1964); Baldwin and Smith (1965)
	Aglycone 4-Aminoazobenzene N-Acetyl-4-aminoazobenzene N-Methylaminoazobenzene N,N-Dimethylaminoazobenzene	Rat, hamster, mouse	A,B	Sato *et al.* (1966)
	2-Acetylaminophenanthrene	Rat	A,B	Miller *et al.* (1966b)

[a] Evidence: A, hydrolysis by β-glucuronidase. B, Chromatographic and spectrophotometric identification of aglycone. C, Isolation and chemical identification of aglycone.

nately, an S_N2 reaction of XXV with guanine residues of the nucleic acids might be facilitated at neutral pH with a displacement of the N-acetyl group of XXV. Hypothetical intermediates involved in the reactions of XXV and XXIV with nucleic acids (guanine residues) are shown in Fig. 7.

B. GLUCURONIDE CONJUGATES WHICH HAVE NOT BEEN FULLY CHARAC-
TERIZED

1. *N-Acetyl-N-arylhydroxylamines*

N-Acetylated arylhydroxylamines are found in urine or bile almost entirely as their O-glucuronide conjugates (XVII, X = aryl group, Y = —$COCH_3$). None, or only traces, of the unconjugated N-acetylaryl-hydroxylamines are excreted. Except for the compounds described in Section III, A, 2, most of these glucuronide conjugates have not been directly characterized. Instead, the aglycone has been identified after treatment of the urine with β-glucuronidase. In many instances, however, the investigators have not reported performing a control experiment in which β-glucuronidase was either omitted or in which the enzymic activity was inhibited by the addition of saccharo(1–4)lactone (Levvy and Conchie, 1966). Observations on the urinary excretion of these O-glu-curonides of N-hydroxylated arylacetamides are summarized in Table IX. The urinary excretion of glucuronides of the N-hydroxy-N-acetylated metabolites of 1-naphthylamine, benzidine, dichlorobenzene, acetanilide, and phenacetin by humans has been reported (Belman *et al.*, 1968). However, the N-hydroxy compounds, obtained after treatment of urine with β-glucuronidase, were not positively identified in these experiments.

2. *Arylhydroxylamines*

Arylhydroxylamines appear to be excreted predominantly in uncon-jugated form. Although there are a few reports in the literature in which investigators have reported the detection of O-glucosiduronic acid con-jugates of arylhydroxylamines (XV, X = —aryl, Y = H) in urine, there is no substantial evidence that this type of conjugate has, in fact, been detected. Kiese *et al.* (1966) reported that a conjugate of N-2-fluorenyl-hydroxylamine (IV) was excreted in the urine of guinea pigs after ad-ministration of 2-aminofluorene (III) and that IV was liberated by treatment of the urine with β-glucuronidase. It was implied that the con-jugate was the glucuronide of IV (XXIV). Weisburger *et al.* (1966) have also suggested that small amounts of XXIV were excreted in the urine of rats given III. It seems doubtful that these investigators detected XXIV, since the compound is extremely unstable in aqueous solutions (Irving and Russell, 1969).

Boyland (1963) claimed to have detected the *O*-glucuronide of 2-naphthylhydroxylamine (XXXIV) as a metabolite of 2-naphthylamine in rat urine. However, in a subsequent detailed study, Boyland and Manson

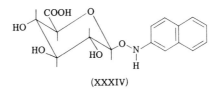

(XXXIV)

(1966) found no evidence for the presence of XXXIV as a metabolite of 2-naphthylamine or 2-naphthylhydroxylamine in the urine of the dog, guinea pig, hamster, rabbit, or rat.

3. Other *N-Hydroxy Compounds*

N-Hydroxyurethane-*N*-glucuronide (XXXV) has been tentatively identified as a metabolite of *N*-hydroxyurethane in the urine of the rat and

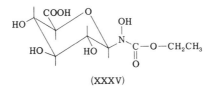

(XXXV)

mouse (Mirvish, 1966). The material obtained by Mirvish from urine of rats given *N*-hydroxyurethane was not pure and was not adequately characterized. The crude product might have been contaminated with some of the *O*-glucuronide of *N*-hydroxyurethane (XXXVI), although no evidence for this was given (Mirvish, 1966). Boyland and Nery (1965) were

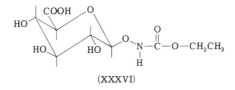

(XXXVI)

unable to detect a glucuronide of *N*-hydroxyurethane as a metabolite of urethane or *N*-hydroxyurethane in the urine of several species, including the rat, when they treated urine with β-glucuronidase and chromatographed the hydrolyzate. However, Mirvish found that a crude fraction, assumed to be mostly XXXV, was not hydrolyzed with β-glucuronidase and was hydrolyzed only with difficulty by hot acid, resembling in this latter respect known *N*-glucuronides of meprobamate (Tsukamoto *et al.*, 1963) and sulfadimethoxine (Bridges *et al.*, 1965). The rapid acid hydrolysis of arylamine *N*-glucuronides presumably depends on the ready protonation of the basic nitrogen atom, which would occur with difficulty with the *N*-glucuronides mentioned above (Mirvish, 1966).

Rabbits excreted about 30% of a dose of *p*-aminopropiophenone as the *N*-hydroxy derivative (von Jagow *et al.*, 1966; von Jagow and Kiese, 1967). The *N*-hydroxy metabolite appeared in the urine primarily as a conjugate which was readily split by incubation of the urine under nitrogen at pH 4.5 and 37°C. However, no increase in free *N*-hydroxy-*p*-aminopropiophenone was observed after incubation of the urine with β-glucuronidase. It does not seem likely that the conjugate was an *O*-glucuronide and the instability in acid solution would seem to exclude the possibility of an *N*-glucosiduronic acid (see above paragraph). The determination of the nature of this conjugate needs further investigation.

C. Biosynthesis

Mechanisms for the biosynthesis of glucuronides have been reviewed by Dutton (1966). Although there are no published studies on the biosynthesis of glucuronides of *N*-hydroxy compounds *in vitro*, it seems likely that the principal route of biosynthesis involves uridine diphosphate (UDP)-glucuronic acid and microsomal glucuronyl transferase. Irving (1964) reported that trace amounts of the glucuronide of *N*-acetyl-*N*-2-fluorenylhydroxylamine (XXV) were formed by incubation of the *N*-hydroxy compound I with the 10,000 × *g* supernatant fraction of rabbit liver homogenate in the absence of exogenous UDP-glucuronic acid. Recent studies (Irving, Veazey, and Russell, in preparation) showed that the glucuronide XXV was formed from I and UDP-glucuronic acid by rat liver microsomes. The glucuronide XXV was separated from the incubation mixture by extraction and chromatography on DEAE-cellulose (Irving *et al.*, 1967b). The glucuronide in the fraction from the DEAE-cellulose column was hydrolyzed with bacterial β-glucuronidase and the aglycone was extracted from the mixture and identified as I by gas–liquid chromatography.

D. Hydrolysis with β-Glucuronidase

Biosynthetic *O*-glucuronides of *N*-acetylated-*N*-arylhydroxylamines are readily cleaved by β-glucuronidase. Treatment of urine collected from animals given a variety of aromatic amines or their N-acyl derivatives has been a necessary step in the identification of the corresponding N-hydroxy-N-acetylated metabolites, since these compounds are excreted almost exclusively as the glucuronide conjugates (Section III, E). Enzymic hydrolysis of biosynthetic sodium (N-acetyl-N-2-fluorenylhydroxylamine β-D-glucosid)uronate (XXV) with bacterial β-glucuronidase gave a quanitative yield of the aglycone I (Hill and Irving, 1967; Hill, 1968). The most detailed studies in this area have been those on the hydrolysis of the glucuronide of N-acetyl-N-phenylhydroxylamine (Ide *et al.*, 1968). Biosynthetic sodium (N-acetyl-N-phenylhydroxylamine β-D-glucosid)uronate (XXVI) was hydrolyzed almost completely by highly purified mouse urinary β-glucuronidase, giving the products, N-acetyl-N-phenylhydroxylamine and glucuronic acid in the expected molar quantities. The hydrolysis of XXVI was inhibited 88% by the addition of saccharo-(1-4)lactone (0.1 mM), a known inhibitor of β-glucuronidase activity (Levvy and Conchie, 1966).

E. EXCRETORY PATHWAYS

Glucuronide conjugates of N-acetylated-N-arylhydroxylamines are excreted in the urine of a variety of species (Section III,A,B). Most available information on the excretion of glucuronide conjugates of N-hydroxy compounds comes from studies on the metabolism of the carcinogens 2-acetylaminofluorene (II) and N-acetyl-N-2-fluorenylhydroxylamine (I). Since a number of reviews are available on this subject (see Miller and Miller, 1969a,b, and references therein; Arcos and Argus, 1968), only some of the findings will be mentioned here. There are marked species and sex differences in the urinary excretion of the glucuronide of N-acetyl-N-2-fluorenylhydroxylamine (XXV). Continuous feeding of II to rats produced a steady increase in the urinary output of XXV until a maximum high level was reached after 12–18 weeks (Fig. 8) amounting to 10–15% of the ingested dose of II (Miller *et al.*, 1960, 1961). Unlike the case with II, continuous feeding of 4-acetylaminobiphenyl or its N-hydroxy derivative to adult male rats for several weeks led to a gradual decrease in the urinary excretion of the O-glucuronide of N-hydroxy-4-acetylaminobiphenyl. Female rats excreted more of XXV than did male rats after administration of I (Weisburger *et al.*, 1964a). Mice excreted XXV in smaller amounts than did rats, and the urine of guinea pigs did not contain detectable amounts of XXV (Miller *et al.*, 1960). XXV is a major urinary metabolite of II in the rabbit; by the third day of a feeding, XXV accounted for approximately 30% of the dose of administered II (Irving, 1962b). The excretion of XXV remained at a level of 20–30% of the dose per day for several weeks.

Frequently, urinary levels of XXV have been equated with endogenous production or gross tissue levels of this metabolite in the rat and other

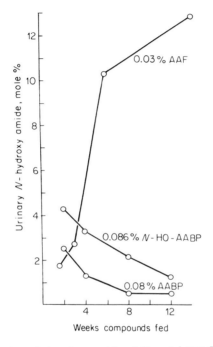

Weeks compounds fed

FIG. 8. The excretion of the glucuronide of N-acetyl-N-2-fluorenylhydroxylamine (XXV) as a function of time of feeding 0.03% 2-acetylaminofluorene (AAF; II) in the diet of rats (from Miller *et al.*, 1960, 1961). Data for the excretion of the glucuronide of N-acetyl-N-4-biphenylylhydroxylamine after feeding N-acetyl-N-4-biphenylylhydroxylamine (N-HO-AABP) or 4-acetylaminobiphenyl (AABP) are given for comparison. Data expressed as percent of the ingested dose of II, N-HO-AABP, or AABP excreted as the N-hydroxy amide (determined after addition of β-glucuronidase to the urine). Each point represents the average excretion for 24 hours for four adult male rats.

species (Lotlikar *et al.*, 1964; Margreth *et al.*, 1964; Shirasu *et al.*, 1966, 1967a,b,c; Weisburger *et al.*, 1964a). Many of the suggestions and conclusions regarding the effects of hepatoxic agents and hormones on the metabolism of 2-acetylaminofluorene (II) or its N-hydroxy derivative (I) *in vitro*, which were based upon measurements of urinary levels of XXV, are not necessarily valid since no consideration was given to the total (urine, bile, and feces) excretion of XXV. Only a small fraction of XXV which is formed by the rat after a single dose of II or I is excreted in the urine (Irving *et al.*, 1967b). Approximately 10% of an oral dose of II or 21% of a single dose of I was excreted by the rat in the bile in 24 hours as XXV, while less than 1–1.5% of the dose of II or I was excreted in the urine of normal rats in 24 hours as the metabolite XXV.

As an example of a mistaken interpretation drawn from consideration of data on urinary excretion alone, earlier experiments on the urinary excretion of XXV after administration of a single dose of II to the rabbit led to the conclusion that the rabbit formed more of XXV *in vivo* than did the rat (Irving, 1962b). However, this conclusion was later shown to be erroneous since the total excretion (bile plus urine) of XXV by the rat and rabbit after a single dose of II was not significantly different (Irving *et al.*, 1967b).

As mentioned above, continued administration of 2-acetylaminofluorene (II) to rats results in a gradually increasing urinary excretion of the glucuronide of the N-hydroxy metabolite (XXV). Miller *et al.* (1960) speculated that this rise was related to the progressive liver damage caused by the carcinogen, resulting in a lower capacity of liver to further metabolize I to nontoxic products. Other explanations for the observed gradual increase in the excretion of XXV (Fig. 8) appear possible. Morris *et al.* (1958) showed that the biliary excretion of Rose Bengal-^{131}I was severely impaired during carcinogenesis in the rat with the closely related carcinogen, N,N-diacetylaminofluorene. Degenerative changes on the surface of the bile canaliculi following the feeding of N,N-diacetylamino-fluorene to rats were later reported (Mikata and Luse, 1964). Since most of the glucuronide XXV formed in the liver of the rat is normally excreted in the bile, it seems likely that with continued feeding of II, resulting in progressive liver damage with a decreased ability of the liver to excrete XXV in the bile, there is a shift in the excretion of XXV to urinary excretion.

There are considerable species differences in the biliary excretion of the glucuronide XXV. The rabbit did not excrete detectable amounts of XXV in the bile after administration of either 2-acetylaminofluorene (II) or its N-hydroxy derivative (I), but did excrete up to 30% of the dose of either of these compounds as XXV in the urine (Irving, 1962b; Irving *et al.*, 1967b).

The metabolic fate of the glucuronide XXV after biliary excretion in the rat has been studied. Most of the conjugate appears to be destroyed, either because of alkaline conditions existing in the intestinal lumen (Section III,A,3) or because of enzymic changes due to bacterial action in the cecum and large intestine. Williams *et al.* (1968) reported that XXV was hydrolyzed by bacterial enzymes in the gut to form the free *N*-hydroxy derivative I, which was then reduced to the amide II by bacterial action (Grantham *et al.*, 1968; Williams *et al.*, 1968). The high level of excretion of XXV in the bile of the rat after administration of II or I, and the subsequent chemical degradation and/or metabolism of XXV in the intestinal lumen, followed by reabsorption of the resulting metabolites, leads to considerable, perhaps unusually prolonged, enterohepatic circulation of metabolites of these carcinogens in the rat. This problem is discussed briefly by Ide *et al.* (1968).

The metabolism of XXV in the rat has been studied by administration of the ^{14}C-labeled conjugate (Irving and Wiseman, 1969). After subcutaneous or intraperitoneal administration of the ^{14}C-labeled XXV, 50% of the radioactivity injected was excreted in the urine and 20% in the feces in 24 hours. Oral administration of ^{14}C-labeled XXV to the rat changed the pattern of excretion; 33% of the radioactivity was excreted in the urine and 38% in the feces in 24 hours. The nature of the urinary metabolites of XXV was also markedly influenced by the route of administration of the conjugate, with the least amount of metabolic change occurring after subcutaneous administration. Following subcutaneous administration of XXV, approximately 55% of the conjugate was excreted unchanged in the bile in 24 hours. In the intestinal lumen, the conjugate was then probably metabolized by bacterial enzymes as discussed above.

IV. *N*- and *O*-Sulfonate Conjugates of *N*-Hydroxy Compounds

A. INTRODUCTION AND DEFINITIONS

Three types of structures resulting from the conjugation of sulfate with *N*-hydroxy compounds are possible: *N*-sulfonic acids (XXXVII) and *O*-sulfonic acids (XXXVIII) of *N*-aliphatic or *N*-arylhydroxylamines, and *O*-sulfonic acids of *N*-acyl derivatives of *N*-aliphatic or *N*-arylhydroxylamines (XXXIX). Compound XXXVII, which is an *N*-substituted derivative of hydroxylamine-*N*-sulfonic acid, will be referred to in this review

$$R{-}N\begin{matrix} OH \\ SO_3H \end{matrix} \qquad R{-}N\begin{matrix} OSO_3H \\ H \end{matrix} \qquad R{-}N\begin{matrix} OSO_3H \\ C{-}R \\ \parallel \\ O \end{matrix}$$

(XXXVII)   (XXXVIII)   (XXXIX)

as an *N*-sulfonic acid (salt form, *N*-sulfonate), whereas, XXXVIII and XXXIX are *N*-substituted (XXXVIII) and *N*-acyl-*N*-substituted (XXXIX) derivatives of hydroxylamine-*O*-sulfonic acid and will be referred to as *O*-sulfonic acids (salt forms, *O*-sulfonates).

There is no direct evidence for the formation *in vivo* of sulfonic acid derivatives of *N*-hydroxy compounds. Several synthetic sulfonates of

TABLE X

Synthetic *O*-Sulfonate Derivatives of Substituted Arylhydroxylamines[a]

$$O\text{-}SO_3^- \text{ —M}^+$$

Aryl-N
$\diagdown$ X

Aryl group	Substitution X	M$^+$	m.p.[b] (°C)	Reference
Phenyl	H	K$^+$	170–174	Boyland and Nery (1962)
Phenyl	—COCH$_3$	K$^+$	151	Boyland and Nery (1962)
Phenyl	—COCH$_3$	NH$_4^+$	122	Boyland and Nery (1962)
Phenyl	—COOCH$_2$C$_6$H$_5$	K$^+$	133	Boyland and Nery (1962)
Phenyl	—COOCH$_2$C$_6$H$_5$	NH$_4^+$	122	Boyland and Nery (1962)
2-Naphthyl	—COCH$_3$	K$^+$	148–150	Boyland and Manson (1966)
2-Naphthyl	—COOCH$_2$C$_6$H$_5$	NH$_4^+$	97–101	Boyland and Manson (1966)

[a] Each of these compounds has been characterized by elemental analysis.
[b] These compounds melt with decomposition.

N-hydroxy compounds have been made and these compounds are very reactive in aqueous systems. If these conjugates were formed *in vivo*, they would probably exist only transiently because of their instability and reactivity, and hence they would not be detectable in urine, bile, or other tissue fluids. On the other hand, by the use of indirect methods, such as trapping the conjugates *in situ* by reaction with nucleophilic compounds, the formation of *O*-sulfonate derivatives of a few *N*-hydroxy compounds has been demonstrated *in vitro* in fortified liver cell fractions. Therefore, the possibility of the formation of these conjugates *in vivo* from *N*-hydroxy compounds must be considered.

B. Synthetic Conjugates

A few *N*- and *O*-sulfonates of *N*-hydroxy compounds have been synthesized; these compounds have been characterized as their potassium or ammonium salts (Tables X and XI). In general, these have been derivatives of phenylhydroxylamine or 2-naphthylhydroxylamine. The derivatives were synthesized by reaction of the arylhydroxylamine or the

TABLE XI

SYNTHETIC N-SULFONATE DERIVATIVES OF SUBSTITUTED ARYLHYDROXYLAMINES[a]

Aryl-N-O-X
|
SO_3^- —M$^+$

Aryl group	Substitution		m.p.[b]	Reference
	X	M$^+$	(°C)	
Phenyl	H	K$^+$	128	Boyland and Nery (1962)
Phenyl	H	NH$_4^+$	156	Boyland and Nery (1962)
Phenyl	—SO$_3^-$—K$^-$	K$^-$	Decomp. >150	Boyland and Nery (1962)
2-Naphthyl	H	K$^+$	170–172	Boyland and Manson (1966)

[a] Each of the compounds has been characterized by elemental analysis.
[b] The compounds melt with decomposition.

N-acylarylhydroxylamine with pyridine-SO$_3$ (Boyland and Nery, 1962). With the arylhydroxylamine, the N-sulfonic acid derivative is obtained, but with more vigorous treatment, the N,O-disulfonic acid derivative can be formed:

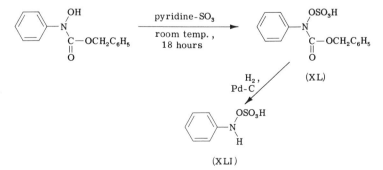

O-Sulfonic acid derivatives of N-acylarylhydroxylamines are obtained under similar conditions:

Ar—N(OH)(C—R, ‖O) pyridine-SO$_3$, 0°, 16 hours → Ar—N(OSO$_3$H)(C—R, ‖O)

R = —CH$_3$ or —OCH$_2$C$_6$H$_5$

Phenylhydroxylamine-O-sulfonic acid (XLI) was prepared indirectly by catalytic hydrogenation of N-benzyloxycarbonylphenylhydroxylamine-O-sulfonic acid (XL):

(XL) pyridine-SO$_3$, room temp., 18 hours

H$_2$, Pd-C

(XLI)

Attempts to prepare 2-naphthylhydroxylamine-O-sulfonic acid (XLIII)
by this method were unsuccessful (Boyland and Manson, 1966). A crude

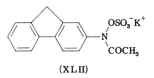

(XLII)

preparation obtained from the reaction of N-acetyl-N-2-fluorenyl-
hydroxylamine (I) with pyridine-SO_3 was reported to contain at least
50% potassium N-acetyl-N-2-fluorenylhydroxylamine-O-sulfonate (XLII)
on the basis of its reactivity with methionine (Section IV, D, 2). However,
XLII has not been obtained in a sufficient state of purity for characterization by elemental analysis (Maher et al., 1968).

C. BIOSYNTHESIS

Direct evidence for the biosynthesis in vivo of N- and O-sulfonates of
N-hydroxy compounds is not available. Because of the instability of these
conjugates in aqueous systems (Section IV, A), it is unlikely that such
conjugates will ever be detected by direct isolation from urine or other
body fluids. DeBaun et al. (1968) indicated that the half-life of XLII in
water was less than 1 minute.

Boyland and Nery (1962) cited unpublished work by Boyland and
Manson in which they reported the detection of 2-naphthylhydroxylamine-
O-sulfonic acid (XLIII) as a metabolite of 2-naphthylamine. Boyland
(1963) also reported the identification and isolation of XLIII as a metabolite of 2-naphthylamine in the urine of rats and dogs. However, in a
later publication (Boyland and Manson, 1966), attempts to synthesize
XLIII were not successful and no evidence was obtained for the formation of XLIII as a metabolite of 2-naphthylamine or of XLIV as a metabolite of N-acetyl-2-naphthylamine in the urine of rats, guinea pigs, rabbits, or hamsters. In the light of these data, it may be questioned whether
XLIII was detected in the earlier experiments quoted by Boyland and
Nery (1962) and by Boyland (1963).

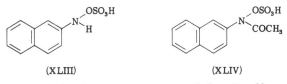

(XLIII)   (XLIV)

The possibility of the formation in vivo of the O-sulfonate of N-hydroxyurethane (XLV) has been mentioned (Nery, 1968); however, XLV

has not been synthesized or detected as a metabolite of urethane or N-hydroxyurethane (Nery, 1968).

$$CH_3CH_2-O-\overset{\displaystyle O}{\underset{\displaystyle \underset{H}{N}}{C}}OSO_3H$$

(XLV)

The formation of adenosine-3'-phospho-5'-phosphosulfate (PAPS) and the role of PAPS in sulfate conjugation has been reviewed by Robbins (1962) and more briefly by Parke (1968) and Hargreaves (1968). The biosynthesis of sulfonates of N-hydroxy compounds would involve transfer of sulfonate from PAPS to the N-hydroxy compounds:

There appear to be at least three different types of sulfonate-transferring enzymes (sulfotransferases or sulfokinases): steroid sulfokinases, which differ from phenol kinases (Nose and Lipmann, 1958), both of which form O-sulfonic acid derivatives, and arylamine sulfokinases (Roy, 1960), which catalyze the formation of N-sulfonic acid derivatives of aromatic amines. Each of these sulfokinases is found in the liver; phenol sulfokinases are found in kidney and intestinal mucosa as well as in liver. Which of these sulfokinases might be involved in the biosynthesis of N- and O-sulfonate conjugates of N-hydroxy compounds of the types described in Section IV, A is not known. However, it seems reasonable that a phenol sulfokinase might effect the biosynthesis of the O-sulfonate conjugates (XXXVIII and XXXIX) and that an arylamine sulfokinase might be involved in the formation of the N-sulfonate conjugates (XXXVII) if, indeed, these conjugates are formed in vivo.

Indirect evidence for the formation and transient existence of the O-sulfonate conjugate of N-acetyl-N-2-fluorenylhydroxylamine (XLII) in vitro has been obtained (King and Phillips, 1968, 1969; DeBaun et al., 1968, 1969; Lotlikar, 1969; Miller and Miller, 1969a). The enzymic synthesis of XLII in vitro in liver preparations has been demonstrated by employing an assay which measures the PAPS-dependent formation of

1- and 3-methylmercapto-2-acetylaminofluorene (XLVI) from N-acetyl-N-2-fluorenylhydroxylamine (I) and methionine:

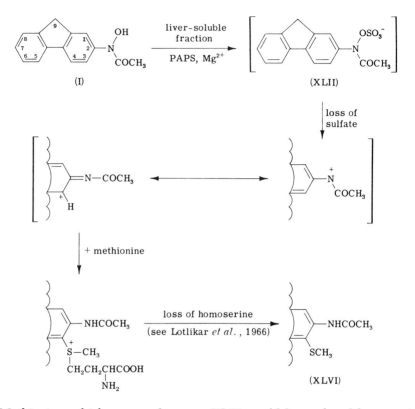

Methionine, which was used to trap XLII, could be replaced by protein or by tRNA or DNA, leading to the binding of some fluorene moiety to these macromolecules. The overall conversion of the N-hydroxy compound I to XLVI was completely dependent upon PAPS, either generated *in situ* from ATP (Table XII) or added (Table XIII). The reactions were stimulated by Mg^{++} which is required for the formation of PAPS from ATP (Robbins, 1962) and for the full activity of several sulfotransferases (Banerjee and Roy, 1966). EDTA inhibited the conversion of I to XLVI when ATP was used to generate PAPS *in situ* (Table XII). Appreciable activity was found in male rat liver; female rat liver and male rabbit liver had considerably lower activities (Table XIV). Only traces of activity were detected in the livers of the hamster, mouse, and guinea pig (Table XIV).

Employing an assay similar to the one described above, N-acetyl-N-4-

TABLE XII

ACTIVATION OF N-ACETYL-N-2-FLUORENYLHYDROXYLAMINE (I) BY THE SOLUBLE
FRACTION OF RAT LIVER AS MEASURED BY REACTION WITH METHIONINE[a]

Additions (final conc.)	o-Methylmercapto-2-acetylaminofluorene (XLVI) formed (μg)
ATP (0.01 M)	1
MgCl$_2$ (0.005 M)	1
ATP + MgCl$_2$	21
ATP + MgCl$_2$ + K$_2$SO$_4$ (0.01 M)	39
ATP + MgCl$_2$ + K$_2$SO$_4$ + EDTA (0.04 M)	4
ATP + MgCl$_2$ + K$_2$SO$_4$ + EDTA + PAPS (0.001 M)	57

[a] Data from DeBaun et al. (1968). Each tube contained, in a final volume of 2 ml, 1 mg of N-acetyl-N-2-fluorenylhydroxylamine (I), 7 mg of L-methionine, and 0.7 ml of 100,000 g (1 hour) supernatant from a 20% rat liver homogenate in 0.05 M tris buffer, pH 7.0. Incubation in air for 2 hours at 37°C.

biphenylylhydroxylamine was converted to 3-methylmercapto-4-acetyl-aminobiphenyl by male rat liver in vitro (Miller and Miller, 1969a). King and Phillips (1968, 1969) and DeBaun et al. (1968) observed that the binding of radioactivity from N-1'-^{14}C-acetyl-N-2-fluorenylhydroxylamine to carrier nucleic acids or protein by a soluble rat liver enzyme in vitro was dependent upon the addition of ATP, sulfate, and either Mn^{++} or Mg^{++}. They inferred that the binding of radioactivity was due to the intermediate formation of the O-sulfonate of the N-hydroxy compound. The combination of ATP, Mg^{++}, and sulfate could be replaced by PAPS (DeBaun et al., 1968).

Thus, the enzymic formation of the O-sulfonate conjugates of at least two N-acetyl-N-arylhydroxylamines has been demonstrated in vitro. Al-

TABLE XIII

REQUIREMENTS FOR THE PAPS-DEPENDENT FORMATION OF o-METHYLMERCAPTO-2-ACETYLAMINOFLUORENE (XLVI) FROM N-ACETYL-N-2-FLUORENYLHYDROXYLAMINE (I) AND METHIONINE BY A SOLUBLE ENZYME FROM RAT LIVER[a]

Additions (final conc.)	o-Methylmercapto-2-acetylaminofluorene formed (μg)
PAPS (0.0006 M)	37
Mg^{++} (0.005 M)	<1.5
PAPS + Mg^{++}	60
PAPS + Mg^{++} + EDTA (0.004 M)	32
PAPS + Mg^{++} (no enzyme)	<1.5

[a] Data from Miller and Miller (1969a). Each tube contained, in a final volume of 1 ml, 0.2 mg of N-acetyl-N-2-fluorenylhydroxylamine (I), 5 mg of L-methionine, 0.1 ml of liver supernatant (20% homogenate centrifuged at 105,000 g for 1 hour; approximately 2.5 mg of protein) and 25 μmoles of tris-HCl buffer, pH 7.0.

though the data raise the possibility that these conjugates may be formed *in vivo*, unequivocal proof that O-sulfonates of N-hydroxy compounds are actually formed *in vivo* is lacking at this time. An important fact to consider in the interpretation of such data is the number and relative

TABLE XIV

COMPARISON OF SULFOTRANSFERASE ACTIVITIES IN RODENT LIVERS *in Vitro* WITH N-ACETYL-N-2-FLUORENYLHYDROXYLAMINE (I) AS SUBSTRATE[a]

Species[b]	μg o-Methylmercapto-2-acetylaminofluorene/ mg protein/30 minutes
Rat (male)	23 ± 2
Rat (female)	4 ± 1
Mouse	0.7
Hamster	<0.5
Guinea pig	0.5
Rabbit[c]	2 ± 0.5

[a] From DeBaun *et al.* (1969). The amount of o-methylmercapto-2-acetylaminofluorene formed on incubation of liver-soluble fraction with ATP, Mg^{++}, sulfate ion, N-acetyl-N-2-fluorenylhydroxylamine (I) and methionine was determined.

[b] All male except for rat as indicated.

[c] The values given for the rabbit are for o-methylmercapto-2-aminofluorene plus o-methylmercapto-2-acetylaminofluorene formed. Deacetylation and destruction of 3-methylmercapto-2-acetylaminofluorene in rabbit liver has been reported (DeBaun *et al.*, 1968).

rates of metabolic reactions which compete for the substrate *in vivo*. These competing reactions may not occur *in vitro* because of the lack of required cofactors. For example, N-acetyl-N-2-fluorenylhydroxylamine (I) is rapidly conjugated with UDP-glucuronic acid, both *in vivo* and *in vitro* (see Section III, C). The N-hydroxy compound I is also deacetylated *in vitro* by liver microsomes (Irving, 1966) and I is also reduced to the amide II (—N(OH)-COCH₃ ——→ —NHCOCH₃) by enzymes in the soluble fraction (Grantham *et al.*, 1965; Lotlikar *et al.*, 1965) (see Section V,D,3). Information on the relative rates of these reactions is not available.

No studies have been reported on the biosynthesis of N-sulfonates of N-hydroxy compounds *in vitro* nor have these conjugates been detected *in vivo* as metabolites of substituted hydroxylamines.

D. REACTIVITY

1. *Rearrangements*

N- and O-Sulfonates of N-hydroxy compounds decompose in aqueous solution; rearrangement products are formed. Potassium N-acetyl-N-2-

naphthylhydroxylamine-O-sulfonate (XLIV, potassium salt) in water, in acetate buffer, pH 7, or in acid solution (2 N hydrochloric acid) at room temperature did *not* give N-acetyl-N-2-naphthylhydroxylamine (XLVII) but yielded 2-acetamido-1-naphthyl hydrogen sulfate (XLVIII) and 2-acetamido-1-naphthol (XLIX) (Boyland and Manson, 1966):

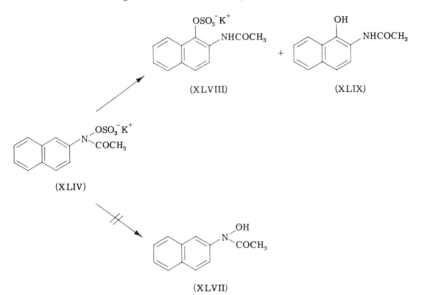

An intramolecular migration of —OSO₃H from nitrogen to carbon was suggested:

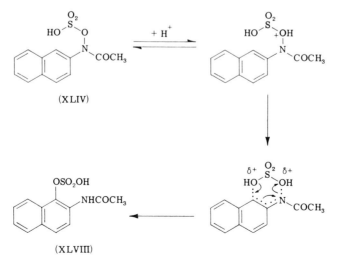

The N- and O-sulfonates of phenylhydroxylamine also undergo rearrangement in acid solution (Boyland and Nery, 1962). Phenylhydroxylamine-N-sulfonic acid (L) in 2 N hydrochloric acid yielded p-aminophenol (LI), while the corresponding O-sulfonic acid gave mainly o-aminophenyl hydrogen sulfate and o-aminophenol. Boyland and Nery (1962) speculated that L was probably hydrolyzed to phenylhydroxylamine (LII) which underwent an intermolecular rearrangement, resulting in the formation of LI:

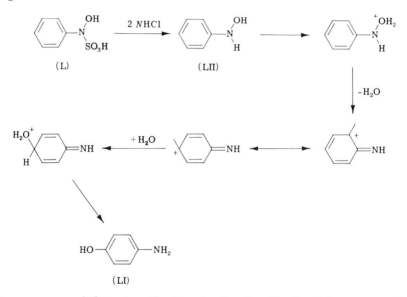

It was proposed that phenylhydroxylamine-O-sulfonic acid rearranged by the intramolecular mechanism shown above for XLIV. Later, Boyland and Manson (1967) reported that the N-sulfonate (L) and 2-naphthylhydroxylamine-N-sulfonic acid in acetone solution were converted to o-aminophenylsulfate (LIII) and 2-amino-1-naphthylsulfate, respectively. An N ⟶ O shift of —SO₃H followed by migration from the nitrogen to the ortho carbon atom was suggested:

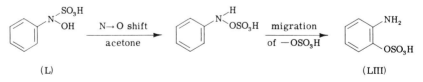

Miller and Miller (1960) presented isotopic evidence that N-acetyl-N-2-fluorenylhydroxylamine (I) rearranged in vivo to the 1-hydroxy isomer LIV, possibly in a manner analogous to the rearrangement of arylhydroxylamines to phenolic amines, as shown above for the acid-catalyzed rearrangement of phenylhydroxylamine (LII). Other studies suggested that 3-hydroxy-4-acetylaminostilbene was derived from the corresponding N-hydroxy metabolite in the rat (Andersen *et al.*, 1964). An NADPH-dependent enzyme in the soluble fraction of rat and rabbit liver which converted several N-acetylarylhydroxylamines, among them I, to the corresponding *o*-amidophenols was later reported (Booth and Boyland, 1964). Gutmann and Erickson (1969) were unable to confirm these results with N-acetyl-N-2-fluorenylhydroxylamine (I). They reported an enzyme system in rat liver capable of conversion of the N-hydroxy compound I to the 1-hydroxy isomer LIV only following pretreatment of rats with 3-methylcholanthrene or other microsomal inducing agents. The inducible component of the isomerase was associated with the microsomal fraction and was dependent for activity on a noninducible factor in the soluble fraction. This isomerase system did not require NADPH. Gutmann and Erickson postulated that the isomerization of I proceeds as a two-step reaction. In the first step, I is dehydroxylated to a positively charged amidonium ion by an enzyme in the soluble fraction of rat liver. The second step, in which hydroxyl ions add to the electrophilic carbon atoms 1 and 3 of the resonance forms of the amidonium ion, yields the *o*-amidofluorenols. The induced microsomal enzyme is presumed to participate in the second step of the reaction.

Because of the ease of rearrangement of O-sulfonates of N-acetylaryl-hydroxylamines (and of synthetic O-acyl derivatives; see Section V,D,1), as illustrated above for the rearrangement of XLIV to XLVIII, the possible role of O-sulfonates in the rearrangement of N-acetylarylhydroxyl-amines *in vivo* should be considered. For example, the N-hydroxy compound I could be converted to the 1-hydroxy isomer as follows:

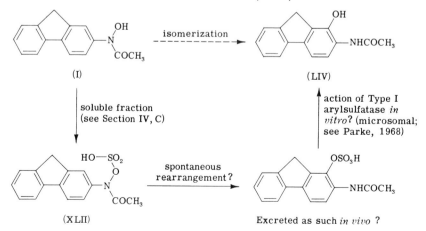

(I)　　　　　　　　　　　　　　　　　　(LIV)

soluble fraction
(see Section IV, C)

action of Type I
arylsulfatase *in*
vitro? (microsomal;
see Parke, 1968)

spontaneous
rearrangement?

(XLII)　　　　　　　　　Excreted as such *in vivo* ?

Despite the fact that conditions were not favorable for the formation of XLII or other conjugates of I in the experiments cited by Gutmann and Erickson (1969), such a mechanism as postulated above could account for the formation of the sulfate ester of 1-hydroxy-2-acetylaminofluorene by the rat *in vivo*. The biochemical mechanism of the rearrangement of *N*-acetyl-*N*-arylhydroxylamines to the corresponding *o*-amidophenols deserves further investigation.

2. Reaction with Nucleophilic Compounds

Synthetic *N*-acylarylhydroxylamine-*O*-sulfonates such as XLII react with a number of nucleophilic compounds, such as methionine, guanosine, or GMP, and proteins or nucleic acids containing these nucleophiles (Miller and Miller, 1969a,b; Maher *et al.*, 1968). Other residues in proteins, such as tryptophan and cysteine residues, or in nucleic acids, such as adenine residues, may also participate in similar reactions. As indicated in Section IV, C, it is by virtue of their reaction with these nucleophiles that evidence for the enzymic formation of *O*-sulfonates of *N*-hydroxy compounds *in vitro* has been obtained. Because of the instability in aqueous solution, in order for appreciable reaction of the synthetic *O*-sulfonates with nucleophilic compounds to be detectable, it is essential to add the solution of the nucleophilic compound to the solid *O*-sulfonate derivative of the *N*-hydroxy compound (Maher *et al.*, 1968). Miller and Miller (1969b) propose that the reaction of the *O*-sulfonate XLII with nucleophiles proceeds by an S_N1 mechanism, but they have not ruled out the possibility of an S_N2 mechanism (Miller and Miller, 1969a). The —OSO_3H group on the *N*-acylarylhydroxylamine-*O*-sulfonate represents an excellent leaving group, resulting in the formation of the amidonium ion (LV) and resonant structures (LVI) which undergo attack by nucleophiles (Fig. 9).

Each of the products of the reaction of XLII with nucleophilic compounds retains the *N*-acetyl group of XLII (Miller and Miller, 1969a). This is in marked contrast to the facile removal of the *N*-acetyl group during the reaction of the glucuronide conjugate of *N*-acetyl-*N*-2-fluorenylhydroxylamine (XXV) with nucleophilic compounds (Section III,A,4).

3. Possible Role of Sulfonates of Arylhydroxylamines as Intermediates in the Formation of Mercapturic Acids of Aromatic Amines

Biochemical mechanisms involved in mercapturic acid formation are reviewed by Wood in Volume II of this treatise. However, because of

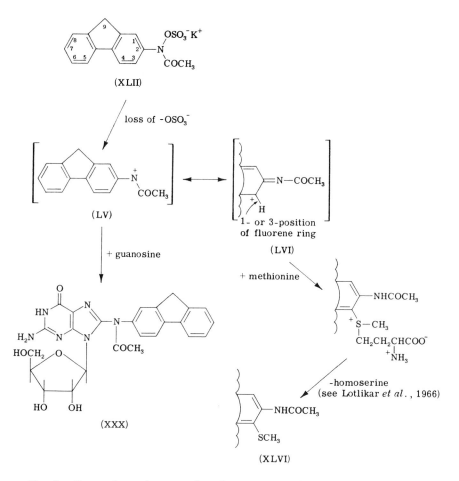

FIG. 9. Proposed mechanisms for the reaction of potassium *N*-acetyl-*N*-2-fluorenylhydroxylamine-*O*-sulfonate (XLII) with guanosine and methionine. (From Miller and Miller, 1969a,b.)

the possible role of sulfonate conjugates of arylhydroxylamines (and perhaps other conjugates of these compounds) as intermediates in the pathway of formation of mercapturic acid derivatives of aromatic amines, this topic will be discussed briefly here. It was found that phenylhydroxylamine reacted with L-cysteine or *N*-acetyl-L-cysteine (LVII) to yield S-aminophenyl-L-cysteine and *o*- and *p*-aminophenylmercapturic acid (LVIII, LIX), respectively (Boyland *et al.*, 1962). However, the *o*- and *p*-aminophenols, which were also formed, were the main products; the para isomer predominated in each case. Reaction of 2-naphthylhy-

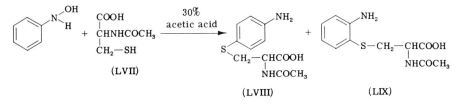

(LVII)   →   (LVIII)   +   (LIX)

droxylamine with LVII under similar conditions gave a product which was characterized as 2-amino-1-naphthylmercapturic acid (LX).

It seemed likely, therefore, that aromatic amines might be excreted by animals as mercapturic acid derivatives. A product identified as S-(2-

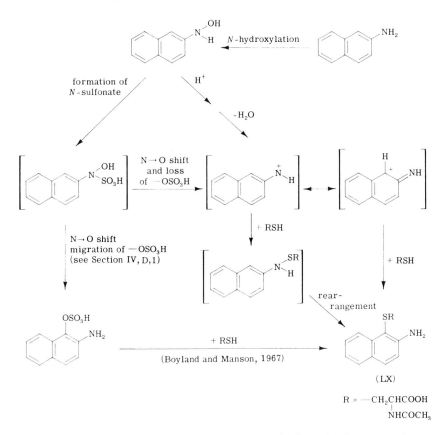

FIG. 10. Possible pathways and mechanisms involved in the formation of mercapturic acid metabolites (LX) of aromatic amines, illustrated with 2-naphthylamine. Compounds and cations in brackets are hypothetical metabolites; the other compounds are known metabolites of 2-naphthylamine.

amino-1-naphthyl)mercapturic acid (LX) was detected in the urine of dogs and rats treated with 2-naphthylamine or 2-napthylhydroxylamine (Boyland *et al.*, 1963; Boyland and Manson, 1966). The urine of rats and rabbits dosed with aniline has also been shown by these investigators to contain aminophenyl- and acetamidophenyl-mercapturic acids. Mercapturic acid metabolites of other aromatic amines have not been reported in the literature.

Several mechanisms, each involving the intermediate formation of the arylhydroxylamine from the corresponding aromatic amine, have been postulated (Boyland *et al.*, 1963), or appear to be possible (Fig. 10).

V. Other Conjugates of N-Hydroxy Compounds

A. INTRODUCTION

There is less evidence for the formation *in vivo* and excretion of other types of conjugates of N-hydroxy compounds. The enzymic formation of the O-methyl derivative of N-acetyl-N-2-fluorenylhydroxylamine *in vitro* (but not *in vivo*) has been demonstrated. Other types of conjugates which would appear to be *possible* metabolites of N-hydroxy compounds are the phosphate and acetate esters.

B. METHYL DERIVATIVES

1. *Introduction*

As with the formation of glucuronides and sulfonates of N-hydroxy compounds, several types of methylated derivatives of N-hydroxy compounds appear to be possible:

$$-N\overset{OCH_3}{\underset{COCH_3}{\diagdown}} \qquad -N\overset{OCH_3}{\underset{H}{\diagdown}} \qquad -N\overset{OH}{\underset{CH_3}{\diagdown}} \qquad -N\overset{OCH_3}{\underset{CH_3}{\diagdown}}$$

Although enzymic transfer of the methyl group from S-adenosylmethionine (SAM) to N-, C-, O- and S-positions of a number of biological and foreign compounds occurs (Williams, 1959; Parke, 1968), there have not been many studies reported on the chemistry or biosynthesis of methylated derivatives of N-hydroxy compounds.

2. *Synthetic Methyl Derivatives and Related Compounds*

Only a few synthetic methylated or related derivatives of N-arylhydroxylamines or their N-acylated derivatives have been reported. N-Methoxy-2-acetylaminofluorene (LXI) has been synthesized (Miller *et al.*, 1964a). However, it has not been demonstrated that LXI exhibits reactivity with nucleophilic compounds shown by the corresponding glucuronide (Section III, A, 4), sulfonate (Section IV, D, 2), or *O*-acetyl derivatives (Section V, D, 1). Ether derivatives of *N*-acyl arylhydroxyl-amines appear to be relatively stable. For example, the trimethylsilyl ether of *N*-acetyl-*N*-2-fluorenylhydroxylamine (LXII) which is readily formed at room temperature by reaction of bis(trimethylsilyl)acetamide with *N*-acetyl-*N*-2-fluorenylhydroxylamine (I) is stable and does not undergo the rearrangement reactions exhibited by the *O*-sulfonate or *O*-acyloxy derivatives. We have used LXII as a derivative for the determination of I by gas-liquid chromatography (3% OV-1 on Gas-Chrom Q at temperatures of 200°–225°C). TMS derivatives of other *N*-acetylarylhy-droxylamines should prove to be quite useful for their quantitative determination.

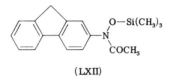

(LXII)

3. Biosynthesis

N-Acetyl-*N*-2-fluorenylhydroxylamine (I) was converted to the *O*-methyl metabolite LXI by an enzyme system in rat liver supernatant fraction (Lotlikar, 1968). The formation of LXI from I was dependent on the presence of SAM and cysteine:

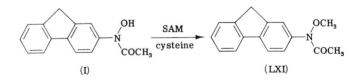

              (I)   (LXI)

N-2-Fluorenylhydroxylamine (IV) also served as a substrate in the system. Although the product of the latter reaction was not characterized, it was suggested to be either the *O*-methyl LXIII or the *N*-methyl LXIV derivative. Some of the product LXIII or LXIV obtained with IV as substrate was also formed in the system when I was used as a substrate. This may have been due to some deacetylation of I by enzymes in the rat liver soluble fraction (Grantham *et al.*, 1965) (see Section V, D,3) resulting in the formation of IV which was then methylated to form LXIII or LXIV.

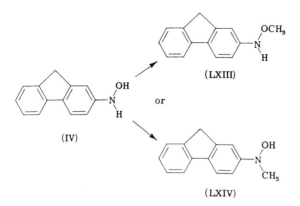

(LXIII)

(IV)   or

(LXIV)

Preliminary studies by Lotlikar (1968) indicate that the formation of LXI from I is catalyzed by an enzyme system different from the catechol-O-methyltransferase described by Axelrod and Tomchick (1958), from iodophenol-O-methyltransferase (Tomita *et al.*, 1964), or from phenol-O-methyltransferase (Axelrod and Daly, 1968).

C. PHOSPHATE ESTERS

Despite the widespread occurrence of phosphate esters in biological systems, the formation of these esters represents a rare mode of conjugation of foreign compounds. One of a few recorded examples involving such conjugation is the excretion of bis(2-amino-1-naphthyl)phosphate (LXV) as a metabolite of 2-naphthylamine in the urine of dogs (Troll *et al.*, 1959; Boyland *et al.*, 1961) and in humans (Troll *et al.*, 1963).

(LXV)   (LXVI)

Later, Troll and Belman (1967) also reported the identification of bis (2-hydroxylamino-1-naphthyl)phosphate (LXVI) as a metabolite of 2-naphthylamine.

Although there is no evidence for the formation or occurrence of phosphate esters of N-hydroxy compounds *in vivo*, two types of phosphorylated conjugates of N-hydroxy compounds might be conceived to be possible: one in which the phosphoryl group is attached to oxygen and

$$-N\overset{OPO_3H_2}{\underset{COCH_3}{}} \qquad\qquad -N\overset{OH}{\underset{PO_3H_2}{}}$$

the other in which the phosphoryl group is attached to nitrogen. Chemical syntheses of these types of derivatives of N-hydroxy compounds have not been recorded in the literature. If such conjugates were to be formed enzymically, their biosynthesis would most likely involve cleavage of the terminal P $\longrightarrow$ 0 bond of ATP (or possibly some other phosphoryl donor) and transfer of the phosphoryl group to the N-hydroxy compounds (see Nordlie and Lardy, 1962).

King and Phillips (1968, 1969) propose that a phosphokinase of rat liver, in the presence of ATP and Mg^{++}, catalyzes the formation of a phosphate ester of N-acetyl-N-2-fluorenylhydroxylamine (LXVII):

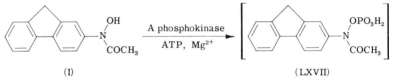

As in the assay system for sulfokinase activity with I as a substrate (Section IV, C), LXVII was not detected but was thought to be trapped by reaction with added tRNA. The ATP-Mg^{++}-dependent formation of o-methylmercapto-2-acetylaminofluorene (XLVI) from methionine and I by rat liver-soluble fraction has also been observed (DeBaun et al., 1968; see Section IV,C, Table XII). However, evidence for the enzymic formation of LXVII in these systems is equivocal because of the difficulties encountered in ensuring that the incubation medium was free of sulfate ion (Miller and Miller, 1969a). Presence of traces of sulfate ion in these systems would result in the generation of PAPS, resulting in the formation of the reactive O-sulfonate derivative (XLII) as described in Section IV,C. King and Phillips (1968), arguing that LXVII might be formed in the system containing ATP and Mg^{++}, indicated that there was less dependence on added Mg^{++} for the formation of the sulfonate XLII than for the phosphate derivative XLVII. Substitution of Mn^{++} for Mg^{++}, in the absence of added sulfate ion, reduced by 98% the ATP-stimulated incorporation of radioactivity from N-1'-^{14}C-acetyl-N-2-fluorenylhydroxylamine into tRNA (Table XV). Replacement of Mg^{++} by Mn^{++} in the presence of ATP and sulfate ion still permitted considerable binding to occur. King and Phillips (1969) have suggested that the phosphokinase and sulfokinase activities in the systems described could also be distinguished on the basis of marked differences in activities at pH 7.4 and pH 9.0.

TABLE XV

REQUIREMENTS FOR THE ATP-DEPENDENT FORMATION OF FLUORENE-BOUND
DERIVATIVES TO tRNA FROM N-1'-^{14}C-ACETYL-N-2-FLUORENYLHYDROXYLAMINE
(I) BY A SOLUBLE ENZYME SYSTEM FROM RAT LIVER[a]

Additions (final conc.)	Fluorene derivative bound to tRNA (mμmoles/mg tRNA)
None	<0.01
$MgCl_2$ (0.004 M)	<0.01
ATP (0.0025 M)	<0.01
$MgCl_2$ + ATP	1.11
$MgCl_2$ + Na_2SO_4 (0.01 M)	<0.01
ATP + Na_2SO_4	0.15
$MgCl_2$ + ATP + Na_2SO_4	4.83
$MnCl_2$ (0.004 M) + ATP	0.02
$MnCl_2$ + Na_2SO_4	<0.01
$MnCl_2$ + ATP + Na_2SO_4	2.17

[a] The equivalent of 50 mg of tissue, 5 mg of yeast tRNA, 0.08 μmole of N-1'-^{14}C-acetyl-N-2-fluorenylhydroxylamine and 110 μmole of tris buffer (pH 7.4) in a final volume of 2.1 ml were incubated for 1 hour in air at 37°C. Data from King and Phillips (1968). Copyright 1968 by the American Association for the Advancement of Science.

Further studies are needed in this area in order to determine if phosphate derivatives of N-hydroxy compounds are formed enzymically.

D. ACETYLATION AND DEACETYLATION

1. *O-Acetylation*

In spite of numerous examples of the biological N-acetylation of nitrogenous compounds, such as aromatic amines and sulfonamides, (Williams, 1959; Parke, 1968), the enzymic O-acetylation of foreign compounds containing hydroxyl groups has never been demonstrated.

Even though the enzymic O-acetylation of N-acetyl-N-arylhydroxylamines has not been demonstrated in mammalian systems, such compounds (represented by structure LXVIII) are of importance as model derivatives of their respective potential glucuronide or sulfonate con-

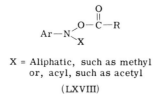

X = Aliphatic, such as methyl
or, acyl, such as acetyl

(LXVIII)

jugates, because they possess interesting chemical properties and because a number of these compounds are highly carcinogenic. As a matter of fact, the initial knowledge of the reactivity and carcinogenicity of these synthetic esters prompted a number of studies on the formation, reactivity, and carcinogenicity of other potential O-conjugates of N-acyl-N-arylhydroxylamines (see Sections III and IV). Miller[2] and his colleagues have been unable to demonstrate the enzymic formation of N,O-diacetyl-N-2-fluorenylhydroxylamine (LXIX) from N-acetyl-N-2-fluorenylhydroxylamine (I) *in vitro* by rat liver in the presence of acetyl CoA; nor has

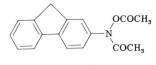

(LXIX)

LXIX been detected as a metabolite of I in intact animals. Several studies on the reactivity of synthetic LXIX have been reported (initial papers by Lotlikar *et al.*, 1966; Miller *et al.*, 1966a; Kriek *et al.*, 1967; subsequent literature reviewed in Miller and Miller, 1969a,b). Compound LXIX reacts with methionine, cysteine, tryptophan, and tyrosine (or proteins containing these amino acids) and with guanosine (or nucleic acids). Reaction of LXIX with methionine and with guanosine yields o-methylmercapto-2-acetylaminofluorene (XLVI) and N-(guanosin-8-yl) acetylaminofluorene (XXX) respectively (see Section III,A,4). Products of the reaction of LXIX with other nucleophilic compounds have not yet been fully characterized. N-Benzoyl-N-methyl-N-p-(phenylazo)phenylhydroxylamine also reacted *in vitro* with these tis-

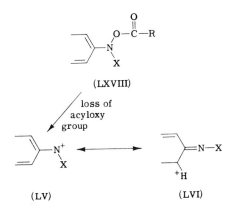

(LXVIII)

loss of
acyloxy
group

(LV)   (LVI)

[2] J. A. Miller, G. H. A. Clowes Memorial Lecture, 60th Annual Meeting of the American Association for Cancer Research, San Francisco, Calif. March 23–25, 1969.

sue nucleophiles (Lotlikar *et al.*, 1966; Poirier *et al.*, 1967). It has been proposed that the reactivity of LXIX and related esters of the type represented by structure LXVIII is due to the generation of the amidonium ion (LV) (or equivalent structure, if X is not an acyl group) or the carbonium ion (LVI) by loss of the acyloxy group (Miller and Miller, 1969a,b). The possibility of the participation of a free radical as an intermediate in the reactions of these compounds (LXVIII) with tissue nucleophiles should be considered.

The diacetyl derivative of 4-hydroxylaminoquinoline-1-oxide has recently been synthesized and shown to react with a variety of nucleophiles, including RNA, DNA, and methionine (Enomoto *et al.*, 1968b). Diacetyl 4-hydroxylaminoquinoline-1-oxide, appears to be an *O,O'*-diacetyl derivative with two *N*-acetoxy groups instead of an *N,O*-diacetyl derivative (Kawazoe and Araki, 1967).

Like the *O*-sulfonate derivatives of *N*-acylarylhydroxylamines (Section IV,D,1), *O*-acyl derivatives of these *N*-hydroxy compounds also undergo rearrangement reactions. Heating *N*-acyloxybenzanilides (such as LXX) resulted in the migration of the acyloxy group from the nitrogen to the ortho position of the ring of the aniline moiety (Horner and Steppan, 1957). The reaction rate was higher the stronger the acid corresponding

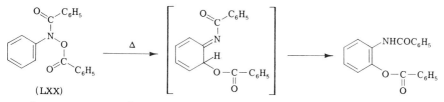

(LXX)

to the migrating acyloxy group. A methoxy group in the para position on the aniline function greatly accelerated the reaction. Tisue *et al.* (1968) have studied the rearrangement of some derivatives of phenylhydroxylamine under the influence of arenesulfonylchlorides. *N*-Benzoylphenylhydroxylamine rearranged upon treatment with *p*-nitrobenzenesulfonylchloride at 0°C; the ortho:para ratio in the product was greater than 50. The oxygen introduced into the aniline ring came exclusively from the sulfonyl group of the *p*-nitrosulfonylchloride, as shown by ^{18}O-labeling. A concerted, cyclic mechanism for the rearrangement of the intermediate *N*-benzoyl-*N*-(*p*-nitrobenzensulfonyloxy)aniline (LXXI) was suggested.

2. *N-Acetylation*

Only a few studies on the *N*-acetylation of *N*-hydroxy compounds have been reported. The occurrence of *N*-acetyl-*N*-hydroxyurethane (LXXII) and *O*-acetyl-*N*-hydroxyurethane (LXXIII) as urinary metabolites of

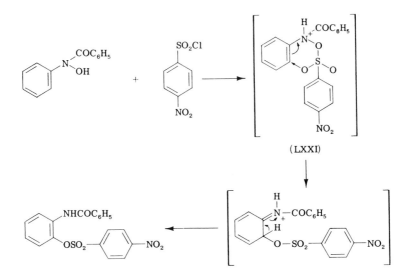

(LXXI)

urethane and N-hydroxyurethane in the rat and rabbit has been demonstrated (Boyland and Nery, 1965). LXXII was assumed to have been formed by the N-acetylation of N-hydroxyurethane, although this has not been confirmed by *in vitro* studies. However, these investigators suggested that LXXIII was formed by the spontaneous rearrangement of LXXII, since authentic LXXII rearranged to LXXIII on standing at room temperature, while LXXIII did not rearrange but was slowly hydrolyzed in water to N-hydroxyurethane. Weisburger *et al.* (1966) reported that

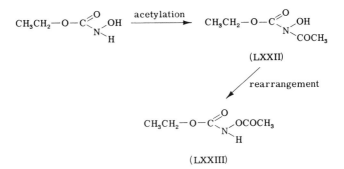

N-2-fluorenylhydroxylamine (IV) underwent acetylation *in vivo* to form N-acetyl-N-2-fluorenylhydroxylamine (I). The *urinary* excretion of I (as the glucuronide conjugate) after administration of IV was cited as evidence for the direct conversion of IV to I (Table XVI). Evidence for the direct acetylation of IV is lacking. Unlike the studies with the N-

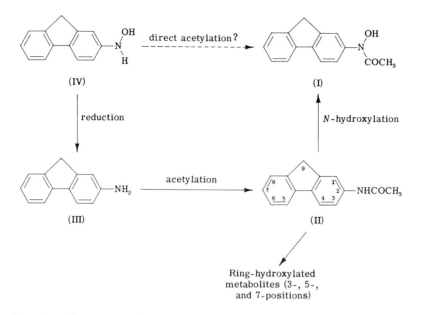

FIG. 11. Alternative pathways for the formation of N-acetyl-N-2-fluorenylhydroxylamine (I) from N-2-fluorenylhydroxylamine (IV) *in vivo*.

acetylation of N-hydroxyurethane, an alternate interpretation of the data of Weisburger *et al.* exists. The possibility must be considered that I could be formed from IV indirectly as shown in Fig. 11.

Compound IV is known to be rapidly reduced to 2-aminofluorene (III) in the rat (Lotlikar *et al.*, 1965) and in other species (Kiese and Wiedemann, 1968). Compound III, in turn, is readily acetylated to form 2-acetylaminofluorene (II) (Peters and Gutmann, 1955) which can then

TABLE XVI

URINARY METABOLITES OF N-2-FLUORENYLHYDROXYLAMINE (IV) IN RATS[a]

Metabolite	Percent of dose, total[b]	
	Male	Female
7-Hydroxy-AAF[c]	19.0	15.5
5-Hydroxy-AAF	5.5	9.2
3-Hydroxy-AAF	1.1	1.7
N-Hydroxy-AAF	1.2	3.2

[a] Data from Weisburger *et al.* (1966); IV was injected intravenously in suspension.
[b] Sum of metabolite excreted as the free compound, the glucuronide conjugate, and the sulfate conjugate.
[c] AAF = 2-Acetylaminofluorene (II).

be N-hydroxylated to form I (see Section II,A). The data showing that the 5- and 7-hydroxy derivatives of II were the *major* metabolites of IV in the rat (Table XVI) would seem to support the view that I could have been formed indirectly from IV, rather than by the direct acetylation of IV.

3. Deacetylation of N-Acetyl-N-arylhydroxylamines

N-Acetyl derivatives of arylhydroxylamines are readily deacetylated by enzymes present in a number of tissues of various species:

$$Ar-N\begin{smallmatrix}OH\\COCH_3\end{smallmatrix} \xrightarrow{\text{deacetylase(s)}} Ar-N\begin{smallmatrix}OH\\H\end{smallmatrix} + CH_3COO^-$$

In most reports, however, the initial products have not been adequately identified and/or characterized. N-Acetyl-N-phenylhydroxylamine (Hustedt and Kiese, 1959) and N-benzoyl-N-phenylhydroxylamine (Kiese and Plattig, 1959) were reported to be deacetylated *in vivo*; these data were confirmed by studies *in vitro* with liver homogenates.

Using indirect methods, Irving (1964) and Grantham *et al.* (1965) found that N-acetyl-N-2-fluorenylhydroxylamine (I) was deacetylated by rat and rabbit liver and by rat brain. The deacetylation of I was inhibited by sodium or potassium fluoride, although later studies (Irving, 1966) revealed that the inhibition by fluoride ion was not complete at concentrations of 0.1 M sodium fluoride. Grantham *et al.* (1965) reported that I was deacetylated by an enzyme present in the 100,000 $\times$ g supernatant fraction of rat liver homogenates but they mentioned that the reaction could possibly have been due to the presence of postmicrosomal elements in their preparation. Grantham *et al.* determined the amount of 2-aminofluorene (III) formed from I as a measure of the extent of deacetylation of I. Although not proven, III was assumed to be formed by deacetylation of I, followed by reduction of the arylhydroxylamine IV to the amine III. However, since I is also reduced to 2-acetylaminofluorene (II) by enzymes present in rat liver soluble fraction (Grantham *et al.*, 1965; Lotlikar *et al.*, 1965) and control experiments using II as a substrate for the deacetylase with rat liver soluble fraction were not reported, the deacetylation of I by rat liver soluble fraction remains to be firmly established.

Arylhydroxylamines react with trisodium pentacyanoamineferrate to form stable-colored complexes (Boyland and Nery, 1964). Irving (1966) found that N-2-fluorenylhydroxylamine (IV) produced enzymically by deacetylation of N-acetyl-N-2-fluorenylhydroxylamine (I) by rat, rabbit, or guinea pig liver microsomes, could be trapped *in situ* with this reagent and that the resulting colored complex could be used as a basis for the quantitative determination of the extent of enzymic deacetylation of I. The stoichiometry of the reaction was established by the direct determination of IV and of the acetate produced. The N-hydroxy compound I was deacetylated by all tissues of the guinea pig which were examined. Guinea pig liver had the highest activity of the species studied and in guinea pig liver, all of the enzymic activity was localized in the microsomal fraction. Since guinea pig liver soluble fraction did not contain detectable deacetylase activity with I as the substrate, it would seem important to clarify the question of deacetylation of I by rat liver soluble fraction (see above), in view of the lack of susceptibility of guinea pig liver to this carcinogen and the high susceptibility of rat liver (Miller *et al.*, 1964b).

Booth and Boyland (1964) have also reported that I and N-acetyl-N-4-biphenylylhydroxylamine (LXXIV) were deacetylated by rabbit liver microsomes. It was claimed that the products (the corresponding arylhydroxylamine) were detected directly by thin-layer chromatography.

An enzyme that transfers the acetyl group from N-acetyl-N-arylhydroxylamines, such as N-acetyl-N-biphenylylhydroxylamine (LXXIV), to arylamines has been found in rat tissues (Booth, 1966). This acetyltransferase was in the soluble fraction of rat liver and required a thiol, such as cysteine or glutathione, for maximum activity. A spectrophotometric method, using 4-aminoazobenzene (LXXV) as acetyl acceptor, was used to measure the enzyme activity.

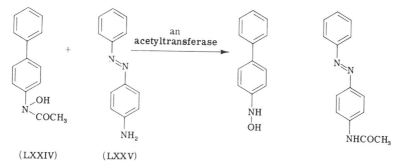

(LXXIV)   (LXXV)

The soluble fractions of various rat tissues showed decreasing activity in the following order: liver, adrenal, kidney, lung, spleen, testes, heart; brain was inactive. With the exception of aniline and aniline derivatives, all of the arylamines tested were effective as acetyl acceptors, but aromatic compounds with side chain amino groups, such as tryptamine or histamine, were inactive. Glucosamine and hydrolyzed yeast RNA did not serve as acetyl acceptors in the system. The N-acetyl derivatives of 2-naphthylhydroxylamine, 4-biphenylylhydroxylamine, and N-2-fluorenyl-hydroxylamine were active acetyl donors, but N-acetylphenylhydroxyl-amine showed only slight activity. Acetyl CoA was not an acetyl donor in this system. The precise role of this acetyl transferase in acetylation–deacetylation reactions of aromatic amines remains to be determined.

VI. O-Conjugates of N-Hydroxy Compounds as Ultimate Carcinogenic Metabolites of Aromatic Amines

Recognition of the carcinogenic properties of the aromatic amines in humans (reviewed in Clayson, 1962; Scott, 1962; see also Boyland, 1969) has led to a vast amount of research on the mechanism of action of these compounds in producing tumors in experimental animals. It has been emphasized that chemical carcinogens, including the carcinogenic aromatic amines, must react in some manner with constituents of the target tissue in order to induce tumors (Miller and Miller, 1969a,b). While such reactions need not result in the formation of covalent bonds between the carcinogen and the cell constituents involved, covalent binding of metabolites of a variety of carcinogens to DNA, RNA, and protein *in vivo* has been demonstrated (reviewed in Miller and Miller, 1966; Miller and Miller, 1967, 1969a,b; Arcos and Argus, 1968). From the fact that tumors are generally not produced at the point of application or the site of administration of carcinogenic aromatic amines, the conclusion was drawn early that the aromatic amines themselves did not act directly on susceptible tissues but that these compounds were enzymically activated in the tissues in which they did induce tumors.

Cramer *et al.* (1960) discovered a new type of metabolic reaction of carcinogenic aromatic amines, namely N-hydroxylation (Section II,A). Although a great deal of subsequent evidence clearly implicated N-hydroxy metabolites of carcinogenic aromatic amines in the mechanism of carcinogenesis by these compounds (reviewed in Miller and Miller, 1969a,b), the fact that these N-hydroxy derivatives showed only limited reaction *in vitro* with biological macromolecules, or components of these, led to the speculation that further metabolic activation of the N-hydroxy metabolite might be necessary. Furthermore, N-hydroxylation in itself does not appear to confer carcinogenicity on aromatic amines or their N-acyl derivatives since phenylhydroxylamine and N-ethylphenylhy-droxylamine (Miller *et al.*, 1966b) and N-hydroxy-N-acetyl-4-aminoazobenzene or N-hydroxy-4-aminoazobenzene (Sato *et al.*, 1966) are not carcinogenic. Gutmann *et al.* (1969) have emphasized that the size and structure of both N-substituents of N-disubstituted arylhydroxylamines greatly influence the carcinogenicity of these compounds.

More recent evidence has indicated that O-conjugates of N-hydroxy metabolites of carcinogenic aromatic amines are the reactive metabolites (see Section III, A, and IV, C, D) and the possible role of these conjugates in the mechanism of action of aromatic amines has been covered

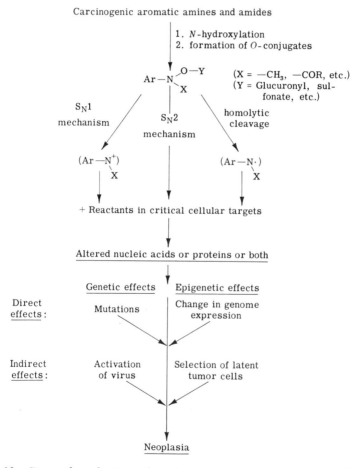

FIG. 12. Proposed mechanisms of carcinogenesis by aromatic amines and amides involving O-conjugates of N-hydroxy compounds as the ultimate (reactive) metabolites which interact with macromolecules in susceptible tissues. (From Miller and Miller, 1969a,b.)

in recent reviews by Miller and Miller (1969a,b). Although the bio-chemical mechanism(s) involved in the activation of carcinogenic aromatic amines, generating chemically reactive species, appear to have been fairly well elucidated, the role of the reactions of these activated metabolites with tissue constituents in the process of carcinogenesis by these compounds is *not* understood in molecular terms. This point has been emphasized (Farber, 1968; Miller and Miller, 1969a,b). Four general hypotheses, summarized in Fig. 12, have been presented as the principal proposals that are now under experimental test.

Any of these mechanisms or others, or combinations of them, may account for the induction of tumors following interaction of the reactive carcinogenic metabolites of aromatic amines with biological macromolecules.

ACKNOWLEDGMENTS

I wish to acknowledge the generous support of my work by the U.S. Veterans Administration. This research has also been supported in part by the National Cancer Institute, Research Grant No. CA-05490. I am grateful to many workers in my laboratory, but special thanks are due to Dr. Jim Hill, Mr. Ralph Wiseman, Mr. Richard Veazey, Mrs. Laretta Russell, and Mrs. Dorothy Bensley. I am also indebted to Dr. H. R. Gutmann, Dr. E. C. Miller, and Dr. J. A. Miller for critical reading of the manuscript and for making helpful suggestions.

REFERENCES

Andersen, R. A., Enomoto, M., Miller, E. C., and Miller, J. A. (1964). *Cancer Res.* 24, 128.

Appel, W., Graffe, W., Kampffmeyer, H., and Kiese, M. (1965). *Arch. Exptl. Pathol. Pharmakol.* 251, 88.

Arcos, J. C., and Argus, M. F. (1968). *Advan. Cancer Res.* 11, 305.

Axelrod, J., and Daly, J. (1968). *Biochim. Biophys. Acta* 159, 472.

Axelrod, J., and Tomchick, R. (1958). *J. Biol. Chem.* 233, 702.

Baldwin, R. W., and Smith, W. R. D. (1965). *Brit. J. Cancer* 19, 433.

Banerjee, R. K., and Roy, A. B. (1966). *Mol. Pharmacol.* 2, 56.

Belman, S., Troll, W., Teebor, G., and Mukai, F. (1968). *Cancer Res.* 28, 535.

Birch, A. J., Massey-Westropp, R. A., and Rickards, R. W. (1956). *J. Chem. Soc.* p. 3717.

Bollenback, G. N., Long, J. W., Benjamin, D. G., and Lindquist, J. A. (1955). *J. Am. Chem. Soc.* 77, 3310.

Booth, J. (1966). *Biochem. J.* 100, 745.

Booth, J., and Boyland, E. (1964). *Biochem. J.* 91, 362.

Boyland, E. (1963). "The Biochemistry of Bladder Cancer." Thomas, Springfield, Illinois.

Boyland, E. (1969). *Progr. Exptl. Tumor Res.* 11, 222.

Boyland, E., and Manson, D. (1966). *Biochem. J.* 101, 84.

Boyland, E., and Manson, D. (1967). *Ann. Rept. Brit. Empire Cancer Campaign* **45**, Pt. II, 20.
Boyland, E., and Nery, R. (1962). *J. Chem. Soc.* p. 5217.
Boyland, E., and Nery, R. (1963). *J. Chem. Soc.* p. 3141.
Boyland, E., and Nery, R. (1964). *Analyst* **89**, 95.
Boyland, E., and Nery, R. (1965). *Biochem. J.* **94**, 198.
Boyland, E., Kinder, C. H., and Manson, D. (1961). *Biochem. J.* **78**, 175.
Boyland, E., Manson, D., and Nery, R. (1962). *J. Chem. Soc.* p. 606.
Boyland, E., Manson, D., and Nery, R. (1963). *Biochem. J.* **86**, 263.
Bridges, J. W., Kibby, M. R., and Williams, R. T. (1965). *Biochem. J.* **96**, 829.
Clayson, D. B. (1962). "Chemical Carcinogenesis." Little, Brown, Boston, Massachusetts.
Conchie, J., Levvy, G. A., and Marsh, C. A. (1957). *Advan. Carbohydrate Chem.* **12**, 157.
Conney, A. H. (1967). *Pharmacol. Rev.* **19**, 317.
Cramer, J. W., Miller, J. A., and Miller, E. C. (1960). *J. Biol. Chem.* **235**, 885.
DeBaun, J. R., Rowley, J. Y., Miller, E. C., and Miller, J. A. (1968). *Proc. Soc. Exptl. Biol. Med.* **129**, 268.
DeBaun, J. R., Miller, E. C., and Miller, J. A. (1969). *Proc. Am. Assoc. Cancer Res.* **10**, 18.
Dutcher, J. D. (1947). *J. Biol. Chem.* **171**, 321.
Dutton, G. J. (1966). *In* "Glucuronic Acid: Free and Combined. Chemistry, Biochemistry, Pharmacology, and Medicine" (G. J. Dutton, ed.), pp. 185–299. Academic Press, New York.
Emery, T. F. (1963). *Biochemistry* **2**, 1041.
Emery, T., and Neilands, J. B. (1961). *J. Am. Chem. Soc.* **83**, 1626.
Enomoto, M., and Sato, K. (1967). *Life Sci.* **6**, 881.
Enomoto, M., Miyake, M., and Sato, K. (1968a). *Gann* **59**, 177.
Enomoto, M., Sato, K., Miller, E. C., and Miller, J. A. (1968b). *Life Sci.* **7**, 1025.
Farber, E. (1968). *Cancer Res.* **28**, 1859.
Fouts, J. R., and Brodie, B. B. (1957). *J. Pharmacol. Exptl. Therap.* **119**, 197.
Gillette, J. R., Kamm, J. J., and Sasame, H. A. (1968). *Mol. Pharmacol.* **4**, 541.
Grantham, P. H., Weisburger, E. K., and Weisburger, J. H. (1965). *Biochim. Biophys. Acta* **107**, 414.
Grantham, P. H., Mohan, L., Yamamoto, R. S., Weisburger, E. K., and Weisburger, J. H. (1968). *Toxicol. Appl. Pharmacol.* **13**, 118.
Gutmann, H. R. (1964). *Experientia* **20**, 128.
Gutmann, H. R., and Erickson, R. R. (1969). *J. Biol. Chem.* **244**, 1729.
Gutmann, H. R., Chen, C. C., and Leaf, D. (1969). *Proc. Am. Assoc. Cancer Res.* **10**, 34.
Hargreaves, T. (1968). "The Liver and Bile Metabolism." Appleton-Century-Crofts, New York.
Hill, J. T. (1968). Biosynthesis and Studies of The Alkaline Sensitivity of the Glucuronide of the Carcinogen N-2-Fluorenylacethydroxamic Acid. Ph.D. Thesis, Univ. of Tennessee, Memphis, Tennessee.
Hill, J. T., and Irving, C. C. (1966). *Federation Proc.* **25**, 743.
Hill, J. T., and Irving, C. C. (1967). *Biochemistry* **6**, 3816.
Horner, L., and Steppan, H. (1957). *Ann. Chem.* **606**, 24.

Hustedt, G., and Kiese, M. (1959). *Arch. Exptl. Pathol. Pharmakol.* **236**, 435.

Ide, H., Green, S., Kato, K., and Fishman, W. H. (1968). *Biochem. J.* **106**, 431.

Irving, C. C. (1962a). *Biochim. Biophys. Acta* **65**, 564.

Irving, C. C. (1962b). *Cancer Res.* **22**, 867.

Irving, C. C. (1964). *J. Biol. Chem.* **239**, 1589.

Irving, C. C. (1965). *J. Biol. Chem.* **240**, 1011.

Irving, C. C. (1966). *Cancer Res.* **26**, 1390.

Irving, C. C., and Russell, L. T. (1969). *158th Natl. Meeting, Am. Chem. Soc., New York,* Abstr. Biol. 102.

Irving, C. C. and Wiseman, R., Jr. (1969). *Cancer Res.* **29**, 812.

Irving, C. C., Veazey, R. A., and Hill, J. T. (1967a). *154th Natl. Meeting, Am. Chem. Soc., Chicago* Abstr. C039.

Irving, C. C., Wiseman, R., Jr., and Hill, J. T. (1967b). *Cancer Res.* **27**, 2309.

Irving, C. C., Veazey, R. A., and Hill, J. T. (1969a). *Biochim. Biophys. Acta* **179**, 189.

Irving, C. C., Veazey, R. A., and Russell, L. T. (1969b). *Proc. Am. Assoc. Cancer Res.* **10**, 43.

Irving, C. C., Veazey, R. A., and Russell, L. T. (1969c). *Chem.-Biol. Interactions* **1**, 19.

Kaczka, E. A., Gitterman, C. O., Dulaney, E. L., and Folkers, K. (1962). *Biochemistry* **1**, 340.

Kampffmeyer, H., and Kiese, M. (1963). *Arch. Exptl. Pathol. Pharmakol.* **244**, 375.

Kampffmeyer, H., and Kiese, M. (1965). *Arch. Exptl. Pathol. Pharmakol.* **250**, 1.

Kato, K., Ide, H., Hirohata, I., and Fishman, W. H. (1967). *Biochem. J.* **103**, 647.

Kato, R. (1966). *J. Biochem. (Tokyo)* **59**, 574.

Kawazoe, Y., and Araki, M. (1967). *Gann* **58**, 485.

Kiese, M. (1966). *Pharmacol. Rev.* **18**, 1091.

Kiese, M., and Plattig, K.-H. (1959). *Arch. Exptl. Pathol. Pharmakol.* **235**, 373.

Kiese, M., and Rauscher, E. (1963). *Biochem. Z.* **338**, 1.

Kiese, M., and Uehleke, H. (1961). *Arch. Exptl. Pathol. Pharmakol.* **242**, 117.

Kiese, M., and Wiedemann, I. (1968). *Biochem. Pharmacol.* **17**, 1151.

Kiese, M., Renner, G., and Wiedemann, I. (1966). *Arch. Exptl. Pathol. Pharmakol.* **252**, 418.

King, C. M., and Phillips, B. (1968). *Science* **159**, 1351.

King, C. M., and Phillips, B. (1969). *Proc. Am. Assoc. Cancer Res.* **10**, 46.

Kluyver, A. J., van der Walt, J. P., and van Triet, A. J. (1953). *Proc. Natl. Acad. Sci. U.S.* **39**, 583.

Kriek, E., Miller, J. A., Juhl, U., and Miller, E. C. (1967). *Biochemistry* **6**, 177.

Lange, G. (1967). *Arch. Pharmakol. Exptl. Pathol.* **257**, 230.

Levvy, G. A., and Conchie, J. (1966). *In* "Glucuronic Acid: Free and Combined. Chemistry, Biochemistry, Pharmacology, and Medicine" (G. J. Dutton, ed.), pp. 301–364. Academic Press, New York.

Lotlikar, P. D. (1968). *Biochim. Biophys. Acta* **170**, 468.

Lotlikar, P. D. (1969). *Proc. Am. Assoc. Cancer Res.* **10**, 52.

Lotlikar, P. D., Enomoto, M., Miller, E. C., and Miller, J. A. (1964). *Cancer Res.* **24**, 1835.

Lotlikar, P. D., Miller, E. C., Miller, J. A., and Margreth, A. (1965). *Cancer Res.* **25**, 1743.

Lotlikar, P. D., Scribner, J. D., Miller, J. A., and Miller, E. C. (1966). *Life Sci.* **5**, 1263.

Lotlikar, P. D., Irving, C. C., Miller, E. C., and Miller, J. A. (1967a). *Proc. Am. Assoc. Cancer Res.* **8**, 42.

Lotlikar, P. D., Enomoto, M., Miller, J. A., and Miller, E. C. (1967b). *Proc. Soc. Exptl. Biol. Med.* **125**, 341.

Maher, V. M., Miller, E. C., Miller, J. A., and Szybalski, W. (1968). *Mol. Pharmacol.* **4**, 411.

Margreth, A., Lotlikar, P. D., Miller, E. C., and Miller, J. A. (1964). *Cancer Res.* **24**, 920.

Marsh, C. A. (1966). *In* "Glucuronic Acid: Free and Combined. Chemistry, Biochemistry, Pharmacology, and Medicine" (G. J. Dutton, ed.), pp. 3–136. Academic Press, New York.

Mason, H. S. (1957). *Science* **125**, 1185.

Matsushima, T., Kobuna, I., Fukuoka, F., and Sugimura, T. (1968). *Gann* **59**, 247.

Mikata, A., and Luse, S. A. (1964). *Am. J. Pathol.* **44**, 455.

Miller, E. C., and Miller, J. A. (1960). *Biochim. Biophys. Acta* **40**, 380.

Miller, E. C., and Miller, J. A. (1966). *Pharmacol. Rev.* **18**, 805.

Miller, E. C., Cooke, C. W., Lotlikar, P. D., and Miller, J. A. (1964a). *Proc. Am. Assoc. Cancer Res.* **5**, 45.

Miller, E. C., Miller, J. A., and Enomoto, M. (1964b). *Cancer Res.* **24**, 2018.

Miller, E. C., Juhl, U., and Miller, J. A. (1966a). *Science* **153**, 1125.

Miller, E. C., Lotlikar, P. D., Pitot, H. C., Fletcher, T. L., and Miller, J. A. (1966b). *Cancer Res.* **26**, 2239.

Miller, E. C., Lotlikar, P. D., Miller, J. A., Butler, B. W., Irving, C. C., and Hill, J. T. (1968). *Mol. Pharmacol.* **4**, 147.

Miller, J. A., and Miller, E. C. (1967). *In* Carcinogenesis: A Broad Critique," pp. 397–420. Williams & Wilkins, Baltimore, Maryland.

Miller, J. A., and Miller, E. C. (1969a). *In* "Physico-Chemical Mechanisms of Carcinogenesis" (E. D. Bergmann and B. Pullman, eds.), pp. 237–261. Academic Press, New York.

Miller, J. A., and Miller, E. C. (1969b). *Progr. Exptl. Tumor Res.* **11**, 273.

Miller, J. A., Cramer, J. W., and Miller, E. C. (1960). *Cancer Res.* **20**, 950.

Miller, J. A., Wyatt, C. S., Miller, E. C., and Hartmann, H. A. (1961). *Cancer Res.* **21**, 1465.

Mirvish, S. S. (1966). *Biochim. Biophys. Acta* **117**, 1.

Morris, H. P., Wagner, B. P., and Lombard, L. S. (1958). *J. Natl. Cancer Inst.* **20**, 1.

Nery, R. (1968). *Biochem. J.* **106**, 1.

Nordlie, R., and Lardy, H. (1962). *In* "The Enzymes" (P. D. Boyer, H. Lardy, and K. Myrbäck, eds.), 2nd ed., Vol. 6, pp. 3–46. Academic Press, New York.

Nose, Y., and Lipmann, F. (1958). *J. Biol. Chem.* **233**, 1348.

Omura, T., Sato, R., Cooper, D. Y., Rosenthal, O., and Estabrook, R. W. (1965). *Federation Proc.* **24**, 1181.

Orrenius, S., and Ernster, L. (1967). *Life Sci.* **6**, 1473.

Otsuka, S. (1961). *J. Biochem. (Tokyo)* **50**, 85.

Parke, D. V. (1968). "The Biochemistry of Foreign Compounds." Macmillan (Pergamon), New York.

Peters, J. H., and Gutmann, H. R. (1955). *J. Biol. Chem.* **216**, 713.

Poirier, L. A., Miller, J. A., Miller, E. C., and Sato, K. (1967). *Cancer Res.* **27**, 1600.

Remmer, H., Estabrook, R. W., Schenkman, J., and Greim, H. (1968). *Arch. Pharmakol. Exptl. Pathol.* **259**, 98.

Robbins, P. W. (1962). *In* "The Enzymes" (P. D. Boyer, H. Lardy, and K. Myrbäck, eds.), 2nd ed., Vol. 6, pp. 363–372. Academic Press, New York.

Roy, A. B. (1960). *Biochem. J.* **74**, 49.

Sato, K., Poirier, L. A., Miller, J. A., and Miller, E. C. (1966). *Cancer Res.* **26**, 1678.

Scott, T. S. (1962). "Carcinogenic and Chronic Toxic Hazards of Aromatic Amines." Elsevier, New York.

Shirasu, Y., Grantham, P. H., Yamamoto, R. S., and Weisburger, J. H. (1966). *Cancer Res.* **26**, 600.

Shirasu, Y., Grantham, P. H., Weisburger, E. K., and Weisburger, J. H. (1967a). *Cancer Res.* **27**, 81.

Shirasu, Y., Grantham, P. H., Weisburger, E. K., and Weisburger, J. H. (1967b). *Cancer Res.* **27**, 865.

Shirasu, Y., Grantham, P. H., and Weisburger, J. H. (1967c). *Intern. J. Cancer* **2**, 59.

Snow, G. A. (1954). *J. Chem. Soc.* p. 2588.

Sugimura, T., Okabe, K., and Nagao, M. (1966). *Cancer Res.* **26**, 1717.

Thauer, R. K., Stoffler, G., and Uehleke, H. (1965). *Arch. Exptl. Pathol. Pharmakol.* **252**, 32.

Tisue, G. T., Grassmann, M., and Lwowski, W. (1968). *Tetrahedron* **24**, 999.

Tomita, K., Cha, C.-J. M., and Lardy, H. A. (1964), *J. Biol. Chem.* **239**, 1202.

Troll, W., and Belman, S. (1967). *In* "Bladder Cancer—A Symposium" (W. B. Deichmann and K. F. Lampe, eds.), pp. 35–44. Aesculapius, Birmingham, Alabama.

Troll, W., Belman, S., and Nelson, N. (1959). *Proc. Soc. Exptl. Biol. Med.* **100**, 121.

Troll, W., Tessler, A. N., and Nelson, N. (1963). *J. Urol.* **89**, 626.

Tsukamoto, H., Yoshimura, H., and Tatsumi, K. (1963). *Chem. Pharm. Bull. (Tokyo)* **11**, 421.

Uehleke, H. (1961). *Experientia* **17**, 557.

Uehleke, H. (1962). *Proc. 1st Intern. Pharmacol. Meeting, Stockholm, 1961* **6**, 31.

Uehleke, H. (1963). *Naturwissenschaften* **8**, 335.

Uehleke, H. (1966). *Life Sci.* **5**, 1489.

Uehleke, H. (1967). *Arch. Pharmakol. Exptl. Pathol.* **259**, 66.

Uehleke, H. (1968). *Experientia* **24**, 108.

Uehleke, H., and Nestel, K. (1967). *Arch. Pharmakol. Exptl. Pathol.* **257**, 151.

Utzinger, G. E. (1944). *Ann. Chem. Liebigs* **556**, 50.

von Jagow, R., and Kiese, M. (1967). *Biochim. Biophys. Acta* **136**, 168.

von Jagow, R., Kiese, M., and Renner, G. (1966). *Biochem. Pharmacol.* **15**, 1899.

Weisburger, E. K., Grantham, P. H., and Weisburger, J. H. (1964a). *Biochemistry* **3**, 808.

Weisburger, J. H., Grantham, P. H., Vanhorn, E., Steigbigel, N. H., Rall, D. P., and Weisburger, E. K. (1964b). *Cancer Res.* **24**, 475.

Weisburger, J. H., Grantham, P. H., and Weisburger, E. K. (1965). *Brit. J. Cancer* **19**, 581.

Weisburger, J. H., Grantham, P. H., and Weisburger, E. K. (1966). *Biochem. Pharmacol.* **15**, 833.

Whistler, R. L., and Wolfrom, M. L., eds. (1963). "Methods in Carbohydrate Chemistry," Vol. II, pp. 326–389. Academic Press, New York.

Williams, J. R., Jr., Grantham, P. H., Weisburger, J. H., and Marsh, H. H., III (1968). *Federation Proc.* **27**, 650.

Williams, R. T. (1959). "Detoxication Mechanisms." Wiley, New York.

Yamamoto, R. S., Pai, S. R., Korzis, J., and Weisburger, J. H. (1968). *Brit. J. Cancer* **22**, 769.

EFFECTS OF CONJUGATED STEROIDS
ON ENZYMES

MERLE MASON

I. Introduction

Studies of the conjugated steroids in recent years have demonstrated
that they cannot properly be regarded as "inactive" excretory products.

121

Partly because of the belated recognition of the possibility of more active roles for these steroid metabolites, there is presently little information available concerning the interaction of steroid conjugates with tissue constituents and the nature of their influence on cellular functions. One purpose of the present review is to assemble the published information dealing with the effects of conjugated steroids and related compounds on enzymes. It is assumed that generalizations concerning their effects on enzyme systems will also be applicable to nonenzymic proteins. This premise is supported to some degree already by the studies to be reviewed.

II. Interaction of Conjugated Steroids with Enzymes

A. PYRIDOXAL ENZYMES

1. *Kynurenine Transaminase*

In studies of a partially purified kynurenine transaminase from rat kidney, Mason (1959) observed that many mono- and dicarboxylic acids were inhibitory at concentrations of 3–6 mM. Dicarboxylic acids with chain lengths similar to that of the keto acid substrate, α-ketoglutarate, were especially potent as inhibitors and appeared to act by interfering with substrate binding. In general, however, the degree of inhibition increased with the size of the hydrocarbon portion of the acids, a trend which led eventually to the testing of high molecular weight anionic hydrocarbons such as the bile acids and conjugated steroids (Mason and Gullekson, 1959, 1960). The estrogen sulfates, especially the disulfate esters of estradiol and diethylstilbestrol (Fig. 1), were exceptionally potent, giving 50% inhibition at levels of approximately 5×10^{-6} M. Free steroids, including estradiol, diethylstilbestrol, estrone, progesterone, pregnanediol, and hydrocortisone, were completely inactive at levels exceeding their solubility in the incubation mixture.

The inhibition was reversed by increasing the concentration of the coenzyme, pyridoxal phosphate, in the incubation mixture and by dialysis. The reversibility suggested that the conjugates interfered with pyridoxal phosphate binding to the apoenzyme, possibly by competing with it for a cationic binding site on the protein. A cationic binding site was assumed because both the inhibitor and the coenzyme are anionic conjugates and because pyridoxal phosphate was known to be bound as a Schiff's base in various pyridoxal enzymes to ε-amino groups of lysyl residues. Such binding has been shown for glycogen phosphorylase (Nolan *et al.*, 1964).

The ability of the conjugates to interfere with the activation of resolved phosphorylase by added pyridoxal phosphate (Mason and Gullekson, 1959) suggested, therefore, that the conjugates interfere with a cationic binding site and possibly with one involved in Schiff's base formation.

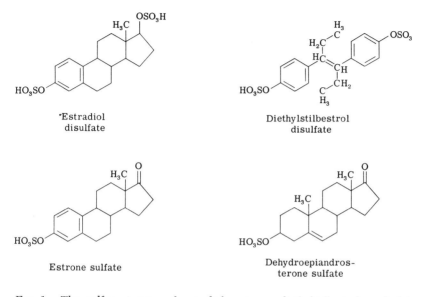

'Estradiol
disulfate

Diethylstilbestrol
disulfate

Estrone sulfate

Dehydroepiandros-
terone sulfate

FIG. 1. The sulfoconjugates of estradiol, estrone, diethylstilbestrol, and dehydroepiandrosterone.

Dialysis of strongly inhibited kynurenine transaminase preparations resulted in enzyme activity substantially greater than that of similarly dialyzed enzyme that had not been treated with the conjugates. This action of the conjugates resulted from their ability to prevent an inactivation process that normally occurred during incubation and storage. The conjugates also protected the enzyme against inactivation by added chymotrypsin, presumably by preventing proteolysis. They did not prevent proteolysis of casein by chymotrypsin, so it was assumed that they protected the transaminase by binding it rather than by inactivating the proteinase. Pyridoxal phosphate also stabilized the transaminase against inactivation by chymotrypsin, as would be expected if the coenzyme and the conjugates bind the same sites on the enzyme.

Both the protective and inhibitory actions of the estrogen disulfates occurred at the same low concentrations (5×10^{-6} M). At the higher concentrations at which estrone sulfate was inhibitory (2.5×10^{-4} M), it was also protective. It should be noted also that inhibitory dicarboxylic acids at relatively high levels, 1 to 6 mM, showed a similar protective action (Mason, 1959). These observations suggested to the authors that the inhibitory and protective actions result from the same interaction of conjugates with the enzyme.

In testing for inhibition, the conjugates were routinely added to the apoenzyme preparations either before or simultaneously with the coenzyme. If apoenzyme, coenzyme, and α-ketoglutarate were incubated together for a short time before the conjugates were added, little or no inhibition occurred. Thus, the inhibitors did not reverse the apoenzyme–coenzyme association in spite of their great effectiveness in preventing association. Later experiments showed in fact that the conjugates *retard* the dissociation of pyridoxal phosphate from the holotransaminase present in freshly prepared homogenates of rat kidney (Ford *et al.*, 1966; Singer *et al.*, 1966). At higher concentrations of conjugate, the inhibitory effects prevailed. Estradiol disulfate and diethylstilbestrol disulfate were again the most potent agents tested; dehydroepiandrosterone sulfate and several carboxylic acids were effective at much higher concentrations.

Scardi *et al.* (1962) examined the inhibition of kynurenine transaminase by estrogen sulfates and estrogen phosphates. By plotting the data according to Dixon, the authors obtained K_i values for both estradiol disulfate and diethylstibestrol disulfate of 0.15 mM (Scardi *et al.*, 1960). The analogous diphosphate esters were considerably more potent (Scardi *et al.*, 1962). Dehydroepiandrosterone sulfate was completely without effect on the reconstitution of kynurenine transaminase at concentrations as high as 2 mM.

2. Glutamate-Aspartate Transaminase

Scardi *et al.* (1962) also examined the effects of estrogen sulfates and estrogen phosphates on the reconstitution of the glutamate-aspartate transaminase of pig heart by added pyridoxal phosphate. The inhibition data gave K_i values of 0.452, 0.12, 0.075, and 0.015 mM for estradiol 3-phosphate, diethylstilbestrol diphosphate, estradiol diphosphate, and estradiol 17-phosphate respectively. The authors concluded that the position of the phosphate groups rather than their number is important in determining the inhibitory potency. Phosphate esterification at C-17 yields the most active compound. The available data with the sulfate esters did not allow such comparisons.

As with the kynurenine holotransaminase (Mason and Gullekson, 1960), the glutamate-aspartate holotransaminase was not inhibited significantly by the conjugates, even at relatively high levels (0.16 M). Prolongation of the incubation of the apoenzyme with pyridoxal phosphate and the conjugates also resulted in reversal of the inhibition. Scardi *et al.* concluded from their kinetic data that the inhibition involved competition between pyridoxal phosphate and the conjugates for the apoenzyme but that the coenzyme eventually became more firmly bound, displacing the conjugates, and thereby reversing the inhibition.

Pulkkinen and Willman (1966) purified a glutamate-aspartate transaminase from normal full-term placentas, partially resolved it, and examined the effects of estrogen sulfates on reconstitution with pyridoxal phosphate. The holoenzyme activity was not influenced by free estrogens or their sulfates. Free estrogens had no effect on reconstitution. The 3-sulfates of estrone, estradiol, and estriol had no effect at 10^{-4} to 10^{-5} M levels. The corresponding 17-sulfated estradiol, on the other hand, inhibited 50% at the 10^{-4} M level. Estradiol 3,17-disulfate and estriol 16,17-disulfate inhibited 75% and 85% at that level, whereas estriol 3,16,17-trisulfate inhibited 95% at 10^{-4} M and 70% at 10^{-5} M.

The authors concluded that the inhibitory potency of the conjugates increases with increasing numbers of sulfate groups. As with the estrogen phosphates studied by Scardi *et al.* (1962), the 17-substituted estrogen was more potent than the 3-substituted one.

3. Pyridoxamine-Oxaloacetate Transaminase

Enzymes which catalyze the reversible transamination between pyridoxamine and oxaloacetic acid were discovered by Wada and Snell (1962) in extracts of *Escherichia coli* and rabbit liver. These enzymes have pH optima near 8.5, require inorganic phosphate for activity, and are inhibited rather than activated by pyridoxal phosphate. Wu and Mason (1964) partially purified an enzyme from rat kidney with very similar properties and examined the effects of estrogen sulfates and diethylstilbestrol diphosphate on enzyme activity. Their kinetic data indicated a competition between pyridoxal phosphate and inorganic phosphate and also between pyridoxal phosphate and pyridoxamine. These results are consistent with the suggestion that the pyridoxamine and inorganic phosphate bind sites on the enzyme that can bind analogous sites of pyridoxal phosphate (Wada and Snell, 1962).

Low concentrations of diethylstilbestrol diphosphate also inhibited the rat kidney enzyme, whereas the sulfate esters of diethylstilbestrol and estradiol, at the same levels, were not inhibitory. Lineweaver-Burk plots of the inhibition data indicated that the estrogen phosphate inhibited by competing with inorganic phosphate for the enzyme and that it did not compete with pyridoxamine. Thus, in this case, the conjugate appears not to have interfered significantly with Schiff's base formation but to have acted at the binding site for the phosphate group. Whether these effects on an atypical transamination reaction are similar to their actions on other pyridoxal enzymes cannot be determined from the available data.

Another enzyme in which inorganic phosphate seems to serve as a substitute for the ester phosphate group of a coenzyme is the microsomal $\triangle^4$-3-ketosteroid reductase of rat liver (Leybold and Staudinger, 1963) which requires NADPH but can use NADH if inorganic phosphate is also present. Wu and Mason (1965) studied testosterone sulfate as a potential inhibitor of the reductase, using NADH and inorganic phosphate. Instead of inhibiting, the conjugate served effectively as a substrate. Diethylstilbestrol diphosphate did not inhibit activation by inorganic phosphate at conjugate levels as high as 10^{-4} M (Wu and Mason, unpublished data).

4. *Tyrosine-α-ketoglutarate Transaminase*

Singer and Mason (1967) examined the effects of a variety of conjugated steroids and carboxylic acids on the tyrosine-α-ketoglutarate transaminase of freshly prepared rat liver homogenates. The homogenates, prepared in ice water, were incubated with and without anions for various periods of time and then assayed for transaminase activity. Much of the activity was lost during incubation. The inactivation was accompanied by the loss of coenzyme but was not completely reversed by the subsequent addition of pyridoxal phosphate. This irreversibility suggests that the enzyme was destroyed under the conditions of incubation.

The coenzyme, the keto acid substrate, and the various anionic steroids and carboxylic acids retarded the inactivation and dissociation; free steroids were inactive at saturation levels. The anionic steroids were effective at 5×10^{-4} to 5×10^{-5} M, aromatic acids at 10^{-2} to 10^{-3} M, 5-hydroxytryptophan at 10^{-3} M, L-glutamate, bicarbonate, and inorganic phosphate at 10^{-2} M. Several other amino acids and NaCl were ineffective at 10^{-2} M.

Many of the *in vitro* stabilizing agents caused elevated levels of hepatic tyrosine transaminase when they were injected into adrenalectomized rats. In general, the most potent stabilizers were the most effective agents in causing elevated enzyme levels. This correlation led to the suggestion that the increase in enzyme activity results from enzyme accumulation that is caused by the stabilization of the transaminase against processes of breakdown, which are known to act at a relatively rapid rate. Such stabilization largely accounts for the accumulation of hepatic tryptophan pyrrolase following the administration of L-tryptophan (Schimke et al., 1965).

In studies of the induction of liver enzymes by steroids, the water-soluble conjugates are often used for convenience in administration, apparently with the assumption that they are converted to the free steroids in the body and act in the same manner as administered free steroids. Although hydrolysis, before entering the liver, appears to be a distinct possibility, it also seems likely from studies of the metabolic fate of inducing levels of hydrocortisone (Litwack, 1967) that the free steroids would be converted again in the liver predominantly to conjugated derivatives. It is interesting, in this connection, that stilbestrol was much less effective in elevating tyrosine transaminase levels in the liver than stilbestrol disulfate. That anionic steroids may be active without hydrolysis was also suggested by the observation that a steroid carboxylic acid, androst-5-ene-3β-o1-17β-oic acid, which cannot be hydrolyzed to yield a neutral steroid, was also fairly effective in elevating the transaminase levels (Singer and Mason, 1967).

It has been suggested (Singer et al., 1966) that anionic compounds impart stability to the transaminases by stabilizing their tertiary or quarternary structure. This view is supported by the observation that stilbestrol disulfate $(5 \times 10^{-4}\ M)$ retarded the inactivation of the enzyme during incubation in 1.1 or 2.2 M urea solutions.

5. Phosphorylases a and b

Early studies of the effects of conjugated steroids on pyridoxal enzymes showed that they interfere with the activation of apophosphorylase a with added pyridoxal phosphate (Mason and Gullekson, 1959). Subsequent studies with highly purified phosphorylase b (Ford and Mason, 1968a,b; Ford, 1968) have demonstrated that the conjugates can modify the binding of pyridoxal phosphate in the holoenzyme. Coenzyme binding and its modification is conveniently monitored by observing changes in the spectral properties of the enzyme. The native phosphorylase dissolved in 0.04 M β-glycerophosphate, pH 7, exhibits an absorption peak with a λ_{max} at 333 mμ which, as reported by Kent et al. (1958), is attributable to the protein-bound pyridoxal phosphate. During incubation of the enzyme with estradiol disulfate (6–7 μmoles/ml) at room temperature, the absorption peak at 333 mμ declined and a new one with a λ_{max} at 420 mμ appeared (Fig. 2). Incubation with diethylstilbestrol disulfate or dehydroepiandrosterone sulfate also caused a spectral shift, but gave a peak

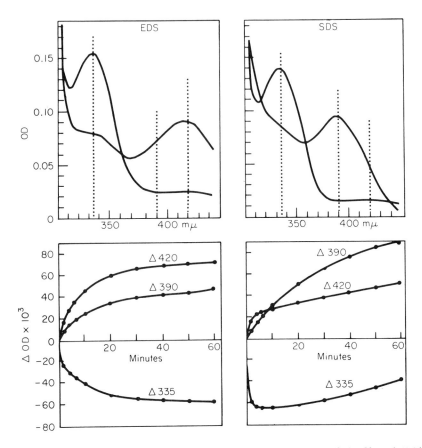

FIG. 2. Effect of estradiol disulfate (EDS) and diethylstilbestrol disulfate (SDS) on the spectral properties of phosphorylase *b*. Three milliliters of a solution of phosphorylase *b* in 0.04 *M* β-glycerophosphate (pH 6.8) were placed in a silica cuvette and a spectrum was taken at room temperature in a Beckman DU spectrophotometer with a blank containing 3 ml of the buffer. Twenty micromoles EDS or SDS were then added to each cuvette in a volume of 0.4 ml and the absorbance changes were measured at 333, 390, and 420 mμ for 1 hour, after which the spectrum was taken again (Ford and Mason, 1968a).

at 390 mμ, apparently formed with the 420 mμ species as an intermediate. These latter shifts were similar to those reported by Kent *et al.* (1958) to occur as a result of incubating the enzyme in solutions of high or low pH or in 7 *M* urea solutions. They proposed that the transient species absorbing above 400 mμ is a protonated Schiff's base of pyridoxal phosphate in which the aldehyde carbon of that coenzyme is linked with the ε-amino group of a lysyl residue of the protein. Formation of the transient species from the 333 mμ form was attributed to cleavage of a covalent bond linking the aldehyde carbon to some other group on the protein. Its further conversion to a species absorbing at 390 mμ was believed to be the result of hydrolysis of the Schiff's base to liberate free pyridoxal phosphate.

Because of the unique ability of estradiol disulfate to convert the enzyme to a stable 420 mμ species, Ford and Mason (1968a,b) studied the actions of that conjugate more extensively. It was observed that the 420 mμ species was formed more rapidly at 4°–6°C than at room temperature. The spectral shift under these conditions was paralleled by a loss of enzyme activity which became almost complete as the spectral conversion approached equilibrium. Approximately 70% of the initial enzyme activity and the original spectrum (333 mμ peak) were restored by dialyzing the treated enzyme against 1000 volumes of 0.005 M β-mercaptoethanol–0.005 M triethanolamine buffer (pH 7.5) overnight at 3°C. This reversibility was taken as evidence that the changes in protein structure caused by the conjugate were somewhat limited. Conformational changes were assumed to account for the ability of the pyridoxal phosphate in the modified enzyme to react with sodium borohydride at neutral pH values and with cysteine, in contrast to the behavior of the unaltered enzyme.

The magnitude and rate of the spectral changes were strongly dependent on the presence of various glycolytic intermediates and other compounds recognized as activators, inhibitors, or substrates of the enzyme. Accompanying increases in the reactivity of the sulfhydryl groups of the enzyme and in the rate of enzyme inactivation were dependent in a similar way on the type of nonsteroid metabolite in the reaction mixture. Of the compounds tested, 3-phosphoglycerate was the most effective agent in facilitating the actions of the conjugates. In contrast, AMP and glucose 6-phosphate caused little change, if any, and instead appeared to prevent the actions of other nonsteroid metabolites.

In ultracentrifuge studies, it was observed that the distribution of the enzyme between the dimeric and tetrameric forms was also modified by the nonsteroid effectors. Low levels of the anionic steroids increased the amount of the tetrameric form. High levels of conjugate caused more extensive aggregation in most cases. In the presence of glucose 6-phosphate, however, even the high concentrations of conjugate did not change the state of aggregation from that present in glucose 6-phosphate buffer alone, i.e., the dimeric state.

The effects of the conjugates on phosphorylase *b* show that anionic steroids can bind the enzyme to cause conformational changes that alter the catalytic action and the reactivity of the coenzyme and the enzymic sulfhydryl groups to environmental agents. The authors proposed that the ability of the conjugates to modify the response of the enzyme to non-steroid effectors may have a physiological counterpart, possibly facilitating or retarding enzymic conformational changes that are important in metabolic regulation.

Since the conjugates caused the pyridoxal phosphate to become more reactive and in some cases appeared to cause its dissociation, it seemed reasonable that they would favor the transfer of that coenzyme to other proteins that can bind it. Such seems to be the case for their effect on phosphorylase *b* in the presence of hepatic tyrosine apotransaminase (Mason *et al.*, 1969). The addition of diethylstilbestrol disulfate to a mixture of these two enzymes resulted in activation of the apotransaminase. Treatment of the phosphorylase with borohydride, which fixes the coenzyme as a reduced derivative to the protein, gave a product that did not activate the apotransaminase. Activation occurred under conditions of pH, ionic strength, and conjugate concentrations that did not bring about spectral shifts in earlier studies, so the transfer of the coenzyme may also be facilitated by protein–protein interactions or by dilution.

6. Other Pyridoxal Enzymes

The observation that steroid conjugates interfere with the cofactor binding by a variety of pyridoxal enzymes, as reviewed above, suggests that these actions may be very general, at least with regard to pyridoxal enzymes. Other pyridoxal enzymes have been tested cursorily, but for the most part the studies were done before the studies of phosphorylase *b*, described above, revealed the degree to which the steroid actions may be dependent on the presence of other effectors.

Davis and Metzler (1962) reported that estradiol disulfate and diethyl-stilbestrol disulfate, at relatively high levels, were without effect on a partially purified serine dehydrase from sheep liver. The coenzyme was very firmly bound; effects on coenzyme association with resolved enzyme were not tested.

In an early survey (Mason and Schirch, 1961) of other pyridoxal enzymes, it was found that the conjugates interfere to varying degrees with the association of pyridoxal phosphate with cysteine sulfinic decarboxylase, dopa decarboxylase, 5-hydroxytryptophan decarboxylase, and serine transhydroxymethylase.

B. DEHYDROGENASES

1. *Glutamate Dehydrogenase*

Tomkins and Maxwell (1963) reviewed studies of the interactions of steroids with various enzymes and proposed that steroids modify enzyme action by causing changes in enzyme conformation. One of the most thoroughly studied enzymes in this respect is glutamate dehydrogenase. Yielding and Tompkins (1960) reported that various steroids and diethylstilbestrol alter the activity of highly purified glutamate dehydrogenase of beef liver and that the activity changes were accompanied by changes in the state of enzyme aggregation. More recently, Willman and Pulkkinen (1967) confirmed that free estrogens and androgens inhibit the glutamate dehydrogenase and also reported that the sulfate esters of these steroids are generally active. Estrone 3-sulfate and estradiol 3-sulfate were approximately as potent as free estrone and estradiol. The 17-sulfate of estradiol was also active but somewhat less potent than the 3-sulfate. Estriol 3-sulfate was less inhibitory than free estriol; estriol 3, 17-disulfate and estriol 3,16,17-trisulfate were ineffective at 10^{-4} M. Estriol 16,17-disulfate at 10^{-4} M, on the other hand, caused a twofold increase in the reaction rate. In studies of these interactions in the analytical ultracentrifuge, the authors observed that diethylstilbestrol and estrone sulfate at 2.5×10^{-4} M did not change the sedimentation behavior under the conditions studied; but that estriol 16,17-disulfate at 6×10^{-4} M caused the enzyme to dissociate. The authors concluded that sulfation of steroids in certain positions may alter or reverse properties as compared to those of free steroids or steroids sulfated in other positions.

2. *17β-Hydroxysteroid Dehydrogenase*

Langer and Engel (1958) first described a 17β-estradiol dehydrogenase localized in the 105,000 g supernatant fraction of human placenta. Talalay and Williams-Ashman (1958) proposed that this enzyme, by virtue of its ability to utilize both NAD and NADP, serves as a transhydrogenase with estradiol acting as indicated in reactions (1) and (2) as a coenzyme:

$$\text{NADPH} + \text{estrone} \rightleftarrows 17\beta\text{-estradiol} + \text{NADP} \qquad (1)$$
$$\text{NAD} + 17\beta\text{-estradiol} \rightleftarrows \text{estrone} + \text{NADH} \qquad (2)$$
$$\overline{\text{Net: NADPH} + \text{NAD} \rightleftarrows \text{NADP} + \text{NADH}}$$

Jarabek *et al.* (1966) reported that this enzyme is cold-inactivated, one of a growing number of enzymes that have been observed to lose activity, usually reversibly, when the temperature at which the enzyme is stored prior to assay is lowered.

With the demonstration that dehydroepiandrosterone sulfate under-goes reduction of the 17-keto group without prior cleavage of the ester linkage (Payne and Mason, 1965), it is reasonable to expect a similar type of specificity in other 17β-dehydrogenases. Crist and Warren (1966) reported that various partially purified preparations of the placental estradiol dehydrogenase can utilize estrone 3-sulfate or estradiol 3-sulfate as activators instead of the free estrogens. They suggested that any important role this enzyme has in metabolic control can be mediated by the estrogen sulfates as well as by the free estrogens.

In a similar study, Pulkkinen and Willman (1967) tested various steroids and their sulfate esters as activators of the transhydrogenation reaction. Estradiol-17β, estrone, and their 3-sulfate esters were equally effective as activators. Androsterone sulfate was also as effective as estradiol-17β. Estradiol 17-sulfate was inactive. No transhydrogenation occurred in the presence of dehydroepiandrosterone, dehydroepiandrosterone sulfate, pregnandiol, or pregnandiol glucuronide.

3. *Malic Dehydrogenase*

Kalman (1952a) reported an inhibitory effect of several water-soluble androgen conjugates on the succinoxidase system of rat liver homogenates. Succinic dehydrogenase appeared to be the target of the inhibitor. As-suming that other dehydrogenases might respond to inhibitors in a similar way, Kalman (1952b) tested the effects of androgen conjugates on rat-liver malic dehydrogenase, which was assayed manometrically through coupling with the electron transport system of the homogenates. Testos-terone sulfate and testosterone 17β-diethylaminoethylcarbonate were in-hibitory at levels of the order of 2×10^{-3} to 2×10^{-5} M. Similar results were found with succinate as substrate.

The author observed that a definite reduction in activity is effected by a given increment of inhibitor and that the inhibition was not reversed by increasing the amount of enzyme. This action was considered to be a possible example of the "irreversible" (or uncompetitive) type of in-hibition described by Ackermann and Potter (1949).

4. *Lactic Dehydrogenase*

The relative amounts of the lactic dehydrogenase isozymes vary in the different tissues. The distribution, as observed in zone electrophoresis, appears to reflect differences in the metabolic patterns of the tissues. Clausen and Gerhardt (1963) observed that diethylstilbestrol diphosphate in the incubation medium selectively inhibited the slowest moving of the isoenzymes of tumor tissue. In studies of these isoenzymes in brain extracts, they observed (Clausen and Gerhardt, 1963) that increasing the level of the conjugate caused inhibition of the isoenzymes in a sequential manner, the slowest-moving band (band 5) being inhibited at the lowest concentration, band 4 at the next higher concentration, etc. They assumed that these actions were a result of the liberation of free diethylstilbestrol, since they could detect the free estrogen in the extracts by infrared analysis.

C. FLAVIN ENZYMES

1. D-*Amino Acid Oxidase*

Hayano and Dorfman (1951) tested the effects of 33 steroids, including several conjugates, on the activity of the D-amino acid oxidase of pig kidney acetone powder. Deoxycorticosterone was the most active inhibitor of the free steroids tested, although it was effective only at relatively high concentrations (1 mg per 3 ml of incubation mixture). Estrone sulfate, estradiol sulfate, and equilen sulfate were slightly more inhibitory than the free steroids. Their greater potency was attributed to their greater solubility in the incubation media rather than to inherently greater activity.

These authors found that the substrate, D-alanine, protected the enzyme against inactivation by deoxycorticosterone. Although the protection was essentially complete at D-alanine levels 10 times the usual substrate level, the addition of these large quantities of substrate 15 minutes after the start of incubation, when inhibition was already established, brought about essentially no relief of inhibition. A similar experiment was carried out with excess flavin adenine dinucleotide additions to test the possibility that the inhibition resulted from competition of the coenzyme and the steroid for the apoenzyme. Again, additional coenzyme present at the beginning of the incubation protected the enzyme from inactivation by the steroid, while additions after the start of incubation brought no relief.

Parallel incubations of the steroid with the coenzyme and of the steroid with the apoenzyme were set up and allowed to proceed for 20 minutes before completion of the systems, with apoenzyme plus alanine in the first instance and with coenzyme and alanine in the second. Under these conditions, only an inhibition occurring in the first 20 minutes would show up in the final assay. Essentially no action of the steroid with the coenzyme was noted. The steroid-apoenzyme incubation, on the other hand, resulted in complete inhibition. It was thus concluded that the apoenzyme was the point of attack of the steroid.

Two attempts were made to reverse the inhibition. Dialysis of the steroid-treated enzyme in a large volume of ice water gave no relief of the inhibition. However, treatment with acetone in which the inhibited enzyme was precipitated, filtered, dried, and reassayed, resulted in almost complete regeneration of enzyme activity. Thus, the authors concluded that the steroid combined reversibly with the apoenzyme, without denaturing it, although reversal of the binding in aqueous media remained to be shown.

D. HEME ENZYMES

1. *Tryptophan Pyrrolase*

Oelkers and Dulce (1964) found that the tryptophan pyrrolase of rat liver was inhibited *in vitro* by estrone sulfate and several other steroid sulfates, but not by cortisol 21-sulfate. Their kinetic data indicated that estrone sulfate competed for the binding of the coenzyme, hematin. The enzyme was also inhibited by free estrogens (Oelkers and Nolten, 1964). The possibility was considered that the inhibition might result from participation of the estrogens in reactions that would change the redox equilibrium of the iron in the coenzyme. Comparisons of the actions of estrone sulfate with those of ferricyanide indicated distinctly different mechanisms of inhibition, however.

E. ENZYMES OF STEROID METABOLISM

One way that conjugated steroids might function as regulators of metabolism is in influencing the activity of enzymes of steroid metabolism. This possibility has received very little attention and, to the author's knowledge, there appears to be little indication in the literature that the conjugates act other than as alternate substrates in the enzymes that act on free steroids.

Raggat and Whitehouse (1966) examined the cholesterol oxidase system of bovine adrenal cortex and found that the various cholesterol esters inhibit oxidation competitively (K_i for cholesterol phosphate, 28 μM; for cholesterol sulfate, 110 μM; for cholesterol acetate, 65 μM). Pregnenolone esters did not inhibit the enzyme system. Pregnenolone, 20α-hydroxycholesterol, and 25-hydroxy-27-norcholesterol inhibited oxidation of cholesterol sulfate by the system but their sulfate esters did not.

In studies of the properties of the steroid alcohol sulfotransferase of human adrenal gland extracts, Adams and Edwards (1968) observed that the substrate and various other factors appeared to influence the association–dissociation equilibrium of the enzyme, causing unusual kinetic behavior. In considering this enzyme as a possible regulatory enzyme, they tested low concentrations (0.01 and 0.02 mM) of steroid hormones (corticosterone, cortisol, cortisone, progesterone, 17β-estradiol, estriol, and estrone sulfate). Only estrone and progesterone had significant effects on activity; they were slightly inhibitory.

Notation and Ungar (1968) reported that the presence of added dehydroepiandrosterone sulfate did not inhibit the conversion of dehydroepiandrosterone to the sulfate ester by extracts of rat testes.

Wengle and Boström (1963) observed that the formation of disulfate esters of various steroids by rat liver extracts showed an optimal substrate concentration at relatively low steroid levels and that formation was strongly decreased at higher steroid levels. Monosulfate formation in the same incubations did not show this behavior. The data do not indicate whether the monosulfate interferes with disulfate formation or the disulfate interferes with its own formation.

F. OTHER ENZYMES

The significance of the very high proportion of the steroids present in fetal tissues in conjugated forms has been the subject of some speculation. One view that has been expressed (Levitz and Dancis, 1963) is that the efficient conversion of steroids to conjugates in fetal tissues protects the fetus against the actions of steroid hormones reaching it from the maternal or placental circulation. Peter et al. (1968) have considered the possibility that estrogen conjugates may suppress the activity or levels of certain hepatic enzymes that are known to be inactive or absent in the fetus until the time of birth. In their study of the liver enzyme that converts methionine to S-adenosyl methionine, they found that relatively high doses of conjugated estrogens resulted in partial, but not complete, inhibition of the neonatal developement of the enzyme level. Such treatment also prevented the increased activity produced by hydrocortisone injections in 4-day-old rats. They were unable to demonstrate any effect of the conjugates at levels of 10^{-4} and 10^{-6} M on the enzyme levels *in vitro*.

Polyestradiol phosphate and analogous nonestrogenic polymers were found by Beling and Diczfalusy (1959) to inhibit seminal acid phosphatase. These polymeric phosphates inhibit other enzymes, including alkaline phosphate and hyaluronidase (Diczfalusy et al., 1953; Fernö et al.,

1958). Beling and Diczfalusy (1959) proposed that this action of the polymers is an example of a widespread and nonspecific type of macro-anion–macrocation interaction.

DeLoecker *et al.* (1966) reported that estradiol disulfate at 3.7 μM stimulated protein synthesis in incubations containing a microsomal fraction from rat skeletal muscle. Estrone sulfate, estradiol 3-sulfate, and estradiol 17-sulfate had marginal effects; free estrogens were inactive. Similar effects were reported for uterine extracts, with estrone sulfate being more active than the disulfates (Brooks and DeLoecker, 1966).

III. Effects of Steroid Conjugates on Nonenzymic Proteins

A. Effects on the Binding of Pyridoxal Phosphate to Albumin

Dempsey and Christensen (1961) observed that equimolar levels of pyridoxal phosphate and serum albumin react to give initially a yellow product absorbing maximally at 415 mμ which is slowly replaced by one absorbing maximally at 332 mμ. These spectral changes resembled those occurring when pyridoxal phosphate is added to apophosphorylase. The authors studied the nature of this reaction as a possible model for the enzymic association reactions.

Following completion of the reaction of equimolar levels of pyridoxal phosphate and serum albumin, when essentially all of the product was in the 332 mμ form, the addition of another equimolar portion of pyridoxal phosphate caused a new rise in absorption at 415 mμ which reached a constant maximal value in 10 minutes with no further change. This second binding was taken as evidence that the first portion of coenzyme added was bound rapidly to a site on the protein (designated Site 2) and subsequently transferred to a second site (Site 1) where it was more stably bound. This would make Site 2 available for binding the second portion of coenzyme. A number of other observations supported this interpretation. To determine the site of binding of the pyridoxal phosphate, the authors treated the various forms of the protein with sodium borohydride, followed by protein hydrolysis. In each case the product was ε-pyridoxyllysine, establishing that the coenzyme was bound to both Site 1 and Site 2 as the Schiff's base derivatives with ε-amino groups of lysyl residues. A number of compounds were found to be inhibitors of the binding of pyridoxal phosphate at Site 1, including several aromatic acids. Estrogen sulfates and bile acids, which were known to interfere with the binding of pyridoxal phosphate to pyridoxal enzymes (Mason and Gullekson, 1959, 1960), were among the most effective agents tested. The disulfates were most potent; free steroids were inactive.

B. Binding of Conjugated Steroids to Plasma Proteins

Puche and Nes (1962) showed that dehydroepiandrosterone sulfate at physiological concentrations is strongly bound to bovine serum albumin. Plager (1965) studied the binding of dehydroepiandrosterone sulfate, androsterone sulfate, and etiocholanolone sulfate to human serum, plasma, and serum albumin with ultracentrifugation and equilibrium dialysis techniques. All three conjugates were bound primarily to the albumin fraction of plasma. Androsterone sulfate and etiocholanolone sulfate appeared to be bound at the same sites of the albumin molecule, with similar affinities. Their binding curves were practically identical and they displaced each other from the sites. Scatchard plots of the binding data indicated one primary site per molecule of albumin and 16 others with less affinity and specificity. Dehydroepiandrosterone sulfate was bound with greater affinity at two sites per molecule albumin and with less affinity to sites similar in number and association constant to those binding the other two steroid conjugates.

Wang and Bulbrook (1967) studied the binding of the sulfate esters of dehydroepiandrosterone, testosterone, and pregnenolone in the plasma of man, rabbit, and rat to determine whether differences in binding might possibly explain differences in the metabolic clearance rates of steroids in these species. In competitive binding studies, they showed that displacement of a radioactively labeled steroid sulfate by the same carrier steroid sulfate was comparable to the displacement with a different carrier steroid sulfate. They suggested, therefore, that the sulfate esters of the three may have common binding sites in the plasma proteins. Since a nonsteroidal sulfate ester (6-benzyl naphthyl sulfate) was almost as effective as carrier dehydroepiandrosterone sulfate in displacing labeled dehydroepiandrosterone sulfate, it was assumed that the sites binding steroid sulfates are not specific for steroid sulfates.

C. Implications of Competitive Binding

Although the phenomenon of competitive binding to plasma proteins is generally recognized, there appears to be relatively little information concerning its relevance to the functions of these proteins. Several examples are known of displacement of one drug by another *in vivo* (Anton, 1961; Kunin, 1965; Thorp, 1964; Keen, 1966). Anton (1961) and

Keen (1966) observed that anionic drugs displaced other anionic drugs from albumin, whereas cationic drugs that were strongly bound to albumin did not displace anionic drugs. This specificity was interpreted as evidence that the drug binding is primarily ionic in nature. One explanation considered for the low order of specificity of anionic drugs in displacing other anionic drugs (Markus and Karush, 1958; Keen, 1966) is that displacement depends not so much on competition for a specific site as upon distortion of the protein structure to alter the binding affinities for the other ligands.

Displacement of endogenous metabolites such as bilirubin and thyroxine from plasma proteins by drugs *in vivo* has also been reported (Harris *et al.*, 1958; Schmid *et al.*, 1965; Ossorio, 1963). Thorp (1964) proposed that displacement from plasma proteins of normally occurring metabolic substrates (fatty acids, keto acids, tryptophan), coenzymes (pyridoxal phosphate), or hormones (thyroxine, steroid sulfates, or glucuronides) may be a factor in the mode of action of drugs with high affinity for the plasma proteins. Such an action was proposed for *p*-chlorophenoxyisobutyrate, a drug capable of decreasing plasma cholesterol and triglyceride levels in hyperlipemic patients. This acid appeared to selectively influence liver metabolism without being taken up by the liver cells. Thorp suggested that a major action of the drug was to displace thyroxine from binding sites in the plasma proteins to make that hormone more accessible to the liver. Many effects on the organism could be explained by such a mechanism. Westerfeld *et al.* (1968) observed with thyroidectomized rats that the drug mimicked thyroxine in causing elevated levels of hepatic α-glycerol phosphate dehydrogenase and malic enzyme, but only if nonstimulating levels of thyroxine were also given. Although these results supported Thorp's proposal, the authors cited several points that discount it. An alternate possibility was proposed that the thyroxine initiates the synthesis of the enzyme and that the drug in some way delays the normal cutoff of the synthetic processes.

Although the consequences of the actions of this particular drug on plasma proteins appear to be unresolved, the suggestion by Thorp (1964) that conjugated steroids and pyridoxal phosphate may be among the metabolites subject to competitive displacement deserves consideration in relation to the biological role of the conjugates.

D. BINDING OF CONJUGATED STEROIDS TO HISTONES

Free steroids have been reported to bind histones (Sekeris and Lang, 1965; Sluyser, 1966a,b; Sunaga and Koide, 1967a,b). Sunaga and Koide (1967c) studied the effect of various experimental conditions on the reaction of arginine-rich histones from calf thymus with corticosteroids. An increase in the degree of interaction was observed with increasing periods of storage of the steroids in phosphate buffer and with the temperature of storage. With cortisol, a derivative was detected but not identified. It differed chromatographically from cortisol 21-phosphate.

Singer (1966) studied the effects of diethylstilbestrol disulfate on the interaction of calf thymus histone with DNA. It was assumed that the anionic steroids and related compounds would bind the cationic groups of the histones and interfere with their association with the DNA and that such actions might be relevant to the role proposed by Bonner and Huang (1963) and Alffrey et al. (1963) for the histones as inhibitors of DNA functions. Singer found that the reconstituted histone-DNA was more resistant than free DNA to digestion by pancreatic DNase. The addition of 6×10^{-4} M diethylstilbestrol disulfate to solutions containing DNA and histone gave a product that was more susceptible to DNase digestion than that present in a similar solution lacking the conjugate. The author suggested that the greater susceptibility reflected the presence of a larger fraction of the DNA in the unbound form.

E. BINDING OF STEROID CONJUGATES TO LIVER PROTEINS

The ability of liver to concentrate corticosteroids has been recorded (Bradlow et al., 1954; Sandberg et al., 1957; Bellamy et al., 1962). Litwack et al. (1963) showed that injected cortisol-^{14}C is rapidly accumulated in rat liver, primarily in the parenchymal cells (Litwack and Baserga, 1967). Accumulation was maximal 45 minutes after intraperitoneal injection. About one-half of the accumulated dose was found in the 100,000 g supernatant fraction, mostly in the form of several anionic metabolites. A portion of the two most anionic metabolites was bound to soluble cytoplasmic proteins (Fiala and Litwack, 1966; Litwack, 1967; Litwack et al., 1965).

Litwack and co-workers (personal communication) have isolated two discrete cortisol binding proteins from the liver cytosol. The larger of the two has a molecular weight in the range of 18,000, an isoelectric point of about 8.4, and contains arginine and lysine residues. It binds a negatively charged cortisol metabolite, believed to be tetrahydrocortisol 3-sulfate. The smaller of the two binding proteins has a molecular weight in the range of 6000 and an isoelectric point of about 6. It binds a more polar metabolite, believed to be a sulfate (possibly a disulfate) derivative. The smaller binding protein also contains in its structure one or more nucleotides.

Although these authors have reported no firm conclusions regarding the functional significance of the binding proteins and the anionic metabolites, they have pointed out (Litwack, 1967) that failure to find any effect of the anionic derivatives on metabolism in the liver would focus attention on the approximately 4% of the accumulated dose that appears in the liver as unchanged steroid.

IV. Effects of Conjugated Steroids on "in Vivo" Functions

Given the fact that the steroid conjugates do react *in vitro* with enzymes and various other proteins, one may ask whether such actions occur *in vivo* and, if so, what effects they may have on metabolic functions. The answer to this question is especially difficult to obtain because it is not readily determined in mammals whether the administration of a steroid causes biological effects by virtue of the actions of the administered compound on cell constituents or by the actions of its metabolic derivatives. This problem is alleviated to some degree in studies of whole cell suspensions or tissue cultures, with which one may more readily compare the effects of free and conjugated steroids on the early responses of the cells. This section describes several *in vivo* systems of various degrees of complexity which respond to conjugated steroids and related compounds or to which the *in vitro* actions of the conjugates appear to be logically related.

A. AMINO ACID TRANSPORT

Pyridoxal and pyridoxal phosphate stimulate the net movement of free amino acids into mammalian cells (Christensen *et al.*, 1954), an action which led to consideration of the possibility that the vitamin B_6 derivatives may serve as carriers in the transfer of the amino acids across the cell membrane. Although these agents do possess the requisite ability to bind the amino acids (as Schiff's base derivatives), a number of observations have been cited as being inconsistent with the carrier role (Christensen, 1960). Among these is the demonstration that several anionic compounds which lack the ability to bind the amino acids also stimulate amino acid transport. Indoleacetate and phenylacetate, for example, intensified amino acid uptake by Ehrlich tumor cells (Christensen *et al.*, 1954), although at rather high levels (5 to 20 mM). Guided by the observations of Mason and Gullekson (1959, 1960) that very low levels of estrogen conjugates influence the binding of pyridoxal phosphate to pyridoxal enzymes, Riggs and Walker (1964) tested the actions of estradiol disulfate, diethylstilbestrol disulfate, and estrone sulfate on glycine uptake by the Ehrlich cells and found that they were as effective at 1 mM levels or lower as 1 mM concentrations of pyridoxal or pyridoxal phosphate.

These results led to a reconsideration (Christensen, 1960) of the features that the various stimulators of amino acid transport have in common and to the generalization that all are capable of binding protonated amino groups. Accordingly, an alternative explanation of the action of pyridoxal phosphate on amino acid transport came under consideration, namely that the coenzyme or the anionic compounds bind protonated amino groups on the cell surface to alter the binding or passage of the transported solutes.

The view that pyridoxal phosphate acts by binding amino groups at the cell surface was supported by the observation that pyridoxal phosphate, in contrast to pyridoxal, is largely excluded from the Ehrlich cells (Pal and Christensen, 1961). The small amount associated with the cells resulted in a distinct yellow color, presumably arising from Schiff's base formation with amine groups on the cell surface (Christensen, 1963). This yellow color was removed instantly following the addition of a small amount of sodium borohydride, an agent which has been used with a number of pyridoxal enzymes to reduce the aldimine double bond of the Schiff's base to give colorless derivatives that are firmly bound to the protein. This fixation did not interfere with the subsequent uptake of amino acids by the cells but did result in the loss of responsiveness to new additions of pyridoxal phosphate and to a lack of stimulation of uptake by estradiol disulfate or diethylstilbestrol disulfate. These observations were considered to be consistent with the view that the steroid conjugates and pyridoxal phosphate bind the same or overlapping sites on the cell surface in exerting their stimulatory effects on amino acid transport.

B. TRANSPORT OF OTHER SOLUTES

Estradiol disulfate was reported also (Christensen and Jones, 1961) to stimulate the mediated uptake of labeled uric acid by erythrocytes *in vitro* and to inhibit the unmediated entry and exit of this compound. These actions were shown to be a secondary result of a slowing of the entry of inorganic phosphate in exchange for cellular chloride ion. Diethylstilbestrol disulfate and free estradiol also inhibited this exchange.

Atkinson *et al.* (1963) reported that erythrocyte potassium ion was released into the plasma when fresh rabbit blood was incubated at 38°C for 15 minutes with a hypnotic steroid, 3α-hydroxy-5β-pregnane-11,20-dione 3-phosphate, at concentrations ranging from 0.9 to 7 mM. This action did not require hydrolysis of the ester linkage. Other active conjugates were 3α-hydroxy-5β-pregnan-20-one 3-phosphate and 3α-hydroxy-16α-methyl-5β-pregnane-11,20-dione 3-phosphate. Of a variety of other conjugates tested, the 21-phosphates and the 3- and 21-hemisuccinates were inactive. Inorganic phosphate was inactive.

Interest in the role of steroid hormones in regulating uterine contractions during pregnancy has led to investigations of the effects of conjugated steroids on the contractions of uterine strips *in vitro* (Robson and Sharaf, 1951, 1952; Jung, 1960; Hempel and Neumann, 1965; Mossman and Conrad, 1967). The uterine response varied with the hormonal status of the donor animals. It is not known whether the steroids act on the uterus in the conjugated forms or as the hydrolyzed derivatives. Their effects on uterine contraction have been interpreted as involving an effect on Na^+ and K^+ distribution (Jung, 1960; Mossman and Conrad, 1967).

C. ANTIINFLAMMATORY ACTION

Known antiinflammatory drugs include both steroid and nonsteroid compounds. Whitehouse and Skidmore (1965) have pointed out that a number of the nonsteroid drugs are aromatic acids and that many of them inhibit enzyme reactions that can logically be related to inflammatory processes. These reactions include (1) oxidative phosphorylation, (2) the formation of histamine by substrate-specific histidine decarboxylase, and (3) the hydrolysis of proteins by nonenteric proteinases. They proposed that these apparently diverse reactions may all be inhibited by the same basic process, namely, the "neutralization" of essential lysyl ε-amino groups on the enzymes by the anionic forms of the drugs.

Although a detailed description of the rationale for the above hypothesis is well beyond the scope of this review, certain similarities in these actions of the drugs to those of the conjugated steroids merit attention. Earlier studies (Gould *et al.*, 1963; Smith *et al.*, 1963; Gould and Smith, 1965b) had shown that decarboxylation of L-glutamate by enzymes from *E. coli* and rat brain was inhibited by relatively high levels of compounds that are congeners of salicylic acid. Inhibition was reduced by preincubation of the enzymes with pyridoxal phosphate but not by increasing the substrate concentration. Other pyridoxal enzymes are sensitive to salicylate (Huggins *et al.*, 1961; Gould and Smith, 1965a) and other antiinflammatory drugs (Hanninen and Hartiala, 1964a,b; Pulver *et al.*, 1956).

Skidmore and Whitehouse (1967) reported that a number of antiinflammatory drugs inhibit histamine formation by the histidine decarboxylases of rat gastric mucosa and rat fetal tissue by competing with pyridoxal phosphate for the coenzyme binding site, presumed to be a lysyl ε-amino group. The drugs also inhibited the binding of pyridoxal phosphate and 2,4,6-trinitrobenzaldehyde to the amino groups of plasma albumin. A generally close parallel between the action of a given drug in inhibiting histamine formation and in displacing pyridoxal phosphate or trinitrobenzaldehyde from albumin was observed.

When more than 90% of the amino groups of albumin were modified either by acetylation or by nitroguanidination, reaction with trinitrobenzaldehyde was strongly inhibited but not abolished. Proteolytic digestion of the native albumin also decreased its reactivity. Polypeptides with very high content of lysyl residues (polylysine, calf thymus histone) reacted similarly to albumin with the reagent. The following substances did not react in the same way: lysine, lysine methyl ester, 6-aminohexoic acid, 6-aminohexoic methyl ester, imidazole, imidazoleacetic acid, and histidine (all at 33.3 mM). Other compounds which reacted in a manner similar to the acetylated albumin included: arginine, arginine methyl ester, N-(α)-benzoylarginine methyl ester (all at 33.3 mM), and salmine sulfate (33.3 mM in arginine side chain, 0.1 mM in lysine side chain). Thus, the typical reaction with trinitrobenzoic acid apparently required an intact polypeptide structure containing lysyl residues.

These authors also observed that the presence of a carbonyl group attached to an aromatic nucleus is not the only structural requirement for reaction with the lysyl amino groups; benzaldehyde, 2-nitrobenzaldehyde, and 2,4-dinitrobenzaldehyde were much less reactive. Similarly, the aromatic acids that interfered with the binding of the reagent required other structural properties as shown by the observation that several acidic analogs of the active drugs were unable to displace the aldehydic reagent and did not inhibit the histidine decarboxylase. It was assumed that the other types of binding were especially important in the actions of the most potent drugs. Urea treatment also interfered with the reaction of trinitrobenzaldehyde with albumin, indicating a dependence of the binding on specific conformations of the protein.

Another facet of the interaction of antiinflammatory drugs with serum proteins was discovered by Gerber et al. (1967). These authors found that the drugs enhanced the rate of reaction of serum protein sulfhydryl groups with 5,5'-dithiobis(2-nitrobenzoic acid). Of 138 compounds tested, consisting of commonly used drugs and commonly occurring biologicals, those that catalyzed the reaction were almost exclusively limited to the nonsteroid compounds possessing antiinflammatory action. Since facilitation of this reaction is usually interpreted to be a result of "unmasking" or exposure of the protein sulfhydryl groups, these results suggest that the reaction with the drugs is accompanied by changes in the protein conformation.

The correlation between antiinflammatory action and ability to interfere with the binding of trinitrobenzaldehyde (or pyridoxal phosphate) with protein amino groups suggested that the latter effects might be useful in screening potential antiinflammatory compounds (Skidmore and Whitehouse, 1965). Phillips *et al.* (1967) have reported, however, that many acids that were effective in the binding tests were not active in the foot edema and pleural effusion tests for antiinflammatory action.

Another *in vitro* screening assay that has been proposed for antiinflammatory drugs is based on the observation by Mizushima (1964) that 1 mM levels of the drugs stabilize a bovine plasma albumin fraction against heat denaturation. Skidmore and Whitehouse (1965) found that trinitrophenyl albumin and N-acetylated albumin could not be protected from heat denaturation in this way and that neither of the modified proteins reacted with trinitrobenzaldehyde. They, therefore, proposed that Mizushima's assay method also affords a measure of the ability of the compounds tested to complex with the lysyl ε-amino groups of the proteins.

The ability of the drugs to stabilize plasma protein fractions *in vitro* raised the question of a similar action of these agents in the circulating blood. Mizushima and Kobayashi (1968) tested their action on biologically active and heat-labile fractions of blood, using nearly undiluted serum and therapeutic levels of the drugs. The drugs did not alter the biological activity of the fractions but did influence their stability to mild heating. Piliero and Columbo (1967) developed a serum turbidity test for blood samples obtained 3 hours after the oral or subcutaneous administration of antiinflammatory drugs to rats. This method gave a good correlation with data obtained by the protein-binding test of Skidmore and Whitehouse and with the clinical antiinflammatory potency of the drugs.

Correlation has also been reported for the potency of the nonsteroidal antiinflammatory drugs and their ability to suppress the release of enzymes from the heavy lysosomes isolated from rat liver (Tanaka and Iizuka, 1968). Several agents capable of releasing hydrolytic enzymes from lysosomes also injure erythrocytes. Brown *et al.* (1967) developed an *in vitro* screening test for nonsteroidal antiinflammatory drugs based on a correlation of their clinical potency with their ability to stabilize canine erythrocytes against heat-induced hemolysis.

D. PORPHYRIN SYNTHESIS

Granick and co-workers have studied extensively the production of porphyrins in chick embryo cells grown in primary culture. The porphyrinogenic action of a number of steroids was noted (Levere et al., 1967; Granick and Kappas, 1967a,b; Kappas and Granick, 1968). For certain natural steroids, porphyrin induction could be detected at concentrations as low as 10^{-6} to 10^{-8} M. This characterized these steroids as having activity equal to or exceeding that of the previously studied porphyria-inducing drugs. The most potent inducers were 5β ring A-saturated steroids, including etiocholanolone, pregnandiol, and pregnanolone. The glucuronide derivatives of even the most active steroids were without activity. The authors concluded that steroid glucuronides are generally devoid of porphyrin-inducing activity. They pointed out, however, that this conclusion does not extend to sulfate conjugates, since the sulfate derivatives of dehydroepiandrosterone and estrone did have detectable, although weak, activity.

The authors have proposed for the actions of steroids the same mechanism suggested earlier for the porphyrinogenic actions of certain drugs. They suggested that the synthesis of heme is limited by a repressor–operator mechanism (Jacob and Monod, 1961) which controls the synthesis of the limiting enzyme, δ-aminolevulinic acid synthetase. They proposed that the steroids may derepress the synthesis of this enzyme by combining with and inactivating a repressor of the δ-aminolevulinic acid synthetase operon.

Uridinediphosphoglucuronate blocked the porphyrin induction by steroids. This effect was assumed to reflect the conversion of the free steroid inducers to inactive glucuronidated derivatives.

E. TRYPTOPHAN METABOLISM

Rose (1966) and Price et al. (1967) reported that the urinary excretion of tryptophan metabolites by normal women, following the ingestion of tryptophan, is markedly changed as a result of the use of birth control pills (estrogen plus progestagen) or of estrogen alone. The excretory pattern resembled that occurring during deficiency of pyridoxine in the diet. The pattern became nearly normal when supplementary pyridoxine was given.

A similar pattern of urinary tryptophan metabolites occurs in pregnant women in response to tryptophan ingestion; this pattern is also corrected by pyridoxine supplementation. Brown *et al.* (1961) suggested that the increased pyridoxine requirement may be caused by the elevated blood levels of estrogens that occur during pregnancy. The discovery (Rose, 1966; Price *et al.*, 1967) that estrogen administration causes similar effects supports this suggestion. The mechanism described by Mason and Gullekson (1959, 1960) in which steroid metabolites interfere with pyridoxal phosphate binding was suggested by these authors as a possible clue to the molecular basis for these actions of estrogens on tryptophan metabolism.

The possibility that the *in vitro* mechanism might explain the effects of estrogens on tryptophan metabolism prompted Mason *et al.* (1969) to compare the effects of estrogen administration and pyridoxine deficiency on kynurenine transaminase levels in rat tissues. Some similarities and some remarkable differences were detected in the effects of the two types of treatment. Both treatments caused a decrease in the kidney supernatant transaminase levels. The decline was associated in the estrogen-treated animals with an *increased* degree of association with the endogenous pyridoxal phosphate relative to the controls, whereas pyridoxine deficiency caused a decreased degree of association. The activity of the mitochondrial transaminase was relatively undisturbed by pyridoxine deficiency but was strongly decreased by estrogen treatment. In both conditions the degree of association of the enzyme with endogenous pyridoxal phosphate was only slightly depressed in this fraction. The enzyme levels and degree of association with coenzyme for the estrogen-treated males were remarkably similar to those of normal females.

Since the studies described above indicate a redistribution of pyridoxal phosphate during treatment with estrogens, they are consistent with the possibility that the estrogens influence the transaminase *in vivo* in a fashion similar to that observed *in vitro*. However, estrogen administration and pyridoxine deficiency do not cause the same pattern of changes in the enzyme so their effects on tryptophan metabolite excretion probably do not reflect the same actions on the enzymes of tryptophan metabolism.

V. Effects of Anionic Ligands on Protein Structure

At this time very little is known about the effects of conjugated steroids on protein structure in spite of the numerous studies that have been made of the binding of a multitude of different ligands to model proteins such as plasma albumin. In many cases these studies do include compounds somewhat similar to the conjugates in that they consist of large hydrocarbon molecules which bear negative charge in solutions at physiological pH values. Such compounds include fatty acids (Teresi and Luck, 1948), anionic dyes (Fredericq, 1956), and detergents (Putnam and Neurath, 1945). Consideration of studies of the binding of such molecules may provide some insight into the interactions to be expected of the conjugated steroids.

Putnam and Neurath (1945) obtained evidence in electrophoretic studies of the interaction of dodecyl sulfate (D) with horse plasma albumin (A) that complexes are formed with the composition AD_n and AD_{2n}, where n is 55. Since the value $2n$ corresponded closely with the number of cationic groups on the protein at neutral pH, these authors concluded that each cationic group binds one detergent anion. Foster (1960), however, suggested that this stoichiometry may be fortuitous. He pointed out that certain other proteins bind detergents considerably in excess of the number of available cationic groups, in which case the binding must depend more on hydrophobic forces.

Yang and Foster (1953) and Pallansch and Briggs (1954) observed that the formation of the AD_n and AD_{2n} complexes was preceded by strong binding of 10 to 12 detergent anions to sites preexisting in the native protein. Both groups of investigators concluded that the higher complexes resulted from cooperative alteration of the protein structure to yield many new, but weaker, binding sites. Pallansch and Briggs (1954) concluded also that the few strong binding sites present in the native albumin are destroyed during the structural change. Foster and Aoki (1958) discovered the N–F transformation of plasma albumin and suggested that it might be the same structural change caused by the detergents. Leonard and Foster (1961) provided strong supporting evidence for this interpretation.

Decker and Foster (1966) studied the binding of alkylbenzenesulfonates to bovine plasma albumin, using equilibrium dialysis and moving boundary electrophoresis. The initial binding involved 11 ± 1 sites and produced only a single electrophoretic boundary. Higher levels of binding gave two additional electrophoretic components, AD_{38} and AD_{76}, similar to those described by Putnam and Neurath (1945) but with a somewhat lower value for n. The composition of the three complexes appeared to be remarkably constant throughout the range of coexistence. Formation of the higher complexes was accompanied by some loss of helical content of the albumin according to optical rotatory dispersion measurements.

The constancy of the composition of the three complexes led Decker and Foster (1966) to conclude that the protein exists in three isomeric forms and that the three forms are each saturated with detergent throughout the range of detergent concentrations in which they coexist. These authors considered one explanation for this all-or-none binding in which it was assumed that the bound detergent anions interact with each other. Such interaction would be analogous to micelle formation, the protein serving as a nucleus for micelle formation and determining its stoichiometry. The protein was assumed to stabilize the micelle because the critical micelle concentrations of the detergents per se were not approached under the conditions of complex formation. The alternative suggestion (Yang and Foster, 1953) that the newly exposed sites bind detergent statistically was also considered. In this case the binding constants must be so strong that formation of the AD_n and AD_{2n} complexes are essentially complete at detergent levels at which the appropriate isomeric forms of the albumin exist in significant concentration. Equations developed for these two theoretical mechanisms were shown (Decker and Foster, 1966) to be equivalent for the conditions of the experiments and, therefore, were unable to distinguish which mechanism is operative.

These authors also discussed another implication of the assumption that the 10–12 binding sites of the native albumin are destroyed and replaced by more numerous and weaker binding sites. Under suitable conditions, low levels of detergent would tend to stabilize the native form and at higher levels to disrupt it. This possibility was pointed out by Foster and Aoki (1958) who showed that low levels of detergent did shift the N–F equilibrium toward the N (native) form. The ability of low levels of detergent to protect albumins against denaturation by urea (Markus *et al.*, 1964) and heat (Foster, 1960) has been attributed to this effect.

Foster and Aoki (1958) developed a mathematical model to describe the effects of detergents on the N–F transition of bovine plasma albumin. A somewhat similar equation was presented by Reynolds *et al.* (1967) which explicitly predicts both the stabilization and the transition of protein structures as a result of ligand association. It also provides a means of determining the equilibrium constant between two different protein states (state 1 and state 2) when the number of binding sites (*n* and *m*, respectively) and the association constants (*K* and *J*, respectively) are known for a particular ligand. They proposed the equation as a general model for the denaturation of proteins that contain masked binding sites for any ligand, including hydrogen ions. According to the formulation, stabilization of state 1 would occur when:

$$K \gg J, \; n \leqslant m$$
$$\text{or}$$
$$K \geqslant J, \; n > m$$

and transition to state 2 would occur when:

$$m > n, \; K \geqslant J$$
$$\text{or}$$
$$m \leqslant n, \; K \ll J$$

The above formulation would be applicable to subunit–subunit interactions as well as to segment–segment interactions and would, therefore, be relevant to the model for allosteric interactions described by Monod et al. (1965).

The binding of other types of ligands has also been shown to stabilize plasma albumin toward heat and urea denaturation (Boyer et al., 1946a, b). The somewhat specific stabilizing actions of antiinflammatory drugs on plasma proteins and their ability to unmask sulfhydryl groups (Section IV, C) suggest that this class of compounds may act, as the detergents do, either to stabilize the native form of the protein or to promote transition to other conformations. Interestingly, dodecyl sulfate, a strong detergent, has been reported to be active in several biological tests for antiinflammatory action (Mizushima, 1964; Mizushima and Kasukawa, 1964). It was moderately effective in comparison with various antiinflammatory drugs, in promoting the unmasking of the sulfhydryl groups of serum proteins (Gerber et al., 1967).

The compounds that influence albumin properties as described above and in Section IV, C are also the ones most effective in displacing pyridoxal phosphate from histidine decarboxylase (Section IV, C). This relationship suggests a similarity in the structural characteristics required of the ligands for the several actions and implies similar mechanisms of action. Since conjugated steroids also interfere with pyridoxal binding to pyridoxal enzymes (Section II, A) and to serum albumin (Section III, A), it seems likely that their actions are fundamentally similar to those of the drugs.

The mechanism by which anionic ligands influence pyridoxal phosphate binding is unsettled. Competitive displacement of the coenzyme is indicated in kinetic studies with kynurenine transaminase (Section II, A) and histidine decarboxylase (Section IV, C) and is implied in the studies of pyridoxal phosphate binding to plasma albumin (Sections III, A and IV, C). On the other hand, the stabilizing effects of conjugates, substrate, and substrate analogs on pyridoxal phosphate binding to transaminases are most readily explained by the assumption (Ford *et al.*, 1966) that these agents stabilize a protein conformation that binds the coenzyme strongly.

Studies of phosphorylase *b* (Section II, A) have not revealed a stabilizing action of steroid conjugates but have shown a strong dependence of coenzyme binding on the enzyme conformation. No indication was found of coenzyme displacement without accompanying changes in conformational state, although, at low conjugate:enzyme ratios, there were changes in the state of enzyme aggregation without noticeable changes in coenzyme binding (Ford and Mason, 1968a).

Whitehouse and Skidmore (1965) suggested that competitive displacement of pyridoxal phosphate may occur most readily in those enzymes to which it is weakly bound. Weak coenzyme binding is indeed characteristic of the two enzymes, kynurenine transaminase and histidine decarboxylase, in which it has been concluded by kinetic criteria (Section II, A and IV, C) that competition occurs. Where the coenzyme binding is strong, as with phosphorylase *b*, the binding may be decreased by conformational changes in the protein prior to any displacement by the conjugates.

Several of the studies reviewed have emphasized the importance of electrostatic binding in the effects of conjugated steroids and aromatic acids on proteins and membranes. By virtue of their anionic groups, the conjugated steroids do possess such binding properties, properties not shared with the free steroids. It is reasonable to expect, however, that many of the binding properties of the conjugates will reflect residual properties of the parent steroid. Such similarities are manifested in the ability of conjugates to serve as substrates in place of the free steroids (Calvin and Liebermann, 1964; Payne and Mason, 1965; Wu and Mason, 1965). Conjugates also apparently serve efficiently in the place of free steroids as coenzymes in the placental transhydrogenation reaction (Section II, B). Both the free and the conjugated steroids inhibit glutamate dehydrogenase (Section II, B), tryptophan pyrrolase (Section II, D), and D-amino acid oxidase (Section II, C).

Free steroids are generally not effective in causing the changes in pyridoxal phosphate binding that have been reported for conjugated steroids (Sections II, A and III, A) and aromatic acids (Sections II, A and IV, C). Specificity for anionic compounds probably relates to the anionic character of the coenzyme and to the positive charge of the site of binding on the ε-amino groups of lysyl residues of the protein.

The important role of the hydrocarbon portion of anionic ligands in binding proteins was emphasized by Ray *et al.* (1966) who compared the binding to serum albumin of *n*-octane and four octane derivatives (*l*-octanol, octyl sulfate ion, and octyl sulfonate ion) and two *n*-dodecane derivatives (*l*-dodecanol and dodecyl sulfate). These results showed that the hydrocarbon moiety of all these substances contributed the major share of the binding energy and that the binding energy increased markedly with the length of the hydrocarbon chain. On the other hand, the binding energies were different for octyl sulfate and octyl sulfonate, so the nature of the anionic group was also important.

Similar forces seem to be involved in the interaction of steroid conjugates with kynurenine transaminase; the important contribution of the hydrocarbon portion of the ligands led in fact to the testing of the effects of conjugated steroids on pyridoxal enzymes (Mason, 1959). The role of the charged groups is also of major importance. With kynurenine transaminase, the disulfates were much more effective both in inhibiting and protecting the enzyme than the steroid monosulfates; free steroids were inactive. The nature of the anionic group was also important. Phosphate esters were reported to be more effective than the sulfates with kynurenine transaminase and glutamate-aspartate transaminase. Diethylstilbestrol diphosphate was much more potent than diethylstilbestrol disulfate in inhibiting the activation of pyridoxamine-oxaloacetate transaminase by inorganic phosphate. This specificity may be related to the specificity of that enzyme for phosphate ion as an activator; sulfate was ineffective (Wu and Mason, 1964).

VI. Biological Relevance

Several years ago, Tomkins and Maxwell (1963) reviewed the literature dealing with the influence of steroids on enzymes. They concluded that steroids can bind proteins reversibly to cause conformational changes and associated changes in biological activity. Most of the current theories concerning the molecular basis of steroid hormone action incorporate this very general mechanism. It is evident from some of the studies reviewed here that the binding of conjugated steroids to proteins can also cause reversible conformational shifts that are related to function. The binding appears also in some cases to favor redistribution of coenzyme and changes in protein stability.

Although the conjugated steroids resemble the free steroids in being able to alter the conformational state of proteins, it is clear that the presence of the charged groups on the conjugated steroids confers a degree of specificity of action. Indeed, the nature of the charged groups, their position, and number also appear to be important in determining the nature or potency of the effects on enzyme action. Theoretically then, conjugation of steroids may decrease or eliminate their action on one protein but increase or initiate an effect on another.

The relevance of these *in vitro* actions to metabolic regulation remains conjectural. As with the free steroids, most of the observed effects of conjugated steroids on enzymes occur at concentrations *in vitro* that are several orders of magnitude greater than the levels to be expected normally in the tissues. While this fact is of obvious importance in assessing the possibility that the interaction of a specific steroid with a specific protein in the tissues occurs, it should not be taken as definite proof that such interaction does not occur. The important role of nonsteroid effectors in determining the effects of estrogen sulfates on phosphorylase *b* emphasizes this point. An effect of differences in the *in vitro* and *in vivo* environments was suggested by Scotto and Scardi (1968) as the reason that levels of estrogen conjugates which did not influence pyridoxal phosphate binding to transaminases in heart homogenates caused a rapid dissociation when perfused through intact heart.

The biological relevance of the *in vitro* studies appears then to rest mainly on their potential for discovering ways that steroids can influence the structure and properties of proteins rather than their ability to locate key sites of action. It seems likely that such information will be useful in testing the properties of proteins that are under investigation as potential sites of hormone action. Such systems would include, for example, the adenyl cyclase system (Sutherland *et al.*, 1968) and binding proteins for estradiol (Shyamada and Gorski, 1969), dihydrotestosterone (Bruchovsky and Wilson, 1968), and cortisol conjugates (Litwack, 1967). The isolation from bacteria of a repressor (Gilbert and Mueller-Hill, 1966) and specific binding proteins for sulfate (Pardee, 1966) and amino acids (Penrose *et al.*, 1968) holds promise of similar achievements with mammalian cells and the possibility that such proteins may be specific sites of steroid action.

VII. Summary

Although conjugated steroids constitute a significant portion of the steroids of various tissues, relatively little is known about their interactions with tissue constituents. In this review, their interactions with various enzymes were described. Comparisons were made with the interactions of the conjugated steroids and various other anionic ligands with model protein systems, such as plasma albumin. Several *in vivo* systems which respond to the conjugated steroids or whose responses appear to be logically related to such actions were described. Although few definitive studies are available, the results of various investigations indicate that the conjugated steroids can reversibly modify protein conformation with accompanying changes in catalytic activity, coenzyme distribution, and enzyme stability.

REFERENCES

Ackermann, W. W., and Potter, V. R. (1949). *Proc. Soc. Exptl. Biol. Med.* **72**, 1.
Adams, J. B., and Edwards, A. M. (1968). *Biochim. Biophys. Acta* **167**, 122.
Allfrey, V. G., Littau, V. C., and Mirsky, A. E. (1963). *Proc. Natl. Acad. Sci. U.S.* **49**, 414.
Anton, A. H. (1961). *J. Pharmacol. Exptl. Therap.* **134**, 291.
Atkinson, R. M., MacGregor, I. G., Pratt, M. A., and Tomich, E. G. (1963). *Biochem. Pharmacol.* **12**, 931.
Beling, C. G., and Diczfalusy, E. (1959). *Biochem. J.* **71**, 229.
Bellamy, D., Phillips, J. G., Jones, I. C., and Leonard, R. A. (1962). *Biochem. J.* **85**, 537.
Bonner, J., and Huang, R. C. (1963). *In* "The Nucleohistones" (J. Bonner and P. O. P. Ts'o, eds.), p. 250. Holden-Day, San Francisco, California.
Boyer, P. D., Lum, F., Ballou, G., and Luck, J. M. (1946a). *J. Biol. Chem.* **162**, 181.
Boyer, P. D., Ballou, G., and Luck, J. M. (1946b). *J. Biol. Chem.* **162**, 199.
Bradlow, H. L., Dobriner, K., and Gallagher, T. F. (1954). *Endocrinology* **54**, 343.
Brooks, S. C., and DeLoecker, W. C. (1966). *152nd Meeting Am. Chem. Soc. New York*, No. 304.
Brown, J. H., Macky, H. K., and Riggilo, D. A. (1967). *Proc. Soc. Exptl. Biol. Med.* **125**, 837.
Brown, R. R., Thornton, M. J., and Price, J. M. (1961). *J. Clin. Invest.* **40**, 617.
Bruchovsky, N., and Wilson, J. D. (1968). *J. Biol. Chem.* **243**, 5953.
Calvin, H. I., and Liebermann, S. (1964). *Biochemistry* **3**, 259.
Christensen, H. N. (1960). *Advan. Protein Chem.* **15**, 239.
Christensen, H. N. (1963). *In* "Chemical and Biological Aspects of Pyridoxal Catalysis" (E. E. Snell, P. M. Fasella, A. Braunstein, and A. Rossi-Fanelli, eds.), p. 533. Macmillan (Pergamon), New York.
Christensen, H. N., and Jones, J. C. (1961). *J. Biol. Chem.* **236**, 76.
Christensen, H. N., Riggs, T. R., and Coyne, B. A. (1954). *J. Biol. Chem.* **209**, 413.
Clausen, J., and Gerhardt, W. (1963). *Acta Psychiat. Scand.* **39**, 305.
Crist, R. D., and Warren, J. C. (1966). *Acta Endocrinol.* **53**, 205.
Davis, L., and Metzler, D. (1962). *J. Biol. Chem.* **237**, 1883.
Decker, R. V., and Foster, J. F. (1966). *Biochemistry* **5**, 1242.
DeLoecker, W. C., Brooks, S. C., and DeWever, F. (1966). *Biochim. Biophys. Acta* **119**, 665.
Dempsey, W., and Christensen, H. N. (1961). *J. Biol. Chem.* **237**, 1113.
Diczfalusy, E., Fernö, O., Fex, H., Högberg, B., Linderot, T., and Rosenberg, T. (1953). *Acta Chem. Scand.* **7**, 913.

Fernö, O., Fex, H., Högberg, B., Linderot, T., Veige, S., and Diczfalusy, E. (1958). *Acta Chem. Scand.* **12**, 1675.

Fiala, E. S., and Litwack, G. (1966). *Biochim. Biophys. Acta* **12**, 1675.

Fischer, E. H., Kent, A. B., Snyder, E. R., and Krebs, E. G. (1958). *J. Am. Chem. Soc.* **80**, 2906.

Ford, J. (1968). Ph. D. Thesis, Univ. of Michigan, Ann Arbor, Michigan.

Ford, J., and Mason, M. (1968a). *Proc. Natl. Acad. Sci. U.S.* **59**, 980.

Ford, J., and Mason, M. (1968b). *Symp. Pyridoxal Enzymes, Nagoya, Japan, 1967* (K. Yamada, N. Katunuma, and H. Wada, eds.), p. 125. Maruzen, Tokyo.

Ford, J., Singer, S., Schirch, L., and Mason, M. (1966). *Abstr. 2nd Intern. Congr. Hormonal Steroids, Milan* p. 143. Excerpta Med. Found., New York.

Foster, J. F. (1960). *In* "The Plasma Proteins. Vol. 1: Isolation, Characterization, and Function" (F. W. Putnam, ed.). Academic Press, New York.

Foster, J. F., and Aoki, K. (1958). *J. Am. Chem. Soc.* **80**, 5215.

Fredericq, E. (1956). *Bull. Soc. Chim. Belges* **65**, 631.

Gerber, D. A., Cohen, N., and Giustra, R. (1967). *Biochem. Pharmacol.* **16**, 115.

Gilbert, W., and Mueller-Hill, B. (1966). *Proc. Natl. Acad. Sci. U.S.* **56**, 1891.

Gould, B. J., and Smith, M. J. H. (1965a). *J. Pharm. Pharmacol.* **17**, 83.

Gould, B. J., and Smith, M. J. H. (1965b). *J. Pharm. Pharmacol.* **17**, 15.

Gould, B. J., Huggins, A. K., and Smith, M. J. H. (1963). *Biochem. J.* **88**, 346.

Granick, S., and Kappas, A. (1967a). *Proc. Natl. Acad. Sci. U.S.* **57**, 1463.

Granick, S., and Kappas, A. (1967b). *J. Biol. Chem.* **242**, 4587.

Hänninen, O., and Hartiala, K. (1964a). *Biochem. J.* **92**, 15.

Hänninen, O., and Hartiala, K. (1964b). *Biochem. Pharmacol.* **14**, 1073.

Harris, R. C., Lucey, J. F., and Maclean, J. (1958). *Pediatrics* **21**, 875.

Hayano, M., and Dorfman, R. I. (1951). *Ann. N.Y. Acad. Sci.* **54**, 608.

Hempel, R., and Neumann, F. (1965). *Acta Endocrinol.* **48**, 656.

Huggins, A. K., Smith, M. J. H., and Moses, V. (1961). *Biochem. J.* **79**, 271.

Jacob, F., and Monod, J. (1961). *J. Mol. Biol.* **3**, 318.

Jarabek, J., Seeds, A. E., Jr., and Talalay, P. (1966). *Biochemistry* **5**, 1269.

Jung, H. (1960). *Acta Endocrinol.* **35**, 49.

Kalman, S. M. (1952a). *Endocrinology* **50**, 361.

Kalman, S. M. (1952b). *Endocrinology* **52**, 73.

Kappas, A., and Granick, S. (1968). *J. Biol. Chem.* **243**, 346.

Keen, P. M. (1966). *Brit. J. Pharmacol.* **26**, 704.

Kent, A. B., Krebs, E. G., and Fischer, E. H. (1958). *J. Biol. Chem.* **232**, 549.

Kunin, C. M. (1965). *J. Lab. Clin. Med.* **65**, 406.

Langer, L. J., and Engel, L. L. (1958). *J. Biol. Chem.* **233**, 583.

Leonard, W. J., and Foster, J. F. (1961). *J. Biol. Chem.* **236**, PC73.

Levere, R. D., Kappas, A., and Granick, S. (1967). *Proc. Natl. Acad. Sci. U.S.* **58**, 985.

Levitz, M., and Dancis, J. (1963). *Clin. Obstet. Gynecol.* **6**, 62.

Leybold, K., and Staudinger, H. (1963). *Biochem. Z.* **337**, 320.

Litwack, G. (1967). *In* "Topics in Medicinal Chemistry" (J. L. Rabinowitz and R. M. Myerson, eds.). Vol. 1, p. 3. Wiley (Interscience), New York.

Litwack, G., and Baserga, R. (1967). *Endocrinology* **80**, 774.

Litwack, G., Sears, M. L., and Diamondstone, T. I. (1963). *J. Biol. Chem.* **238**, 302.

Litwack, G., Fiala, E. S., and Filosa, R. J. (1965). *Biochim. Biophys. Acta* **111**, 569.

Markus, G., and Karush, F. (1958). *J. Am. Chem. Soc.* **80**, 89.

Markus, G., Lowe, R., and Wissler, F. (1964). *J. Biol. Chem.* **239**, 3687.
Mason, M. (1959). *J. Biol. Chem.* **234**, 2770.
Mason, M., and Gullekson, E. H. (1959). *J. Am. Chem. Soc.* **81**, 1517.
Mason, M., and Gullekson, E. H. (1960). *J. Biol. Chem.* **235**, 1312.
Mason, M., and Schirch, L. (1961). *Federation Proc.* **20**, 200.
Mason, M., Ford, J., and Wu, H. L. C. (1969). *Ann. N.Y. Acad. Sci.* (in press).
Mizushima, Y. (1964). *Arch Intern. Pharmacodyn.* **149**, 1.
Mizushima, Y., and Kasukawa, R. (1964). *Intern. Arch. Allergy* **24**, 100.
Mizushima, Y., and Kobayashi, M. (1968). *J. Pharm. Pharmacol.* **20**, 169.
Monod, J., Wyman, J., and Changeux, J. P. (1965). *J. Mol. Biol.* **12**, 88.
Mossman, R. G., and Conrad, J. T. (1967). *Am. J. Obstet. Gynecol.* **99**, 539.
Nolan, C., Novoa, W. B., Krebs, E. G., and Fischer, E. H. (1964). *Biochemistry* **3**, 542.
Notation, A. D., and Ungar, F. (1968). *Can J. Biochem.* **46**, 1185.
Oelkers, W., and Dulce, H. J. (1964). *Z. Physiol. Chem.* **337**, 150.
Oelkers, W., and Nolten, W. (1964). *Z. Physiol. Chem.* **338**, 105.
Ossorio, C. (1963). *In* "Salicylates" (A. S. J. Dixon, ed.), p. 82. Churchill, London.
Pal, P. R., and Christensen, H. N. (1961). *J. Biol. Chem.* **236**, 894.
Pallansch, M. J., and Briggs, D. R. (1954). *J. Am. Chem. Soc.* **76**, 1396.
Pardee, A. B. (1966). *J. Biol. Chem.* **241**, 5886.
Payne, A. H., and Mason, M. (1965). *Steroids* **6**, 323.
Penrose, W. R., Nichoalds, G. E., Piperno, J. R., and Oxender, D. L. (1968). *J. Biol. Chem.* **243**, 5921.
Peter, C. H., Volpe, J. J., and Laster, L. (1968). *J. Clin. Invest.* **47**, 2099.
Phillips, B. M., Sancilio, L. F., and Kurchacova, F. (1967). *J. Pharm. Pharmacol.* **19**, 697.
Piliero, S. J., and Colombo, C. (1967). *J. Clin. Pharmacol. J. New Drugs* **7**, 198.
Plager, J. E. (1965). *J. Clin. Invest.* **44**, 1234.
Price, J. M., Brown, R. R., and Thornton, M. J. (1967). *Am. J. Clin. Nutr.* **20**, 452.
Puche, R. C., and Nes, W. R. (1962). *Endocrinology* **70**, 857.
Pulkkinen, M. O., and Willman, K. (1966). *Steroids* **8**, 51.
Pulkkinen, M. O., and Willman, K. (1967). *Acta Obstet. Gynecol. Scand.* **46**, 494.
Pulver, R., Exer, B., and Herrmann, B. (1956). *Schweiz. Med. Wochschr.* **86**, 1080.
Putnam, F. W., and Neurath, H. (1945). *J. Biol. Chem.* **159**, 195.
Raggat, P. R., and Whitehouse, M. W. (1966). *Biochem. J.* **101**, 819.
Ray, A., Reynolds, J. A., Polet, H., and Steinhardt, J. (1966). *Biochemistry* **5**, 2606.
Reynolds, J. A., Herbert, S., Polet, H., and Steinhardt, J. (1967). *Biochemistry* **6**, 937.
Riggs, T. R., and Walker, L. M. (1964). *Endocrinology* **74**, 483.
Robson, J. M., and Sharaf, A. A. (1951). *J. Endocrinol.* **7**, 223.
Robson, J. M., and Sharaf, A. A. (1952). *J. Endocrinol.* **8**, 133.
Rose, D. P. (1966). *Clin. Sci.* **31**, 265.
Sandberg, A. A., Slaunwhite, W. R., Jr., and Antoniades, H. N. (1957). *Recent Progr. Hormone Res.* **13**, 209.
Scardi, V., Magno, S., and Scarano, E. (1960). *Boll. Soc. Ital. Biol. Sper.* **36**, 1719.
Scardi, V., Iaccarino, M., and Scarano, E. (1962). *Biochem. J.* **83**, 443.
Schimke, R. T., Sweeny, E. W., and Berlin, C. M. (1965). *J. Biol. Chem.* **240**, 4609.
Schmid, R., Diamond, I., Hammaker, L., and Gundersen, C. B. (1965). *Nature* **206**, 1041.

Scotto, P., and Scardi, V. (1968). *Life Sci.* **7**, 1121.
Sekeris, C. E., and Lang, N. (1965). *Z. Physiol. Chem.* **340**, 92.
Shyamada, G., and Gorski, J. (1969). *J. Biol. Chem.* **244**, 1097.
Singer, S. (1966). Ph.D. Thesis, Univ. of Michigan, Ann Arbor, Michigan.
Singer, S., and Mason, M. (1967). *Biochim. Biophys. Acta* **146**, 452.
Singer, S., Ford, J., Schirch, L., and Mason, M. (1966). *Life Sci.* **5**, 837.
Skidmore, I. F., and Whitehouse, M. W. (1965). *J. Pharm. Pharmacol.* **17**, 671.
Skidmore, I. F., and Whitehouse, M. W. (1967). *Biochem. Pharmacol.* **16**, 737.
Sluyser, M. (1966a). *J. Mol. Biol.* **19**, 591.
Sluyser, M. (1966b). *J. Mol. Biol.* **22**, 411.
Smith, M. J. H., Gould, B. J., and Huggins, A. K. (1963). *Biochem. Pharmacol.* **12**, 917.
Sunaga, K., and Koide, S. S. (1967a). *Biochem. Biophys. Res. Commun.* **26**, 342.
Sunaga, K., and Koide, S. S. (1967b). *Steroids* **9**, 451.
Sunaga, K., and Koide, S. S. (1967c). *Arch. Biochem. Biophys.* **122**, 670.
Sutherland, E. W., Robison, G. A., and Butcher, R. W. (1968). *Circulation* **37**, 279.
Talalay, P., and Williams-Ashman, H. G. (1958). *Proc. Natl. Acad. Sci. U.S.* **44**, 15.
Tanaka, K., and Iizuka, Y. (1968). *Biochem. Pharmacol.* **17**, 2023.
Teresi, J. D., and Luck, J. M. (1948). *J. Biol. Chem.* **174**, 653.
Thorp, J. M. (1964). *In* "Absorption and Distribution of Drugs" (T. B. Binns, ed.), p. 64. Livingstone, Edinburgh and London.
Tomkins, G. M., and Maxwell, E. S. (1963). *Ann. Rev. Biochem.* **32**, 677.
Wada, H., and Snell, E. E. (1962). *J. Biol. Chem.* **237**, 127.
Wang, D. Y., and Bulbrook, R. D. (1967). *J. Endocrinol.* **39**, 405.
Wengle, B., and Boström, H. (1963). *Acta Chem. Scand.* **17**, 1203.
Westerfeld, W. W., Richert, D. A., and Ruegamer, W. R. (1968). *Biochem. Pharmacol.* **17**, 1003.
Whitehouse, M. W., and Skidmore, I. F. (1965). *J. Pharm. Pharmacol.* **17**, 668.
Willman, K., and Pulkkinen, M. O. (1967). *J. European Steroides* **2**, 77.
Wu, H. L. C., and Mason, M. (1964). *J. Biol. Chem.* **239**, 1492.
Wu, H. L. C., and Mason, M. (1965). *Steroids* **5**, 45.
Yang, J. T., and Foster, J. F. (1953). *J. Am. Chem. Soc.* **75**, 5560.
Yielding, K. L., and Tomkins, G. M. (1960). *Proc. Natl. Acad. Sci. U.S.* **46**, 1483.

GLUCURONIC ACID PATHWAY

TATU A. MIETTINEN
and ERKKI LESKINEN

I. Introduction

Metabolic transformation of foreign compounds and of many endogenous substances in the animal organism consists of reduction, oxidation,
hydrolysis, or conjugation with some other compound. The latter process
may take place directly, provided that the compound to be conjugated
possesses one or more functional groups, such as hydroxyl-, carboxyl-,
amino-, imino- or sulfhydryl groups, or it may follow after such groups
have been formed by one of the other detoxication processes. Accordingly, the various detoxication reactions are more or less coupled with
each other. Of the conjugation reactions, glucuronidation, as compared
with sulfate, cysteine, and glycine conjugation, methylation, and acetylation, appears to be the most important, particularly because glucuronic
acid reacts with a great number of different compounds. The enzyme
activities needed for the conjugation occur in many organs, primarily the
liver, kidney, and intestines, and they normally have a strong glucuronidation capacity. Furthermore, the substrate for the conjugation, glucuronic acid as UDP-glucuronic acid (UDPGA), is easily generated from
glucose via UDP-glucose (UDPG).

But glucuronidation is not the only reaction utilizing glucuronic acid.
The latter is supplied in the form of UDPGA for the synthesis of mucopolysaccharides. Furthermore, glucuronic acid is released from UDPGA
via direct hydrolysis and the conjugates formed from UDPGA and endogenous or exogenous aglycones by the microsomal enzyme, glucuronyl
transferase, are also cleaved. Since free glucuronic acid is further metabolized, in addition to glucarate and ascorbate, via L-xylulose into the
pentose phosphate cycle, the process as a whole forms an additional
route, the glucuronic acid pathway, for the catabolism of glucose.

The significance of glucuronic acid conjugation in the detoxication process is well known, and glucuronidation was actually one of the first
processes, in addition to glycine and sulfate conjugations, recognized to
detoxicate foreign compounds within the body (cf. Neumeister, 1895).
In many respects, less is known about the significance of the hydrolytic
product, free glucuronic acid, of the pathway and the regulation of the

further metabolism of this acid. That the latter phenomenon is also associated with the detoxication process is indicated by the finding that many drugs, the so-called "inducer" drugs, which stimulate the microsomal drug-metabolizing enzyme system (cf. Conney, 1967), also stimulate the glucuronic acid pathway without themselves being necessarily conjugated with glucuronic acid.

Many extensive review articles have been published dealing with one or more aspects of the glucuronic acid pathway (Hollmann, 1964; Marsh, 1966; Dutton, 1966; Levvy and Conchie, 1966; Burns and Conney, 1966; Smith and Williams, 1966; Schmid and Lester, 1966; Jayle and Pasqualini, 1966; Hänninen, 1966, 1968; Dohrmann, 1967). In the present account the pathway will be treated as a whole, while we attempt to explore primarily the factors which regulate glucuronidation and the release of glucuronic acid from UDPGA *in vivo*, and the relationship and significance of the further metabolism of glucuronic acid to the glucuronidation process itself and to the general metabolism. With these objectives in mind, changes in glucuronidation and the metabolism of free glucuronic acid via the xylulose, ascorbate, and glucarate routes, as well as enzyme activities and the availability of the nucleotides and substrates involved in these processes, are related to various experimental and clinical conditions which have an influence on general metabolism or on one or several steps of the glucuronic acid pathway.

II. Biosynthesis of Glucuronides

Investigations carried out in many laboratories have led to a rather good understanding of the reaction mechanisms by which glucuronic acid is formed from glucose. The reactions, including glucuronidation and the release of free glucuronic acid, are summarized schematically in Fig. 1. (See p. 173.) A more detailed presentation of the initial reactions is given in Section II.

A. Biosynthesis of UDPG

UDPG biosynthesis proceeds by way of the following reaction:

$$\text{D-Glucose-1-P} + \text{UTP} \rightleftharpoons \text{UDP-D-glucose} + \text{PP} \qquad (1)$$

The corresponding enzyme, UDPG pyrophosphorylase, was first discovered in yeast (Munch-Petersen *et al.*, 1953). UDPG itself was first found in the liver (Caputto *et al.*, 1950; Rutter and Hansen, 1953) and mammary gland of the rat and the guinea pig (Smith and Mills, 1954a).

UDPG pyrophosphorylase activity has been shown in the liver (Smith et al., 1953), the lactating mammary gland of the rat (Smith and Mills, 1955), and in epiphyseal cartilage homogenate (Castellani et al., 1957). The enzyme has been crystallized from a calf liver preparation (Albrecht et al., 1966). Intracellular synthesis of UDPG was first demonstrated in the nuclear fraction (Smith et al., 1953; Mills et al., 1954; Smith and Mills, 1954b) though Reid (1959) reported that the liver supernatant fraction is the main site of UDPG synthesis. Cautious about the intracellular localization of UDPG pyrophosphorylase, Reid later (1961a) suggested that the nucleus may be the site of the enzyme, which may be released into the supernatant fraction during preparation.

Thyroxine treatment depresses the activity of UDPG pyrophosphorylase, resulting in a reduced hepatic UDPG level (Reid, 1961a,b). UDPG pyrophosphorylase activity is moderately reduced in homozygote Gunn rats (Halac and Frank, 1960). The physiological significance of the enzyme is twofold:

(1) Biosynthesis of UDPG, an essential intermediate in glycogen synthesis. During induced glycogen synthesis (with lactate or hydrocortisone), decreased levels of UDPG and glucose-6-P have been demonstrated in the rat liver (Hernbrook et al., 1965). UDPG may regulate its own rate of synthesis, because rat liver UDPG pyrophosphorylase is inhibited by AMP and UDPG (Kornfeld, 1965). UDPG is also consumed in the formation of UDP-glucuronic acid (Storey and Dutton, 1955).

(2) The reverse reaction may have significance as the last step in the utilization of D-galactose in the animal organism. There are two ways by which UDPG can participate in the metabolism of galactose:

(a) According to the following reaction:

$$\text{Gal-1-P} + \text{UDPG} \rightleftharpoons \text{UDP-Gal} + \text{G 1-P} \qquad (2)$$

The enzyme catalyzing the reaction is called Gal-1-P uridyl transferase.

(b) The other reaction is:

$$\text{UDPG} \rightleftharpoons \text{UDP-Gal} \qquad (3)$$

The enzyme catalyzing the reaction is called—UDP-Gal-4-epimerase (galactowaldenase). The balance of the reaction is attained with 75% UDPG and 25% UDP-Gal. These reactions demonstrate that galactose may enter the glucuronic acid pathway directly without being diluted by the large glucose pool.

B. BIOSYNTHESIS OF UDPGA

1. *Biosynthesis via UDPG Dehydrogenase*

When Dutton and Storey (1951, 1953, 1954) found that glucuronide synthesis requires a thermostable factor, this was soon identified as UDPGA (Smith and Mills, 1954b; Storey and Dutton, 1955). Because UDPG and UDPGA are both present in the liver, it seemed likely that UDPG was the precursor of UDPGA. It was found (Strominger *et al.*, 1954) that UDPGA is formed in liver tissue from UDPG by a specific enzyme, UDPG-dehydrogenase, which is NAD^+- specific, according to the following reaction:

$$\text{UDP-D-glucose} + 2\ NAD^+ \rightarrow \text{UDP-glucuronic acid} + 2\ NADH_2 \quad (4)$$

Strominger *et al.* (1957) soon purified the enzyme from calf acetone powder up to 400-fold. Since then it has been further purified (700-fold) and its molecular weight is estimated to be 3×10^5 (Wilson, 1965). The enzyme is located in the supernatant fraction, has K_m for UDPG $= 2 \times 10^{-5} M$ and for NAD^+ of the order $\times 10^{-4}$ (Maxwell *et al.*, 1956). The K_m for NAD^+ is consequently of the same order as that of the other dehydrogenases. One of the important rate-determining factors in this reaction seems to be oxidized NAD^+ (Mills *et al.*, 1958). Accumulation of the reaction products UDPGA and NADH appears to inhibit the reactions *in vitro* (Zalitis and Feingold, 1968). The formation of UDPGA and glucuronides by the isolated enzyme system is also dependent upon a constant supply of ATP because 2 moles of ATP are used per mole of UDPG (1 mole of ATP is consumed for glucose-6-P synthesis and 1 mole of ATP is used for rephosphorylation of UDP to UTP) and upon glucose or mainly glycogen (Mills *et al.*, 1958). UDPG dehydrogenase activity occurs not only in the liver but also in the kidney cortex and gastric mucosa (Dutton and Stevenson, 1959; Stevenson and Dutton, 1962), intestine (Miettinen and Leskinen, 1963), skin extract (Jacobson and Davidson, 1962a), epiphyseal cartilage homogenate (Castellani *et al.*, 1957), synovial tissue, and connective tissue (Bollet *et al.*, 1959). Dutton and Greig (1957) did not find UDPGA in the intestinal homogenate, probably owing to high UDPGA-destroying pyrophosphatase activity. In rat liver homogenates UDPG dehydrogenase can be inhibited with UDP-D-xylose (Neufeld and Hall, 1965). It appears possible that this compound, which is a product of UDPGA decarboxylation, controls the entrance of hexose units into the pathway of D-xylose synthesis.

UDP-glucuronic acid decarboxylase has not yet been demonstrated in the animal organism, although its presence seems very likely.

Marchi *et al.* (1964) have shown that hydrocortisone depressed the activity of rat liver UDPG dehydrogenase by about 40%. Because hydrocortisone also reduces the level of liver UDPG (Hernbrook *et al.*, 1965, see Section II, A), this hormone may reduce UDPGA synthesis, and this, in turn, may affect the synthesis of such substances as mucopolysaccharides, and/or glucuronides and free glucuronic acid. This suggestion is supported by the work of Beck *et al.* (1964), who demonstrated retarded conjugation and excretion of N-acetyl-p-aminophenol glucuronide in the rabbit immediately after administration of glucocorticoids. They consider that a plausible explanation for this phenomenon might be a decreased $NAD^+/NADH_2$ ratio (Hübener, 1962) and, owing to that, decreased activity of UDPG dehydrogenase. A reduced $NAD^+/NADH_2$ ratio was also considered by Strömme (1965) to cause a decreased rate of glucuronidation of diethyldithiocarbamate (disulfiram) after administration of ethanol. An ethanol-induced shift in the $NAD^+/NADH_2$ ratio (Forsander *et al.*, 1958) lowers the capacity for the formation of UDPGA from UDPG.

2. Biosynthesis via UDP-Glucuronic Acid-5-Epimerase

UDP-glucuronic acid can also be synthesized from UDP-iduronic acid by the enzyme UDP-glucuronic acid-5-epimerase. This enzyme has been demonstrated in rabbit skin; it requires a catalytic amount of NAD^+ and is inhibited by $NADH_2$ (Jacobson and Davidson, 1962b, 1963). It is logical, however, to suppose that the reverse reaction, i.e. synthesis of UDP-L-iduronic acid, is more important.

3. Biosynthesis from Glucuronic Acid-1-Phosphate

An alternative pathway of UDP-glucuronic acid formation in plants is phosphorylation of glucuronic acid to glucuronic acid-1-phosphate by ATP. In the presence of UDP, glucuronic acid pyrophosphorylase, glucuronic acid-1-phosphate and UTP produce UDP-glucuronic acid and pyrophosphate. There is no direct evidence that glucuronic acid can be phosphorylated or that UDP-glucuronic acid pyrophosphorylase occurs in animal tissues (Smith and Mills, 1954b; Storey and Dutton, 1955). However, there are some investigations which have indirectly thrown some light on this reaction. Shirai and Ohkubo (1954b) were able to demonstrate that α-glucuronic acid-1-phosphate showed high activity of glucuronide formation with anthranilic acid by rat liver slices. The β-isomer of glucuronic acid-1-phosphate was without effect on glucuronidation. Earlier, Touster and Reynolds (1952) had also observed that β-D-glucuronic acid-1-phosphate did not stimulate o-aminophenyl glucuronide formation by mouse liver slices. Arias *et al.* (1958) presented evidence that 4-methylumbelliferone can be conjugated with glucuronic acid in the presence of ATP, UTP and the soluble fraction of rat liver homogenate. α-Glucuronic acid-1-phosphate can act as a substitute for glucuronic acid and ATP. In liver, kidney, and gastric mucosal preparations o-aminophenol glucuronide synthesis has been obtained from ATP, UTP, and glucuronate (or glucurone) (Stevenson and Dutton, 1962).

C. GLUCURONIDE SYNTHESIS

1. Reaction Mechanisms

Glucuronic acid conjugation takes place with: (1) hydroxyl groups of phenols, enols, and primary, secondary, and tertiary alcohols, (2) carboxyl groups of aromatic and certain aliphatic compounds, (3) amino and imino groups of some aliphatic, heterocyclic, and aromatic compounds, (4) sulfhydryl groups of certain sulfur compounds, and (5) hydroxyl groups of carbohydrates (cf. Dutton, 1962; Marsh, 1966; Smith and Williams, 1966). The conjugation is carried out primarily by glucuronyl transferase, though other mechanisms also appear to exist. Since the transferase is located within the microsomes, it is segregated by a lipid barrier which can be penetrated only by substances with some degree of lipid solubility (Gaudette and Brodie, 1959; Brodie, 1964). Steroid hormones (Jayle and Pasqualini, 1966; Schriefers, 1967), bilirubin (Schmid and Lester, 1966), and vitamin A derivatives (Lippel and Olson, 1968) are easily conjugated with glucuronic acid, while glucuronidation of cholesterol seems to be negligible.

Glucuronic acid is transferred from UDPGA to various aglycones by this microsomal enzyme according to reaction 5:

$$\text{UDP-}\alpha\text{-D-glucuronic acid} + \text{ROH} \rightarrow \text{RO-}\beta\text{-D-glucuronic acid} + \text{UDP} \quad (5)$$

Attempts to reverse the reaction have so far been unsuccessful (Storey and Dutton, 1955; Isselbacher and Axelrod, 1955). This reaction has been demonstrated to form glucuronides with the compounds of the first four groups presented above. S-Glucuronide formation will be discussed first.

2. S-Glucuronides

The observation of Parke (1952) revealed that administration of thiophenol increased the excretion of total glucuronides in rabbit urine.

Later Storey (1964b) demonstrated that the same substance competitively inhibits glucuronidation of *p*-nitrophenol but not that of *o*-aminophenol *in vitro*.

One of few S-glucuronides is the glucuronide of 2-mercaptobenzothiazole, which has been isolated from the urine of dogs to which 2-benzothiazolesulfonamide has been administered intravenously (Clapp, 1956). Kamil *et al.* (1953b) detected that rabbits receiving antabuse (tetraethylthiuramdisulfide) excreted extra amounts of glucuronic acid. Kaslander (1963) administered antabuse to man and isolated from the urine the corresponding S-glucuronide. The amount of the glucuronide was less than 1% of the dose administered. Recently Dutton and Illing (1969) demonstrated the formation of an S-glucuronide by a UDP-glucuronyltransferase in mouse-liver homogenate and showed that this glucuronide was hydrolyzed by β-glucuronidase.

3. N-Glucuronides

N-Glucuronide synthesis is interesting because in physiological conditions it can proceed enzymically or nonenzymically. Bridges and Williams (1962), using free glucuronic acid and *p*-choloroaniline, aniline, *p*-toluidine and sulfanilamide as substrates, demonstrated that N-glucuronide formation was nonenzymic *in vitro* and *in vivo*. N^1-Glucuronide of sulfadimethoxine was synthesized enzymically in monkey liver, while N^4-glucuronide was formed spontaneously when the drug was added to human urine (Bridges *et al.*, 1968). N^1-Glucuronide was a major urinary metabolite in man and rhesus monkey, but not in dog, guinea pig, rat or rabbit, after administration of sulfadimethoxine.

There are four types of N-glucuronides: (1) those with aliphatic amino groups, (2) those with aromatic amino groups, (3) those with sulfonimido or a carbamoyl group, and (4) those with heterocyclic nitrogen (Williams, 1967). The first two glucuronides are extremely labile compounds in acidic solution (Bridges and Williams, 1962), while those pertaining to the third group are relatively stable (Williams, 1967).

N-Glucuronide synthesis appears to proceed enzymically, too. Axelrod *et al.* (1957, 1958) showed that several aromatic compounds formed N-glucuronides in the presence of UDPGA and UDPGA transferase of guinea pig liver microsomes. This observation has since been confirmed by Arias (1961) and Isselbacher *et al.* (1962). It has been observed that Gunn rats, which have a defective transferase activity (Schmid *et al.*, 1958), are also able to synthesize N-glucuronide (aniline) *in vitro* and *in vivo* (Arias, 1961). Leventer et al. (1965) have solubilized the N-glucuronyltransferase (aniline as aglycone) from guinea pig liver microsomes.

Japanese workers (cf. Fishman 1959; Takanashi *et al.*, 1964) have postulated that there are at least three possible pathways by which *N*-glucuronide is synthesized *in vitro:* (1) via microsomal glucuronyltransferase, (2) nonenzymically, and (3) via transglucuronidation by β-glucuronidase. Apart from *N*-glucuronide formation, the other types of glucuronides are not directly formed from free glucuronic acid. This had been shown already by Butler and Packham (1955) and Eisenberg *et al.* (1955). Although exogenous glucuronolactone lowers the serum bilirubin level (Danoff *et al.*, 1958) and increases the excretion of bound morphine in rabbit urine (Hosoya and Otobe, 1959), this does not mean that glucuronic acid is directly incorporated into glucuronides. It might enhance glucuronidation indirectly, possibly after transformation to glucose, or after conversion to glucaric acid which inhibits hydrolysis of the newly formed glucuronides by β-glucuronidase. N-OH glucuronides are the subject of a chapter of this volume.

4. *Transglucuronylation by β-Glucuronidase*

Glucuronide synthesis can also occur via β-glucuronidase transfer activity, a mechanism originally proposed by Fishman (1940).

This worker considered that the enhanced β-glucuronidase activity found in various tissues after administration of glucuronogenic compounds was the reason for the increased conjugation. The works of Florkin *et al.* (1942) concerning bornylglucuronide synthesis with β-glucuronidase, with free glucuronic acid and borneol as substrates, seemed to support this view. *In vitro* studies have shown that it is actually possible to transfer the β-glucuronosyl residue from aryl and alicyclic glucuronides to aliphatic alcohols and glycols, but not to phenols or alicyclic alcohols (Fishman and Green, 1956, 1957; Levvy and Marsh, 1960; Takanashi *et al.*, 1964).

Saccharo-1,4-lactone, an inhibitor of β-glucuronidase, in a concentration which causes almost complete inhibition of β-glucuronidase in liver and kidney extracts, has no effect on glucuronide synthesis by mouse liver slices or suspension (Karunairatnam *et al.*, 1949; Karunairatnam and Levvy, 1949; Levvy, 1952; Lathe and Walker, 1958c). Takanashi *et al.* (1964), on the other hand, were able to demonstrate inhibition of *N*-glucuronide formation by β-glucuronidase in the presence of saccharo-1,4-lactone. Touster and Reynolds (1952) could not observe *o*-aminophenolglucuronide formation by β-glucuronidase when D-glucuronic acid-1-phosphate was used as substrate.

It is apparent that the direct role of β-glucuronidase in glucuronidation is negligible *in vivo;* indirectly, however, it may have a marked significance in the hydrolysis of the glucuronides formed.

5. Inhibition of UDPGA Transferase(s) in Vitro

From the very early studies of Dutton and Storey (1954; Storey and Dutton, 1955) and from those of Isselbacher (1956; Isselbacher and Axelrod, 1955), it became evident that glucuronyltransferase was a very labile enzyme. This and its difficult solubilization make purification difficult. Using heat-treated preparations of *Trimeresurus flavoviridis* venom Isselbacher *et al.* (1962) solubilized and partially purified this enzyme from rabbit liver microsomes. The solubilized and partially purified transferase is heat-labile and very unstable on storage, has a broad pH optimum with a maximum at pH 7.4 and is inhibited by Ca^{++} and Mg^{++}. The Mg^{++} requirement may differ between various glucuronyltransferase enzymes (Storey, 1965b), a finding taken as evidence that the enzyme is multiple.

The transferase activity in intestinal slices of guinea pig is strongly inhibited by sodium azide, 2,4-dinitrophenol, sodium cyanide and sodium arsenate, and by substances which uncouple oxidative phosphorylation (Schachter *et al.*, 1959). But tetrahydrocortisone glucuronidation in guinea pig liver microsomes, although *not* inhibited by cyanide, fluoride, or heavy metals, is blocked by glucuronolactone or glucuronate (Isselbacher, 1956). This variable cyanide effect may have been due to the different tissue or enzyme preparations.

Sulfhydryl blocking agents (*p*-chloromercuribenzoate, phenylmercuriacetate, 2-chlorovinylarsenoxide, etc.) were able to inhibit the *p*-nitrophenol and *o*-aminophenol glucuronidation both in homogenates of mouse liver (Storey, 1964a,b, 1965a) and in the purified enzyme system (Isselbacher *et al.*, 1962). Glutathione and cysteine do not affect the enzyme but can protect it against blocking agents. Experiments show that SH-groups are essential for the activity of the enzyme.

EDTA in certain experimental conditions is inhibitory to the glucuronyltransferase. Such a situation exists, for example, if UDPGA pyrophosphatase is already inhibited by excess ATP and UDP-N-acetylglucosamine (Pogell and Leloir, 1961) or if the UDPGA pyrophosphatase activity is endogenously low, as in guinea pig and rabbit liver microsomes (Pogell and Leloir, 1961; Halac and Bonevardi, 1963), or if the UDPGA concentration is high enough to guarantee substrate saturation (Huttunen and Miettinen, 1965). The purified enzyme from rabbit liver microsomes is inhibited by EDTA *in vitro,* but is activated by dialysis at pH 9 with or without EDTA.

Of the drugs tested, synkavit and novobiocin (Hsia *et al.*, 1963a) and chloromycetin and streptomycin *in vitro* inhibit transferase (Waters *et al.*, 1958).

Studies concerning competitive inhibition should fulfill two conditions: (1) preferentially solubilized (purified) enzyme preparations should be used, and (2) UDPGA must be present in excess. Since most of the studies only occasionally fulfill these conditions, the conclusions drawn concerning the multiplicity or singularity of glucuronyltransferase vary greatly. Isselbacher *et al.* (1962) found that anthranilic acid functioned as an effective competitive inhibitor of *p*-nitrophenol glucuronide formation in the presence of UDPGA excess (8 m*M*). They were unable to separate the ester and ether glucuronide-forming enzyme(s) from each other, whereas the enzyme synthesizing *N*-glucuonides was distinctly different. *p*-Nitrophenol as the aglycone for ether-type glucuronides could be replaced by tetrahydrocortisone and *o*-aminophenol, and anthranilic acid for ester-type glucuronides by bilirubin.

The activity differences in the presence of thiol group blocking agents between the glucuronidation systems, according to whether *p*-nitrophenol or *o*-aminophenol is used as acceptor, seems to indicate that two different enzymes catalyze the transferase reaction (Storey, 1965a). In other experiments, phenol, menthol, and benzoic acid were shown competitively to inhibit *o*-aminophenol- and *p*-nitrophenolglucuronide formation; thiophenol, which probably itself forms glucuronides, acted as very strong competitive inhibitor of *p*-nitrophenol glucuronidation, but not at all in the *o*-aminophenol system.

The following studies, on the other hand, suggest that the various aglycones are conjugated by the same enzyme. In the experiments of Karunairatnam *et al.* (1949), menthol caused a 73% drop in *o*-aminophenol glucuronidation in mouse liver slices (UDPGA was not present). Dutton (1966) found that benzoate or α-ethylhexanoate competes with *o*-aminophenol (UDPGA 0.05 m*M*), and that aniline glucuronide formation was inhibited by phenolic and carboxylic acceptors in guinea pig liver microsomes (Axelrod *et al.*, 1957). Borneol inhibited the production of "direct-reacting" bilirubin in rat liver homogenates (Grodsky and Carbone, 1957) and bilirubin salicyl glucuronide biosynthesis in intestinal slice preparations of guinea pig (Schachter *et al.*, 1959). More recently, bilirubin has been demonstrated to cause a very strong inhibition of 4-methylumbelliferone glucuronidation in the presence of excess UDPGA (1.6 m*M* final concentration) in human liver homogenate (Frezza *et al.*, 1968). The finding that anthranilic acid, in the presence of excess of UDPGA, competitively inhibits *p*-nitrophenylglucuronide formation also supports the single enzyme hypothesis (Isselbacher *et al.*, 1962).

Glucuronyl transferase is also inhibited by some natural inhibitors. The heat-labile inhibitor found by Dutton and Greig (1957) in liver, gut, and kidney homogenates of various animals, seems to have been UDPGA pyrophosphatase. The deficient glucuronide-producing enzyme system found in Gunn rats is not dependent on the presence of natural inhibitors in the liver microsomal fraction of that rat strain (Schmid *et al.*, 1958) though addition of diethylnitrosamine increases the activity to a level obtained in normal Wistar rats (Stevenson *et al.*, 1968). Such substances as cholesterol and cortisone do not inhibit glucuronyltransferase (Hsia *et al.*, 1963b). But the progestational hormones studied by Hsia's group have been considered as competitive inhibitors (see also Lathe and Walker, 1958a,b). In considering the bilirubin-conjugating capacity of the newborn, it has become apparent that human pregnancy serum contains an inhibitor of the bilirubin-conjugating system (Lathe and Walker, 1958a,b). The inhibitory effect could be demonstrated in the breast milk of some women whose infants showed prolonged mild hyperbilirubinemia (Arias *et al.*, 1967).

6. Activation of UDPGA Transferase in Vitro

The activating effect of some substances on *in vitro* glucuronidation is brought about by several mechanisms, some of which are unknown.

ATP and UDP-N-acetylglucosamine stimulate glucuronidation by inhibiting the breakdown of UDPGA by pyrophosphatase in rat liver microsomes (Pogell and Leloir, 1961). As mentioned earlier in this chapter, these substances are devoid of stimulatory action in guinea pig and rabbit liver microsomes. In boiled extract of mouse liver Taketa (1962a,e) found an endogenous activator of glucuronide formation, which operated as an activator, not as a substrate. The purification and chemical properties suggested that it was UDP-N-acetylglucosamine. EDTA also stimulates glucuronidation in rat tissue homogenates by inhibiting UDPGA breakdown (Pogell and Leloir, 1961; Miettinen and Leskinen, 1962, 1963). EDTA seems to have at least the following effects on glucuronidation *in vitro:* (1) It inhibits UDPGA pyrophosphatase, leading via better substrate saturation to increased glucuronidation. (2) It protects certain conjugates against deconjugation (Miettinen and Leskinen, 1963: Airaksinen *et al.*, 1965). (3) It increases glucuronidation by solubilizing transferase (Halac and Bonevardi, 1963). (4) It counteracts the inhibitory effect of Ca^{++} and Mg^{++} on transferase (Isselbacher *et al.*, 1962). (5) It inhibits transferase under certain circumstances (Pogell and Leloir, 1961). (6) It activates transferase by converting rough endoplasmic reticulum into smooth membranes (Halac and Weiss, 1967).

The activation of UDP-glucuronyltransferase by serum albumin and by the dialyzed supernatant demonstrated by Pogell and Leloir (1961) may be regarded as some kind of protective action against transferase enzyme or microsomes.

Greenwood and Stevenson (1965) and Stevenson *et al.* (1968) have demonstrated that diethylnitrosamine stimulates glucuronidation in both *in vivo* and *in vitro* experiments using rat liver homogenate, or microsomal or solubilized preparations as enzyme source. The mechanism of action seems to differ from that of ATP and UDP-N-acetylglucosamine, because they could not demonstrate reduction of UDPGA destruction. This effect was specific for the substrate, being demonstrable with *o*-aminophenol and paracetamol, but not with *p*-nitrophenol, phenolphthalein, or menthol as substrates. It was also species-specific. The normal Wistar and Gunn rats exhibited activation to the same level of enzyme activity. No stimulation was observed with liver preparations from species other than the rat.

D. DISTRIBUTION OF GLUCURONIDE FORMATION IN TISSUES

1. *Liver*

Owing to the great amount of liver tissue and high glucuronyltransferase activity per unit weight, the liver appears to be the main site of glucuronide production in the animal organism. However, the hepatic glucuronidation capacity is dependent on the glycogen content, because the latter substance is the ultimate precursor of UDPGA. Glucose as such is less effective because of low hexokinase activity and the position of phosphoglucomutase equilibrium (Shirai and Ohkubo, 1954a; Mills *et al.*, 1958; Schachter *et al.*, 1959).

The cat is the only known mammal whose liver cannot produce glucuronides to any appreciable extent. Hartiala first pointed out this phenomenon in 1955. Since then, Robinson and Williams (1958) have demonstrated that the cat cannot enhance its glucuronide excretion in response to several glucuronidogenic substances and that even corticoids are not excreted as glucuronides by this animal as they are by others (Borrell, 1958). But bilirubin glucuronide is stated to occur in cat bile (Sawada, 1959). The formation of the latter conjugate by cat liver preparations has been reported by Lathe and Walker (1958c), but the conjugate was not formed by liver microsomes fortified with UDPGA (Sawada, 1959).

2. Kidney

Renal tissue is able to synthesize glucuronides (Lipschitz and Bueding, 1939; Storey, 1950) and it contains an enzyme mechanism for UDPGA synthesis. Crude UDPGA preparations have been isolated from kidneys and transfer of glucuronic acid from UDPGA to aglycones has been demonstrated (Dutton and Stevenson, 1959; Stevenson and Dutton, 1962).

An inhibitor which was found in rat kidney homogenates (Dutton and Greig, 1957) has been found to be active in UDPGA pyrophosphatase (Conney and Burns, 1961; Stevenson and Dutton, 1962).

The relative physiological significance of the glucuronidation process in the kidney awaits further elucidation, but there are some studies which show that the kidney can, to some extent, compensate for the absence of liver. Thus in hepatectomized dogs Bollman and Mendez (1955) noted a slow conversion of bilirubin into "pigment I" (= monoglucuronide), perhaps by the kidney; in rat kidney homogenate Grodsky and Carbone (1957) observed conversion of bilirubin into "direct-reacting bilirubin to an appreciable extent"; and in hepatectomized rats Béraud and Vannotti (1960) observed that the kidney could carry out glucuronidation of triiodothyronine (T_3).

3. Gastrointestinal Tract

The gastrointestinal tract is another important site of glucuronide formation (Shirai and Ohkubo, 1954a). Among the aglycones which have been studied in this respect are o-aminophenol, cinchophen, phenolphthalein, estradiol, estriol, salicylic acid, and testosterone (Hartiala, 1954, 1955, 1961; Hartiala et al., 1957, 1958; Lehtinen et al., 1958a–c; Schachter et al., 1959; Dahm and Breuer, 1966a,b; Dahm et al., 1966). Glucuronyltransferase activity is detectable throughout the gastrointestinal tract and appears to be localized in the mucosa, where most of the activity is contained in the microsomal fraction (Stevenson and Dutton, 1962). Glucuronidation and transferase activity decreases linearly from the oral to the aboral end of the small intestine (Hänninen et al., 1968). This organ forms glucuronide more efficiently in in vitro studies if the medium contains glucose or if the experimental animals are fed carbohydrates before conjugation studies (Stevenson and Dutton, 1962; Schachter et al., 1959).

Dutton and Greig (1957) failed to demonstrate UDPGA or glucuronyltransferase activity in gut homogenates of various animals. This was due to the very active UDPGA pyrophosphatase activity, which degraded UDPGA. From the studies of Dutton and Stevenson (1959, 1962) the mucosa is known to synthesize UDPGA from UDPG.

The glucuronyltransferase of the intestine seems to resemble that of the liver and kidney but in their developmental enzyme pattern these organs are dissimilar (Dutton, 1959; Dutton and Stevenson, 1959). In fetal tissue the gastrointestinal tract is perhaps the most active site of glucuronide synthesis (Dutton 1959). Gut preparations from Gunn rats or from the cat have only low activity or are totally devoid of glucuronide-producing capacity (Schachter *et al.*, 1959; Arias *et al.*, 1963a).

An interesting finding is the presence of a soluble glucuronyltransferase in the human intestine. The human gastric mucosa contains both a soluble and a microsomal glucuronyltransferase. These enzymes seem especially to attack estrogens (Dahm and Breuer, 1966a,b; Hoffmann and Breuer, 1968).

4. *Skin and Connective Tissue*

An inducible glucuronyltransferase activity has been demonstrated in mouse and guinea pig skin (Stevenson and Dutton, 1960; Dutton and Stevenson, 1962). The skin also seems to be capable of UDPGA synthesis (Jacobson and Davidson, 1962a), although in *in vitro* experiments high concentrations of UDPGA are required for glucuronide synthesis owing to the highly active pyrophosphatase (Pogell and Krisman, 1960). UDPG dehydrogenase and glucuronyltransferase activities have been demonstrated in epiphyseal cartilage homogenate (Castellani *et al.*, 1957) and in extracts of human synovial tissue and connective tissue of the guinea pig (Bollet *et al.*, 1959).

E. INTRACELLULAR DISTRIBUTION OF GLUCURONYLTRANSFERASE ACTIVITY

Dutton and Storey (1954) and Strominger *et al.* (1954) first observed that glucuronyltransferase activity is connected "with the larger (5000 g for 20 min) and smaller cytoplasmic granules." Its localization to the microsomal fraction has been demonstrated in the liver (Isselbacher and Axelrod, 1955; Isselbacher, 1956; Dutton and Spencer, 1956), the stomach and intestines (Stevenson and Dutton, 1962; Breuer and Dahm, 1967; Dahm *et al.*, 1966; Hoffmann and Breuer, 1968), and the kidney (Stevenson and Dutton, 1962).

Transferase activity, like that of many other drug-metabolizing enzymes (Fouts, 1962), seems in some unknown way to be connected with and dependent on the liver glycogen content. *In vitro* studies indicate that the highest specific activity is attached to the particulate glycogen fraction (Halac and Frank, 1960). In the rat liver depleted of glycogen by total fast, the activity remains almost unchanged, however, while in the human liver transferase activity appears to be reduced (Miettinen and Leskinen, 1963, 1968).

Solubilization and purification studies also point to the intracellular origin of glucuronyltransferase. Isselbacher (1956; Isselbacher and Axelrod, 1955) was the first to attempt to solubilize glucuronyltransferase activity from microsomes, but with little success. Since then it has been tried by Pogell and Leloir (1961) and the best preparations so far have been made by Isselbacher (1961) and Isselbacher *et al.* (1962) from rabbit liver microsomes with the aid of a heat-treated preparation of *Trimeresurus flavoviridis* venom. Their preparations were devoid of N-glucuronyltransferase activity. Leventer *et al.* (1965), on the other hand, were able to solubilize N-glucuronyltransferase activity from guinea pig liver microsomes.

Studies concerning the intramicrosomal distribution of glucuronyltransferase activity with various substrates seem to indicate that the transferase(s) may be further differentiated on the basis of submicrosomal distribution (Gram *et al.*, 1968).

The first information concerning the so-called soluble (localized in the cytoplasm) glucuronyltransferase came from studies by Dahm and Breuer (1966a,b). A partial purification of soluble glucuronyltransferase from human intestinal mucosa was achieved with differential centrifugation and protein precipitation. It was later found that human gastric mucosa also has a soluble UDP-glucuronyltransferase (Hoffmann and Breuer, 1968). These soluble enzymes seem to carry out glucuronyl transfer from UDPGA to estrogens.

III. Biosynthesis of Free Glucuronic Acid

Free glucuronic acid could be formed in the animal organism, as shown in Fig. 1, via direct hydrolysis of UDPGA by pyrophosphatase, via the transferase route associated with glucuronide formation, via oxidation of myoinositol and via hydrolysis of mucopolysaccharides.

A. Formation via the Glucuronyltransferase–β-glucuronidase Route

The possible role of the β-glucuronidase route in the release of free glucuronic acid has been the subject of many discussions (Pogell and Leloir, 1961; York *et al.*, 1961; Hollmann and Touster, 1962; Touster and Shaw, 1962). This enzyme rapidly hydrolyzes β-glucuronides and because it is also present in the microsomal fraction, a part of the newly formed glucuronides synthesized with some endogenous or exogenous aglycones might already be hydrolyzed before they are released into the cytoplasm or from the cell. Glucuronidogenic drugs, such as borneol and menthol, enhance the excretion of glucuronides and either D-glucuronic acid, L-ascorbic acid, or L-xylulose (Burns *et al.*, 1957; Enklewitz and Lasker, 1935; Touster and Shaw, 1962; Longenecker *et al.*, 1939). These

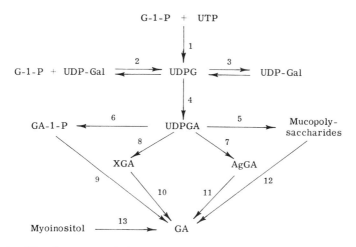

Fig. 1. The formation of free glucuronic acid. The enzymes catalyzing the various steps are indicated in the figure by numbers. The meaning of the numbers is given in the legend to Fig. 2.

Abbreviations: G-1-P, glucose 1-phosphate; UTP, uridine 5-triphosphate; UDPG, uridine 5-diphosphate glucose; UDP-Gal, uridine 5-diphosphate galactose; UDPGA, uridine 5-diphosphate glucuronic acid; GA-1-P, glucuronic acid 1-phosphate; X, aglycone (endogenous); Ag, aglycone (exogenous); GA, glucuronic acid.

compounds also enhance the hepatic activity of β-glucuronidase (Fishman, 1940). The increased excretion of free glucuronic acid found in these experiments can be explained not only by increased hepatic β-glucuronidase activity but also by postulating that a labile glucuronide (Pogell and Leloir, 1961), formed and hydrolyzed in increased amounts under these conditions, is a precursor of glucuronic acid (Hollmann and Touster, 1962) or that β-glucuronidase exerts its effect extracellularly.

Glucuronidation and β-glucuronidase activity correlate poorly with each other, although a fairly close parallelism exists between them in the stomach after irradiation (Hartiala *et al.*, 1961; Hartiala, 1961). Feeding of saccharolactone, an inhibitor of β-glucuronidase, brings about (1) depression of β-glucuronidase activity in the liver and kidney (Kiyomoto *et al.*, 1963), (2) no change in the normal urinary ascorbate level (Hollmann and Touster, 1962), but (3) has a protective action against gastric ulcer induced by administration of cinchophen (Hartiala, 1961). Saccharolactone added to the incubation medium *in vitro* did not usually increase glucuronidation, suggesting that β-glucuronidase did not hydrolyze the glucuronides formed. However, the uptake of the polar acid by the lipid material of the microsomes may be negligible and intramicrosomal β-glucuronidase may not be inhibited. In some studies (Miettinen and Leskinen, 1962, 1963) saccharolactone has actually caused a significantly increased glucuronidation in liver homogenates but not in renal or intestinal homogenates of normally fed or fasted rats, even in the presence of EDTA.

On the other hand, enhancement of glucose catabolism through the glucuronic acid pathway did not require the formation of glucuronide with exogenous compounds, since aminopyrine (Touster *et al.*, 1955) and barbital (Burns *et al.*, 1957), which are the most potent "inducers" of the glucuronic acid pathway (Burns, 1961), are not excreted as glucuronides. Since UDPGA pyrophosphatase activity is inhibited by these drugs, the channeling of glucose to the pathway takes place primarily via the transferase route, probably with a labile glucuronide as intermediate (Touster and Hollmann, 1961b). That this inhibition has some significance in the glucuronic acid pathway is indicated by the findings that blocking of UDPGA pyrophosphatase by ATP, EDTA, and *N*-acetylglucosamine *in vitro* increases glucuronidation, owing to better substrate concentration in the liver, kidney, intestinal, and skin preparations (Stevenson and Dutton, 1960; Pogell and Leloir, 1961; Miettinen and Leskinen, 1962; Taketa, 1962a,e). 3,4-Benzpyrene, which has a very profound effect on glucuronic acid pathway activity, enhances not only glucuronyltransferase but also β-glucuronidase activity (Pulkkinen and Hartiala, 1965).

B. FORMATION VIA THE PYROPHOSPHATASE ROUTE

An alternative pathway from UDP-glucuronic acid to free glucuronic acid is via glucuronic acid 1-phosphate (GA-1-P) and subsequent hydrolysis of GA-1-P to free acid.

The first reaction, UDPGA hydrolysis to GA-1-P and UMP, is catalyzed by an enzyme called UDPGA pyrophosphatase. It has a rather broad substrate specificity and is probably a nucleotide pyrophosphatase (Ogawa *et al.*, 1966). The second reaction is catalyzed by an enzyme, GA-1-P phosphatase, which seems to be the same enzyme as nonspecific alkaline phosphatase (Takanashi et al., 1966). UDPGA pyrophosphatase was first found in rat kidney microsomes (Ginsburg et al., 1958) and later in rat skin (Pogell and Krisman, 1960) and in rat liver microsomes (Conney and Burns, 1961). GA-1-P phosphatase activity has been found in the rat kidney (Ginsburg et al., 1958) and skin (Pogell and Krisman, 1960) and in very low activity in the liver also (Conney and Burns, 1961). More recently, this enzyme activity has been reported by Touster (1966) to be present in the liver also, though Hänninen (1968) was unable to detect it in that organ. There is evidence that the pyrophosphatase route may play some role in the formation of free glucuronic acid for the synthesis of ascorbic acid (Ginsburg et al., 1958) and pentoses, whereas its influence on the transferase reaction in vivo (possibly via competition with transferase from UDPGA) is unknown. It seems probable however, that in the liver the transferase pathway plays the major role, as GA-1-P phosphatase activity is low. In the kidney and gastrointestinal tract its activity is high, but the significance of these tissues for the production of free glucuronic acid is unknown.

C. FORMATION FROM MYOINOSITOL

The synthesis of free glucuronic acid from myoinositol has been shown by in vitro and in vivo experiments using radioactive inositol as substrate (Charalampous and Lyras, 1957; Richardson and Axelrod, 1958, 1959; Burns et al., 1959). Myoinositol catabolism seems to take place mainly in the kidney (Richardson and Axelrod, 1959; Howard and Anderson, 1967), although very low activity has been observed in the rat liver (Hänninen, 1968). This oxygenation has also been demonstrated in chicken and hog kidney preparations but not in the rat pancreas or diaphragm (Richardson and Axelrod, 1958).

In the kidney there are two myoinositol-catabolizing systems, one of which is localized in the supernatant (56,000 g) fraction and is inhibited by glucuronate, glucuronolactone, gulonate, or gulonolactone, while the other is found in the whole kidney homogenate fraction and is not inhibited by the abovementioned substances. It is probable that the latter system is mainly responsible for myoinositol catabolism. The free glucuronic acid which is synthesized via this route is further metabolized to L-xylulose and via the pentose phosphate cycle to glucose and glycogen (Anderson and Coots, 1958). The enzyme system synthesizing D-glucuronate has been partially purified and separated from that synthesizing L-glucuronate.

Although inositol is the precursor of D-glucuronic acid and L-gulonic acid, it is not converted to L-ascorbic acid (Burns et al., 1958, 1959). The latter is readily formed from glucuronate and gulonate in the liver homogenate or from the respective lactones by intact liver cells. Thus, glucuronic acid, which is formed in vivo from inositol in the kidney, is not transported to the liver or taken up by the liver cells, probably because the acid is not in lactone form owing to the lack of aldonolactonase in the kidney (Burns et al., 1959). Richardson and Axelrod (1958) have found gulonolactonase in the rat kidney, however.

D. RELEASE FROM MUCOPOLYSACCHARIDES

Because the mass of uronic acid-containing mucopolysaccharides is rather large in the body and because these compounds show a considerable turnover, the daily release of glucuronic acid from these compounds may be considerable. But the degradation of mucopolysaccharides in vivo is a poorly understood process (cf. Silbert, 1966). Hyaluronic acid is degraded by hyaluronidase, activity of which is found in many animal tissues (Bollet et al., 1963). Testicular hyaluronidase has been studied extensively and is known to digest hyaluronic acid to oligosaccharides which contain considerable quantities of tetrasaccharides and a small amount of a disaccharide, hyalobiuronic acid (Weissmann et al., 1954). Oligosaccharides are cleaved to monosaccharides by the sequential action of β-glucuronidase and β-N-acetylglucosaminidase. Chondroitin sulfate A and C are also split to oligosaccharides by chondroitinase, which appears to be identical with hyaluronidase (Matthews and Dorfman, 1954). The split products are further hydrolyzed by β-glucuronidase but not by β-N-acetylgalactosaminidase (Linker et al., 1955). Owing to the presence of iduronic acid, chondroitin sulfate B is resistant to hyaluronidase.

The release of glucuronic acid during the in vivo degradation of mucopolysaccharides takes place primarily extracellularly, suggesting that the acid may be excreted into the urine as such, because the uptake of free acid by the cells is known to be poor. Glucuronate may also be released during the synthesis of mucopolysaccharides from UDPGA because the skin exhibits transferase activity, and the enzymes of the pyrophosphatase route have also been found in that organ (see Section III, B).

IV. Further Metabolism of Free Glucuronic Acid

The metabolism of glucuronate and of its lactone appear to differ from each other in vivo (Hollmann, 1964). Thus, the latter, when administered to human subjects or experimental animals, is catabolized almost as readily as glucose, whereas administered glucuronic acid is mainly excreted as such into the urine. In homogenates, however, the two compounds are metabolized similarly, suggesting that in intact cells there is a block in the uptake of glucuronate. Free glucuronic acid formed *in vitro* within a cell or added to a tissue homogenate can be further metabolized along three different pathways, viz., (1) via L-xylulose to D-xylulose phosphate, (2) to ascorbic acid and (3) to glucaric acid (Burns and Ashwell, 1960; Hollmann, 1964; Burns and Conney, 1966; Levvy and Conchie, 1966). The relative importance of these pathways appears to vary in each organ and considerable variation is found in different species. Thus, the ascorbic acid pathway in man and particularly in the guinea pig and monkey is so little used that this compound must be obtained as a vitamin from outside the body.

Each of these pathways appears to have an important biological significance, suggesting that the free glucuronic acid formed through metabolic hydrolysis during the glucuronidation process is not simply a product which must be degraded in some way or other in order to avoid its accumulation. Thus, the glucuronic acid–xylulose pathway serves not only as a possible route for glucose metabolism (reported to be important in adipose tissue; Winegrad and Shaw, 1964) but, according to more recent studies, as a link in the hydrogen transport cycle between the intra- and extramitochondrial compartments and also between NADPH and NAD in the transhydrogenation process (Arsenis *et al.*, 1968). In most species the ascorbic acid route produces the vitamin which is necessary in many biological processes (Burns and Ashwell, 1960), particularly in the metabolism of connective tissue, and especially in the hydroxylation of proline and lysine during collagen synthesis (Hutton *et al.*, 1967; Kivirikko and Prockop, 1967). Finally, glucaric acid appears to regulate the activity of β-glucuronidase which, on the other hand, can regulate the formation of free glucuronic acid (Levvy and Conchie, 1966). A more detailed description of these pathways is given in the following sections; outlines of the reactions are found in Fig. 2.

A. THE GLUCURONATE–XYLULOSE PATHWAY

Free glucuronic acid is reversibly converted to L-gulonic acid by a NADP-linked dehydrogenase (ul-Hassan and Lehninger, 1956; Ishikawa and Noguchi, 1957; Ashwell *et al.*, 1961; Mano *et al.*, 1961), which is widely distributed in various organs (Mano *et al.*, 1961) independently of their glucuronidation capacity. It has been suggested that since this enzyme, which has been purified up to 500-fold, reduces a number of

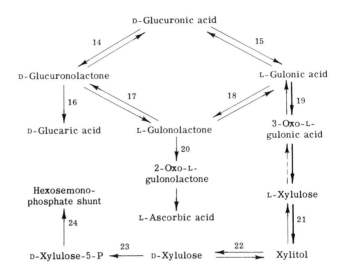

different uronic acids, it should be called NADP-L-hexonate reductase (Mano *et al.*, 1961). The enzyme acts not only on glucuronic and gulonic acids but also on their lactones (Mano *et al.*, 1961). Accordingly, parenterally administered glucuronolactone, when taken up by cells, is converted directly to gulonolactone by this enzyme. Glucuronolactone appears to be formed *in vivo* through a reversible reaction from free acid by a specific lactonase, uronolactonase, found in the microsomal fraction of the liver but not of any other tissues (Winkelman and Lehninger, 1958; Yamada *et al.*, 1959). However, there is another lactonase, which is found in the soluble fraction of the liver and kidney but not in intestinal homogenate (Hänninen, 1968), and which synthesizes and splits a number of 5–7 carbon aldonolactones, including gulonolactone (ul-Hassan and Lehninger, 1956; Winkelman and Lehninger, 1958; Yamada *et al.*, 1959; Yamada, 1959). Accordingly, the presence of these lactonases and of glucuronic acid reductase in the liver and kidney renders these organs capable of metabolizing glucuronic acid to gulonic acid also via glucurone and gulonolactones.

The gulonic acid synthesized within cells is released in small amounts into the extracellular space and finally excreted into the urine; its amount is elevated after administration of glucuronolactone or after stimulation of the glucuronic acid pathway by inducer drugs (Burns, 1957b). However, the bulk of the gulonic acid formed in cells is converted to L-xylulose by a two-step enzymic reaction (Hollmann, 1964). In the first step L-gulonic acid is converted to β-ketogulonic acid (Ashwell *et al.*, 1959,

Fig. 2. The further metabolism of free glucuronic acid. The enzymes catalyzing the various steps are indicated in the figure by numbers. The meaning of the numbers is given below.

Number	Systematic name	Trivial name
1	UTP:α-D-glucose-1-phosphate uridyltransferase	UDPG pyrophosphorylase
2	UDP-glucose:α-D-galactose-1-phosphate uridyltransferase	Uridyl transferase
3	UDP-glucose 4-epimerase	UDP-glucose epimerase
4	UDP-glucose:NAD oxidoreductase	UDPG dehydrogenase
5	—	Mucopolysaccharide synthesizing enzymes
6	Dinucleotide nucleotido-hydrolase	Nucleotide pyrophosphatase (UDPGA pyrophosphatase)
7 and 8	UDP glucuronate glucuronyl-transferase	UDP glucuronyltransferase
9	Alkaline phosphatase	GA-1-P phosphatase
10 and 11	β-D-Glucuronide glucurono-hydrolase	β-Glucuronidase
12	—	Mucopolysaccharides catabolizing enzymes
13	—	Myoinositol oxygenase
14	D-Glucurone-δ-lactone hydrolase	Uronolactonase
15	L-Gulonate:NADP oxidoreductase	Glucuronate reductase
16	D-Glucuronolactone:NAD oxidoreductase	Glucuronolactone dehydrogenase
17	L-Gulonolactone:NADP oxidoreductase	Glucuronolactone reductase
18	D (or L)-Gulono-γ-lactone hydrolase	Aldonolactonase
19	L-3-Hydroxy acid:NAD oxidoreductase	3-Hydroxyacid dehydrogenase
20	L-Gulono-γ-lactone: oxygen oxidoreductase	L-Gulonolactone oxidase
21	Xylitol:NADP oxidoreductase (L-xylulose forming)	L-Xylulose reductase
22	Xylitol:NAD oxidoreductase (D-xylulose forming)	D-Xylulose reductase
23	ATP:D-Xylulose 5-phosphotransferase	Xylulokinase
24	Sedoheptulose-7-phosphate: D-glyceraldehyde-3-phosphate glycolaldehydetransferase	Transketolase

1961; Smiley and Ashwell, 1961), which is decarboxylated in the second step to L-xylulose. The enzyme of the first step, L-gulonic acid oxidase, is NAD-dependent, and has been purified from pig kidney and guinea pig liver (Ashwell et al., 1959; Ishikawa, 1959). It has a very broad substrate specificity and has therefore been called a 3-hydroxy acid dehydrogenase (Smiley and Ashwell, 1961). During the second step β-keto gulonate is converted to L-xylulose by a decarboxylase which has been purified about 20-fold from an acetone powder of guinea pig liver (Winkelman and Ashwell, 1961). Evidence has been presented suggesting that the formation of L-xylulose from L-gulonate is a reversible phenomenon (Bublitz et al., 1958).

Since the conversion of glucuronic acid or its lactone to gulonic acid results in oxidation of NADPH and since during the next step, i.e., during the conversion of gulonic acid to β-keto gulonic acid, NAD is reduced, these reactions serve as a transhydrogenation process between reduced NADPH and NAD.

The mechanism of conversion of L-xylulose via xylitol to D-xylulose has been extensively studied (Hollmann, 1964). Mammalian liver contains two mitochondrial xylitol dehydrogenases, of which the specific NADP-linked one is located in the core of the mitochondrion and reversibly reduces xylitol to L-xylulose. The relatively unspecific NAD-linked dehydrogenase appears to be attached weakly to the outer mitochondrial membrane and reversibly reduces xylitol to D-xylulose (Arsenis et al., 1968). The latter enzyme closely resembles the unspecific xylitol dehydrogenase which is found in liver cytoplasm and which also reduces sorbitol to fructose. It is apparent that the L-xylulose, which is formed from L-gulonate, must be transported from the cytoplasm into the mitochondrion and there reduced by the NADP-linked dehydrogenase to xylitol, which is then oxidized in outer mitochondrial compartments and/or in the cytoplasm to D-xylulose. This reaction sequence could again serve as a transhydrogenation process between NADPH and NAD when xylitol transfers hydrogen from the mitochondria. On the other hand, a cooperative function of the cytoplasmic and mitochondrial NAD-linked dehydrogenases could result in a transfer of hydrogen into the mitochondria in the form of xylitol (Arsenis et al., 1968).

It is conceivable that a change in the nucleotide status, i.e., in the redox state, would interfere with xylulose metabolism. Evidence that this is the case has been obtained from studies with ethanol, which showed that serum and urinary xylitol increase in man after ethanol loading (Pitkänen and Sahlström, 1968). The latter is known to increase the NADH/NAD ratio in the liver (Forsander et al., 1958). Since the ethanol, though slightly glucuronidized (Kamil *et al.*, 1953a,b), is not known to be an "inducer" of the glucuronic acid pathway, it is reasonable to assume that the increment of xylitol is not due to stimulation of the activity of the glucuronate route, but that more probably the change in the redox state shifts the xylitol–D-xylulose system, analogously with the other redox couples, toward the more reduced state, with the result that the reduced components, including xylitol, accumulate in the body. Production of L-xylulose might even be decreased under conditions associated with an increased NADH/NAD ratio, because 3 moles of oxidized NAD are reduced for the synthesis of 1 mole of L-xylulose from UDPG. Accordingly, conversion of the latter to UDPGA and of gulonate to L-xylulose may be inhibited. However, the excess of reducing equivalents might also increase the NADPH/NADP ratio, suggesting that during ethanol oxidation the equilibrium of the NADP-linked reactions of the glucuronic acid pathway, i.e., glucuronate $\rightleftarrows$ gulonate and L-xylulose $\rightleftarrows$ xylitol, lies far over to the right. That L-xylulose excretion may even have been decreased simultaneously with increased xylitol output after ethanol consumption (Pitkänen and Sahlström, 1968) is in agreement with this view.

The extent to which glucuronic acid is metabolized *in vivo* via the xylulose route by the different organs is not known, although the high activity of glucuronidation in the liver, kidney, and intestinal mucosa (cf. Dutton, 1966; see also Section II,D) suggests that the formation of free glucuronic acid and its conversion to D-xylulose takes place primarily in these organs. Studies by Hollmann (Hollmann and Wille, 1952; Hollmann, 1954a) suggested that the liver metabolizes the bulk of exogenous oral glucurone; the role of the intestinal mucosa may have been considerable, however. *In vitro* studies have shown that many tissues are able to utilize glucurone to some extent (Hollmann, 1954b) probably owing to the wide distribution of unspecific glucuronate reductase (Mano *et al.*, 1961). Glucuronate reductase and β-hydroxy acid dehydrogenase activities show rather similar patterns in the liver, kidney, and intestinal mucosa (Hänninen, 1968). Animal experiments have indicated that xylitol, which is frequently used in therapeutic measures, particularly in diabetes, as an antiketotic substance (Bässler *et al.*, 1962; Yamagata *et al.*, 1965), is mainly utilized by the liver (Müller *et al.*, 1967). Serum xylulose, which is increased in diabetes, has also been assumed to originate from the liver, for its concentration is higher in the hepatic vein than in the vena cava (Winegrad and Burden, 1965, 1966). However, xylulose, possibly formed in the intestinal mucosa, is transported via the portal blood to the liver, and may contribute to hepatic vein xylulose.

The tissue distribution of the two xylitol dehydrogenases has also been studied (Hollmann, 1964), although the activity in the intestine is not known. The activity of the NADP-dependent dehydrogenase is highest in the liver and kidneys (Hollmann and Laumann, 1967). However, a relatively low activity is also present in lungs, spleen, adrenals, brains, and heart muscle, although these organs are not able to perform glucuronidation (cf. Dutton, 1966). Skeletal muscle, which oxidizes glucose extensively, is unable to effect glucuronidation or utilization of L-xylulose. It is interesting to note that the hepatic and renal activities of L-xylulose reductase are low in the cat (Hollmann and Laumann, 1967), which is little able to synthesize and excrete glucuronides and which, in contrast to other species, is unable to increase ascorbate excretion in response to administration of an "inducer" drug (Hollmann and Touster, 1962).

Adipose tissue has been suggested to metabolize a relatively high percentage of glucose through the glucuronic acid pathway (Winegrad and Shaw, 1964), although more recent studies indicate that this may be only a few percent (Landau et al., 1966). No L-xylulose reductase activity is demonstrable (Kumahara et al., 1961), although xylitol is known to be utilized by rat adipose tissue (Mori et al., 1967). NAD-linked xylitol reductase shows highest activity in the liver, kidney, and prostate. In the skeletal muscle the activity is only about 2% of that in the liver (Hollmann, 1964).

Evidence that stimulation of the glucuronic acid pathway could actually result in increased metabolism of glucuronate into the pentoses is to be found in the earlier studies by Enklewitz and Lasker (1935) on subjects with pentosuria. Administration of the "inducer" drugs aminopyrine and antipyrine to these subjects, who are unable to metabolize L-xylulose further, obviously owing to enzyme deficiency (cf. Hiatt, 1966), was followed by a marked elevation in urinary L-xylulose excretion. Later on, the conversion of D-glucuronolactone-1-^{13}C to L-xylulose-5-^{13}C was demonstrated in a pentosuric (Touster et al., 1957).

The information on the relationship of the urinary pentoses to the activity of the glucuronidation process in experimental animals and in normal man is markedly more meager than that relating to ascorbate in the animals which are able to synthesize this vitamin. This is obviously mainly owing to the difficulty of determining xylulose and xylitol. Small amounts of the latter compounds are normally present in the urine and serum of man and experimental animals (Coover et al., 1950; Roe and Coover, 1950; Touster et al., 1954, 1955; Pitkänen and Sahlström, 1968). Administration of glucuronolactone significantly increases urinary excretion and blood levels of these pentoses (Touster et al., 1955; Hiatt, 1958;

Winegrad and Burden, 1965, 1966; Pitkänen and Sahlström, 1968). It is thus conceivable that any state which activates the glucuronidation process or enhances the formation of UDPGA might result in increased endogenous production of free glucuronic acid and consequently in increased pentose excretion. The latter, on the other hand, need not indicate that the pathway as a whole is stimulated, for the further metabolism of pentoses may have been inhibited, as already demonstrated by enzyme deficiency in pentosuria and probably also during ethanol consumption by the changed nucleotide status. A high serum xylulose level in diabetics has been assumed to be due to a hyperactive glucuronic acid pathway (Winegrad and Burden, 1965, 1966). However, more recent studies have indicated that the activity of glucuronidation is reduced in diabetic rats (Müller-Oerlingenhausen et al., 1967b). On the other hand, increased β-glucuronidase (cf. Dohrmann, 1968) may enhance the formation of free glucuronic acid and thus stimulate the synthesis of xylulose. Administration of insulin reduces the level of serum xylulose, while epinephrine and growth hormone increase it (Winegrad and Burden, 1966). Urinary xylulose excretion is elevated in experimental hyperthyroidism, and reduced in hypothyroidism (Coover et al., 1950; Roe and Coover, 1950) and its formation is increased in fasting (Stirpe and Comporti, 1965; Stirpe et al., 1965) in spite of the fact that the glucuronidation process and glucuronide excretion are stimulated in hyper- and hypothyroidism (Miettinen and Leskinen, 1967) and inhibited in fasting (Miettinen and Leskinen, 1963, 1968).

The proportion of the total daily glucose oxidation of the body taking place via the xylulose route is apparently very low. Thus, on the basis of the xylulose excretion of pentosurics, it has been calculated that 5–15 gm of glucose per day is catabolized via this pathway (Hollmann, 1964). Since xylitol is primarily metabolized by the liver (Müller et al., 1967), it is quite possible that the bulk of that glucose is also catabolized by the liver, and that in that particular organ the route may even have a metabolic significance despite the fact that pentosurics do well without being able to convert L-xylulose further to D-xylulose. Energy production via this route may not be important, for the amount of glucose utilized is small and 1 mole of glucose converted via L-xylulose to D-xylulose phosphate requires 3 moles of ATP, while only 6 moles of ATP are formed simultaneously from the two net moles of NADH formed, because the other two NADH-generating steps are transhydrogenations. From the theoretical point of view, the significance may lie in factors which are more or less dependent on each other, viz., (1) transhydrogenation, (2) hydrogen transport in and out of the mitochondrion, (3) regulation of the redox state, which rise to (4) increased NADH production. The latter could cause (5) a "fat-sparing" effect, at least during xylitol infusion, by shifting the hydroxyacetone phosphate-glycerophosphate system to the more reduced site (Bässler and Stein, 1967). Consequently, hepatic triglyceride synthesis increases and the acyl CoA content of the liver decreases when acetone body production is also reduced. This has been claimed to explain, partly at least, the well-known antiketogenic effect of xylitol infusion in diabetes (Bässler and Stein, 1967). In addition, (6) xylitol and other intermediates of the glucuronate pathway are potent gluconeogenic compounds (Hollmann, 1964; Krebs *et al.*, 1966), xylitol (7) stimulates insulin secretion in the dog (Kuzuya *et al.*, 1966), but only slightly in man (Geser *et al.*, 1967; Kuzuya *et al.*, 1967), and also (8) reduces the plasma free fatty acid concentration (Yamagata *et al.*, 1965) by its fat-sparing effect, by inhibition of lipolysis through its metabolism in adipose tissue or by stimulation of insulin secretion. It should be noted that the glucuronate-xylulose cycle consumes NADPH and could thus interfere with (9) hepatic fatty acid synthesis.

It is already seen from these observations that theoretically, at least, the synthesis and hydrolysis of UDPGA or glucuronides, apart from the detoxication process, and presumably also the release of glucuronic acid from mucopolysaccharides or its synthesis from myoinositol have a great number of metabolic consequences that not only influence the metabolism of carbohydrates but are also reflected in fat and energy metabolism. The future will show whether any of these consequences are of significance.

B. The Ascorbic Acid Pathway

L-Gulonolactone, which could be formed either from glucuronolactone by glucuronate reductase after lactonization of glucuronic acid or from the gulonic acid of the glucuronate-L-xylulose route, has been shown to be the actual precursor of ascorbic acid in all other species except man, monkey, and guinea pig (Hollmann, 1964). These species are unable to synthesize ascorbate, for their livers are almost totally devoid of the soluble aldonolactonase necessary for L-gulonolactone formation and particularly of the microsomal L-gulonolactone oxidase which converts L-gulonolactone to 2-keto-L-gulonolactone (Chatterjee *et al.*, 1961). The latter enolizes nonenzymically to L-ascorbate. The inability of man, monkey, and guinea pig to synthesize ascorbate appears actually to be related solely to the lack of the oxidase, because the liver of these species is unable to convert L-gulonolactone to ascorbate (Burns, 1957a) and because soluble hepatic lactonase did not appear to be totally absent (Yamada *et al.*, 1959).

Though human subjects have been suggested by Baker *et al.* (1962) to convert almost one-fourth of labeled glucuronolactone to labeled ascorbic acid, more recent studies with unlabeled glucuronolactone have once again shown the inability of man to convert this precursor to ascorbate *in vivo* (Chandrasekhara *et al.*, 1968). Similarly, human fetal and newborn liver microsomes lack the capacity to synthesize ascorbate from L-gulonolactone (Hollmann and Neubaur, 1966). It has also been suggested that the labeled ascorbic acid that Baker *et al.* (1962) isolated was actually β-ketogulonic acid (Hollmann and Neubaur, 1966), which has been found normally in urine and the amount of which increases after administration of glucuronolactone (see Section IV, A).

The liver, but not other organs, of most mammals (except man, guinea pig and monkey) and some birds possesses the aldonolactonase–gulonolactone oxidase system, and so the liver of these species is the only organ capable of synthesizing ascorbic acid. However, in the lower animals, most birds, reptiles and amphibians, this capacity has been retained solely by the kidney, although in chick embryos the oxidase activity is still also found in other organs up to hatching (Farbo and Rinaldini, 1965).

Ascorbic acid is further metabolized through dehydroascorbate to 2, 3-diketogulonate, oxalate, L-xylose, L-lyxonate and L-xylonate (Hollmann, 1964; Burns and Conney, 1966). The latter is also converted to L-erythroascorbate. Of the labeled ascorbate, 5% is oxidized to CO_2 by man, 19% by the rat, and 25% by the guinea pig. The rest appears in urine as ascorbate, diketo-L-gulonate and oxalate. Labeled ascorbic acid is also converted to glycogen, probably through the 5-carbon compounds and trioses. Accordingly, in animals capable of synthesizing ascorbate there appears to be an additional cycle (glucose–ascorbate–glucose) for the metabolism of sugars.

In view of the obligatory role of ascorbic acid in many biochemical processes of the animal organism, the disappearance of the ascorbate pathway during the evolution of some species is surprising and the reason for this remains unknown. It is hard to imagine that the two remaining pathways of glucuronic acid metabolism (the xylulose and glucaric acid routes) or even the reaction products of these pathways could play an even more important role than the major end product of the ascorbic acid route. Apart from its vitamin-producing role, the latter pathway also contributes to the degradation of the hydrolysis product of the glucuronidation process (glucuronic acid), a function carried out solely by the xylulose and glucaric acid routes in man, monkey, and guinea pig. Though increased intracellular release of free glucuronic acid, e.g. by phenobarbitone (Longenecker *et al.*, 1940; Burns *et al.*, 1957), results in increased ascorbate production in the rat, the increment may not be due to an increased demand for the vitamin but more probably reflects increased catabolism of glucuronate, because in man and the guinea pig no ascorbate is formed and yet catabolism of glucuronate is now stimulated by the drug solely through the other two pathways (Enklewitz and Lasker, 1935; Burns *et al.*, 1957). It is true that ascorbic acid-deficient animals metabolize many drugs ineffectively because the vitamin is required in the hydroxylation of these compounds (Conney and Burns, 1962). Even in the rat the bulk of the extra glucuronic acid formed during the stimulation of the glucuronic acid pathway may have been metabolized through the other two routes. It could be speculated that during the course of evolution the equilibrium lay so much in favor of these two routes that the ascorbate pathway finally disappeared from man, monkey, and guinea pig. At least *in vitro*, ascorbic acid synthesis from glucuronolactone is negligible in the rat liver presumably because the xylulose route degrades glucuronolactone more efficiently. Blocking of gulonate dehydrogenase with cyanide results in a net ascorbic acid synthesis (Chatterjee *et al.*, 1957; Ishikawa, 1959).

C. The Glucaric Acid Pathway

β-Glucuronidase is known to be inhibited by glucaric acid and particularly by its 1,4-lactone (cf. Levvy and Conchie, 1966). Although the compound was originally used for experimental purposes as a nonphysiological substance, Marsh pointed out that it is present in normal urine (Marsh, 1963a) and that the animal organism is able to synthesize it from glucuronolactone (Marsh, 1963b,c, 1966). This discovery not only demonstrated the existence of a third pathway for the further metabolism of glucuronic acid, in addition to the ascorbate and L-xylulose routes, but also suggested that the animal organism actually possesses a feedback mechanism between glucaric acid synthesis and β-glucuronidase-induced intracellular release of free glucuronic acid, and that the extent to which this inhibitor is synthesized and released into the body fluids controls the hydrolysis of various glucuronides by β-glucuronidase.

For the time being, glucaric acid is known to be synthesized from glucuronolactone—from the free acid only after lactonization—by the irreversible action of NAD-dependent cytoplasmic glucuronolactone dehydrogenase (Marsh, 1963b,c). It is obvious that D-glucaro-1,4-6,3-dilactone and its nonenzymic conversion products, the 3,6-1,4-lactones, are the intermediates. Administration of these lactones to human subjects or to experimental animals results in increased excretion of glucaric acid into the urine (Ishidate et al., 1965; Matsui et al., 1969a). The enzyme, though originally assumed to be specific for glucuronolactone (Marsh, 1963c), also oxidizes mannuronolactone to the corresponding mannaric acid (Sadahiro et al., 1966; Matsui et al., 1965, 1969b). The enzyme has been purified from the liver, where, with the kidney and testis, its activity is highest among the tissues tested (Marsh, 1963b). The high testicular activity is surprising, for this organ is not capable of any significant glucuronidation; the release of glucuronic acid may take place primarily from mucopolysaccharides or from UDPGA formed for the synthesis of the latter, although even then the lactonization is a problem, because uronolactonase exists only in the liver and aldonolactonase in both the liver and the kidney. The enzyme appears to be present in the intestinal mucosa (Hänninen, 1968), which has an active glucuronidation pathway but no lactonase for glucuronolactone formation. The equilibrium between glucuronic acid and its lactone lies far over to the right at physiological pH values. In the presence of a trapping agent, however, lactonization occurs spontaneously (Yamada, 1959).

The hepatic activity of the enzyme is of the same magnitude in man and many other mammals (Marsh, 1963b), is low in the fetus and in young animals, and in rats appears to be higher in males than in females, and is particularly high at about the time of sexual maturation (Marsh and Carr, 1965). The enzyme activity was absent in an experimental liver tumor, and increases in the liver during pregnancy (Marsh and Carr, 1965), but is not activated by inducer drugs of the glucuronic acid pathway (Marsh and Reid, 1963).

Glucaric acid, which can be measured either by an indirect enzymic method (Marsh, 1963a) or by direct chemical methods (Ishidate et al., 1965), is excreted in small amounts into the urine (Marsh, 1963a; Ishidate et al., 1965; Aarts, 1965) of man (10–20 mg/day) and other species, and is also found in the bile (Matsushiro, 1965). Administration of its precursor glucuronolactone increases urinary excretion in man (independently of the dose) by 16–20% of the dose (Marsh, 1963a; Ishidate et al., 1965; Matsui et al., 1969a). The simultaneous increment of uronic acid excretion is about 30% of the dose (Marsh, 1963a). Since the as-

corbic acid pathway is absent in man, the rest, about 50%, of the dose
is metabolized through the L-xylulose route, though the results of Matsui
et al. (1969a) show that a sizable portion of glucaric acid may have been
further metabolized. Similar results have been obtained in the guinea
pig, which is also unable to produce ascorbate, while the rat, which is
able to do so, excretes only a small percentage of the dose into the urine
as uronic and glucaric acids, the bulk being catabolized, obviously
through the ascorbate and L-xylulose routes (Marsh, 1963a; Matsui *et al.*,
1969a). Glucuronolactone administration increases the plasma glucaric
acid concentration also (Okada *et al.*, 1964), thus providing a reasonable
explanation for the earlier finding by Fishman *et al.* (1951) that glucuronolactone administration is followed by a marked reduction of serum
β-glucuronidase activity. A similar fall is also found in the liver after administration of both glucuronolactone (Marsh, 1966; Kiyomoto *et al.*,
1963) and glucaro-1,4-lactone. The fall appears to be due to inhibition
of the enzyme and not to reduction of the enzyme protein synthesis
(Marsh, 1966), although glucaric acid could conceivably inhibit the
synthesis of β-glucuronidase, at least under certain conditions. These
observations clearly demonstrate that *in vivo* glucaric acid is able to
control β-glucuronidase activity and that determinations of the latter are
influenced both by the amount of enzyme present and by its inhibition.

Free glucuronic acid, in contrast to its lactone, promotes only a
negligible increase in urinary glucaric acid output, probably because it
is not taken up by the cells. Accordingly, any increase of the intracellular
glucuronic acid level would result in increased glucaric acid synthesis.
Thus, administration of inducer drugs such as barbiturates increases the
serum glucaric acid level (Okada, 1969) and urinary glucarate excretion
(Marsh and Reid, 1963; Aarts, 1965, 1966). Salicylic acid, which is
markedly glucuronized, did not stimulate glucarate excretion in man
(Aarts, 1965). Inducer drugs, however, are unable to stimulate the conversion of administered glucuronolactone to glucaric acid, an observation
which is in agreement with unchanged glucuronolactone reductase activity. This suggests that both the uptake of exogenous glucuronolactone by
cells and its further relative and absolute metabolism via the glucarate,
xylulose, and ascorbate routes remain unchanged; stimulation of basal
glucarate excretion by inducers is presumably caused by enhanced intracellular release of free glucuronic acid from the UDPGA formed in increased amounts from UDPG, i.e., in a way suggested for the stimulation
of the ascorbate route (Hollmann and Touster, 1962). Since β-glucuronidase is obviously involved in the release mechanism of glucuronate,

there should be a correlation between glucaric acid and β-glucuronidase activity. As a matter of fact, serum β-glucuronidase activity is elevated in pregnancy (cf. Levvy and Conchie, 1966), diabetes (cf. Dohrmann, 1968), and barbiturate-treated epilepsy (Okada, 1969), i.e., in states that exhibit elevated glucarate serum concentration and urinary excretion (Okada, 1969). On the other hand, hepatic β-glucuronidase activity is high in newborn animals, which exhibit a poor glucuronidation capacity, low glucuronolactone dehydrogenase activity (cf. Levvy and Conchie, 1966), and thus probably also decreased glucarate formation.

Though conversion of glucuronic acid (or its lactone) to gulonic acid (or its lactone) appears to be a reversible reaction in at least some species (see Section IV, A), administration of gulonolactone to human subjects is not followed by increased glucarate formation (Matsui et al., 1969a). Accordingly, although administration of this acid results in its increased metabolism through the ascorbate and xylulose routes (see Section IV, A and B), the reverse reaction back to glucuronolactone appears to be negligible in man in vivo. Nor is myoinositol, which is converted to glucuronic acid (see Section III, C), metabolized further through the glucaric acid pathway in vivo (Matsui et al., 1969a), suggesting that the kidneys produce negligible amounts of glucaric acid.

The significance of the glucaric acid route in vivo appears to be negligible from the quantitative point of view as far as glucose catabolism is concerned. It could be hypothesized that its primary role in metabolism is to regulate the release of free glucuronic acid by inhibiting β-glucuronidase, which could then have a great many different consequences, namely, (1) controlling the L-xylulose pathway and particularly pentose production and the nucleotide status of the route, (2) regulating ascorbic acid synthesis in animals able to produce this vitamin (saccharo-1,4-lactone given to barbital-treated rats actually reduces ascorbate excretion; Marsh and Reid, 1963), (3) regulating the synthesis of glucaric acid itself, (4) determining the rate of detoxication through varying inhibition either of UDPGA destruction via unknown aglyconeglucuronide or hydrolysis of newly formed glucuronides of compounds to be conjugated, (5) participating in the control of mucopolysaccharide synthesis (through degradation of UDPGA) or catabolism (by inhibiting the β-glucuronidase effect), (6) controlling the release of biologically active substances, e.g., hormones, from their inactive glucuronides and (7) controlling the synthesis of β-glucuronidase itself. In view of the greatly varying β-glucuronidase activity in different organs and in subcellular compartments, and of the poorly understood role of this enzyme in the glucuronate pathway, it is difficult for the time being to conclude what is the actual relationship between the β-glucuronidase activity and the amount of glucaric acid in different sites of the body.

V. Regulation of the Glucuronic Acid Pathway by Nucleotides and Substrates

The rate at which glucose is metabolized via the glucuronic acid pathway is dependent not only on the activity of various enzymes but also on the availability of glucose or other sugars, nucleotides, and foreign compounds which need not even be conjugated with glucuronic acid. Since availability of glucose appears to be one of the very primary factors in the initial reactions of the pathway, its role will be discussed first.

A. AVAILABILITY OF CARBOHYDRATES

Administration of glucose to fasting frogs improves the glucuronidation capacity of these animals, as pointed out long ago by Schmid (1936). Subsequently, it has been demonstrated repeatedly that administration of glucose and fructose either *in vivo* or *in vitro* enhances the glucuronidation capacity (Lipschitz and Bueding, 1939; Südhof, 1952, 1954a,b; Heyde and Wieland, 1960; Beck and Richter, 1962b). The exact mode of action is unknown, though increase of G-1-P could markedly increase the formation of UDPG, which in turn could lead to augmented synthesis of UDPGA. These reactions may actually be markedly stimulated *in vivo*, because a glucose load is normally followed by a pronounced increase of plasma insulin, which facilitates the synthesis of hepatic G-1-P, UDPG, and glycogen. Administration of insulin to normally fed animals actually increases the hepatic UDPGA concentration and stimulates glucuronidation *in vitro* (Müller-Oerlingenhausen *et al.*, 1968), while alloxan diabetes or diabetes provoked by insulin antibodies is followed by impaired glucuronidation, reduced hepatic UDPGA concentration, reduced UDPG-dehydrogenase and unchanged UDPGA-transferase activities (Müller-Oerlingenhausen *et al.*, 1967b). That the availability of UDPG affects glucuronidation is also indicated by apparently improved bilirubin conjugation after administration of UDPG to newborn children, in whom the amounts of endogenous UDPG available may be decreased (Careddu and Marini, 1968).

Galactose is a better precursor of L-ascorbate than glucose. Since galactose is converted directly from UDPGal to UDPG without entering into the glucose pool, it is not diluted to the same extent as glucose itself and results in effective UDPGA synthesis and subsequent L-ascorbate formation (see Section II, A; Evans et al., 1959, 1960).

Though myoinositol is converted to glucuronate and gulonate in the kidney (see Section III, C), it is unable to promote ascorbate excretion, because obviously renal glucuronate and gulonate are not transported to the liver or, as free acids, they are not taken up by liver cells for ascorbate synthesis. Nor does myoinositol give rise to glucarate excretion (Matsui et al., 1969a), suggesting that the formation of glucarate by the kidney, as compared to that by the liver, is negligible, even though the renal tissue is able to synthesize glucaric acid in vitro.

B. THE ROLE OF NAD AND NADP NUCLEOTIDES

As appears from Fig. 3, there are four reactions in the pathway which are mediated by NAD-linked enzymes and function primarily in the cytoplasmic compartment of the cell. It is thus conceivable that if the level of NAD or perhaps only the ratio NAD/NADH increases in the cytoplasm this, as such, might result in enhanced channeling of glucose to the pathway, which would increase glucuronidation or perhaps only the release of free glucuronic acid and its further metabolism in the three pathways degrading glucuronic acid. This consideration is supported by the finding of Mills et al. (1958) that the activity of UDPG-dehydrogenase is largely dependent on the availability of oxidized NAD. Unfortunately there are no studies dealing with the effect of changing NAD-nucleotide status in vivo on the rate of glucose metabolism via the glucuronic acid pathway. Dietary and endocrinological manipulations, such as fasting and diabetes, are known to change the ratio NAD/NADH cf. Krebs, 1967) and also appear to change the activity of the pathway. The change of nucleotide status under those circumstances might be secondary to the lack of normal metabolism of glucose.

Administration of nicotinamide (NA) has been demonstrated primarily to enhance the NAD level—and to a lesser extent that of NADH—in the liver and many other tissues (Bonsignore and Ricci, 1957; Kaplan et al., 1956). Though this compound might serve as an "inducer" drug without itself being glucuronized and might, in addition, have direct metabolic consequences, its effect as a potent NAD precursor on the glucuronic acid pathway is of special interest. Since the authors have carried out some studies on this line, they will be briefly reviewed in the following sections (Tables I and II).

Administration of NA to rats resulted in marked elevation of urinary

ascorbic acid excretion, suggesting that either metabolism of ascorbate
was decreased, washed out of the tissues in increased amounts, or that
synthesis of the vitamin was actually stimulated via increased glucuronic
acid release or inhibition of the L-xylulose and glucarate routes. The

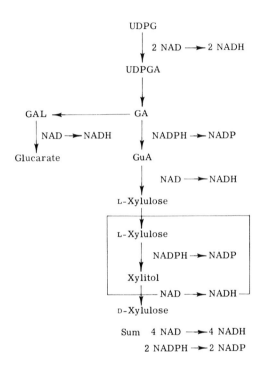

Fɪɢ. 3. Schematic illustration of NAD- and NADP-nucleotide requirement by
the glucuronic acid pathway. Mitochondrial reactions are in the box. UDPG, UDP-
glucose; UDPGA, UDP-glucuronic acid; GA, glucuronic acid; GAL, glucuronic acid
lactone; GuA, gulonic acid.

effect was also found in rats whose liver glycogen was depleted and
gluconeogenesis stimulated by a 48-hour fast. Unfortunately, end products of the glucaric or L-xylulose routes were not measured. A menthol
load which enhanced glucuronate and ascorbate excretion in normal rats
further increased the output of the latter in both fed and fasted animals
pretreated with NA. The latter compound had no effect on basal or
menthol-induced glucuronide excretion, suggesting that NA did not

stimulate glucuronidation itself. Glucuronolactone increased urinary glucuronide excretion (sum of free and bound) slightly less in the NA-
treated animals and enhanced ascorbate output more as compared to
controls, an additional menthol load being followed by a nonsignificant
further elevation of ascorbate excretion in both groups. Finally, administration of L-gulonolactone increased ascorbate excretion under basal
conditions and during the menthol load up to two-fold, the respective
increment in NA-treated animals being fourfold.

TABLE I

EFFECT OF NICOTINAMIDE AND METHYLCHOLANTHRENE ON GLUCURONIDE
AND ASCORBATE EXCRETION IN THE RAT

Compound given[a]	Glucuronides		Ascorbic acid	
NA[b]	− 4	(NS)[c]	+251	(S)[d]
Me[b]	+179	(S)	+ 40	(S)
Me (NA)[e]	+198	(S)	+ 55	(S)
GAL[b]	+980	(S)	+ 65	(S)
GAL (NA)[e]	+772	(S)	+206	(S)
GuAL[b]	—		+ 32	(S)
GuAL (NA)[e]	—		+280	(S)
MC[b]	+ 25	(NS)	+660	(S)
Me (MC)[e]	+ 40	(S)	+220	(S)

[a] Abbreviations: NA, nicotinamide; Me, menthol; MC, methylcholanthrene; GAL,
D-glucuronic acid lactone; GuAL, L-gulonic acid-γ-lactone.

[b] Percent increment of nontreated controls.

[c] NS, statistically nonsignificant change.

[d] S, statistically significant change.

[e] Percent increment of NA- or MC-pretreated controls.

Though other explanations are possible, these experiments suggest
that NA enhances the release of free intracellular glucuronic acid without
affecting the net glucuronidation, subsequently leading to increased
ascorbate synthesis and excretion, that NA stimulates the further metabolism of free glucuronic acid, the amount of which was increased either
by augmented glucuronidation (e.g., by a menthol load) or by administration of exogenous glucuronolactone and that NA in some way or other
increases the conversion of endogenous and exogenous L-gulonolactone
to ascorbate. It is interesting to note that although drugs did not accelerate the conversion of administered glucuronolactone or gulonolactone
to ascorbic acid (Burns and Evans, 1956), NA appeared to be able to do
this, probably through altered nucleotide status.

In vitro experiments revealed that 12 hours after administration of NA the hepatic NAD concentration was elevated (Table II), the effect being less significant at the NADP level. However, the ratio NAD/NADH was decreased from 7 to 4, indicating that, in contrast to earlier findings, the amount of reducing equivalents was increased relatively more than that of oxidized ones. It is possible that after administration of NA the nucleotide primarily formed is NAD, which is then converted in reduced form by reactions generating reducing equivalents, including NAD-linked

TABLE II

EFFECT[a] OF NICOTINAMIDE AND 3,4-BENZPYRENE ON LIVER PYRIDINE NUCLEOTIDES IN THE RAT

Compound given[b]	NAD		NADH		NAD/ NADH		NADP	NADPH		NADP/ NADPH	
NA	+190	(S)[c]	+430	(S)	−46	(S)	+40	+7		+30	(S)
3,4-BP	−38		−45		+11		−24	−38	(S)	+23	(S)

[a] Change, percent of nontreated controls.
[b] Abbreviations: NA, nicotinamide; 3,4-BP, 3,4-benzpyrene.
[c] S, statistically significant change.

reactions of the glucuronate pathway. It should also be borne in mind that the method used measures total nucleotide levels, so that errors caused by free and bound nucleotides, and different subcellular compartmentation of nucleotides could have significantly changed the ratio (cf. Krebs, 1967) in the cytoplasm.

Since addition of NAD to the liver homogenate in the absence of UDP-nucleotides markedly stimulated glucuronidation and because the stimulation of the latter *in vivo* may have accounted for the increased ascorbate excretion during NA treatment, the enzyme activities involved with glucuronidation were measured in the liver of the NA-pretreated animals. Moderately increased (50%) UDPG dehydrogenase activity and UDPGA concentration (50%) were found in the liver, while the rest of the enzymes determined, including glucuronyltransferase, UDPGA pyrophosphatase, and glucuronate reductase, did not show any change in activity. Accordingly, either NA-induced change in NAD status without any change in enzyme protein synthesis, or NA-(or NAD) induced stimulation of the latter enhanced UDPG dehydrogenase activity, and so led to increased UDPGA formation. The latter may have been hydrolyzed in increased amounts to free glucuronate via reaction(s) which did not result in net glucuronide formation. Increased ascorbate excretion might thus be a reflection of increased hydrolysis of UDPGA initiated by increased synthesis of the latter via NA-induced activation of UDPG dehydrogenase. If the latter is ultimately caused by an initially increased concentration of oxidized NAD, then the free glucuronate may have been metabolized at an even higher rate via the xylulose and glucarate pathways than via the ascorbate route, because the two former are NAD-linked reactions.

Though the level of NADP nucleotides remained unchanged after NA administration, the ratio NADP/NADPH tended to increase a finding which is to be expected if the reactions from free glucuronic acid shown in Fig. 3 are enhanced. It should be borne in mind, however, that quantitatively the glucuronate pathway may contribute little to the oxidoreduction of NADP or NAD couples.

Evidence presented above that it is primarily the change in the nucleotide status which stimulates the glucuronic acid pathway after NA administration is far from conclusive, and the possibility still exists that NA as such functions by stimulating the drug-metabolizing enzyme system as a whole, including the glucuronate pathway. The alteration found in the enzyme pattern of the route after administration of NA resembles that observed following administration of "inducer" drug, i.e., UDPG dehydrogenase is activated (see Table III). On the other hand, in hyperlipidemic patients large doses of nicotinic acid for long periods of time are not accompanied by any significant proliferation of endoplasmic reticulum (Miettinen et al., 1968), a change caused by many inducer drugs (Remmer and Merker, 1963a,b). Nicotinic acid also increases urinary ascorbate excretion (Longenecker et al., 1940) and impairs glucose utilization, probably in the liver via an alteration of nucleotide status (Miettinen et al., 1969). As a matter of fact, nicotinic acid may stimulate gluconeogenesis (Miettinen et al., 1969), and this, like nicotinamide, might stimulate channeling of glucose to the pentose cycle via the glucuronate pathway.

It is possible that the "inducer" drugs exert their effect, partly at least, by changing hepatic nucleotide status, for hydroxylation of many foreign compounds requires reduced NADP (cf. Conney, 1967). Furthermore, the NADP-linked reactions of the glucuronate pathway are stimulated by increased NADPH level. Determination of the liver nucleotides after 3,4-benzpyrene induction, however, did not change (Table II) their levels dramatically, even though NADPH appeared to decrease. The ratio NAD/NADH now remained unchanged, whereas that of NADP/NADPH increased slightly, perhaps being a sign of increased utilization of NADPH by the hydroxylation reactions or the glucuronate–xylulose route. However, the changes are so negligible that, in view of the methodological

problems in their determination from different compartments, no definitive conclusions could be drawn on their role in benzpyrene-induced
stimulation of the glucuronic acid pathway.

The effect of an ethanol-induced change in the NAD/NADH ratio on
the glucuronate pathway has been discussed in Sections II, B, 1, and IV, A.
In diabetes and fasting the ratio NAD/NADH decreases (Krebs, 1967),
which could inactivate the glucuronate pathway. As a matter of fact,
glucuronidation of o-aminophenol is decreased in diabetes, apparently
owing to reduced UDPG dehydrogenase activity of the liver (Müller-
Oerlingenhausen *et al.*, 1967b) and the development of glucuronidation
in newborn animals is retarded by starvation (Flint *et al.*, 1963). On the
other hand, serum L-xylulose concentration and increased glucaric acid
excretion by diabetics (Winegrad and Burden, 1965, 1966; Okada, 1969)
and increased urinary ascorbate output by diabetic rats (Straumfjord
and West, 1957) suggest that the pathway may even be overactive *in
vivo* at least as far as glucose metabolism, though not necessarily the
detoxication process, via this pathway is concerned. Since the glucuronate-xylulose cycle leads to the pentose phosphate shunt, the route may
be associated with the enhanced gluconeogenesis of diabetes and fasting.

Since ATP stimulates glucuronidation *in vitro* by inhibiting degradation of UDPGA (Pogell and Leloir, 1961), it is possible that conditions
which reduce the level of this high-energy compound could impair the
glucuronidation process. Ethionine-induced inhibition of the drug-pro-
duced stimulation of the pathway has been related to the initial decrease
of ATP content (see Section V, C). However, administration of ATP to
experimental animals has shown that ascorbate excretion decreases
markedly in both normal and "inducer"-treated animals, possibly as a
sign of reduced metabolism via the glucuronate pathway (Ganguli *et al.*,
1956; Conney *et al.*, 1961). UDP-N-acetyl glucosamine stimulates glucuronidation *in vitro* by inhibiting UDPGA pyrophosphatase; the compound has been suggested to be active *in vivo* also (Taketa, 1962e).

C. Effect of Exogenous Compounds

As far as the glucuronate pathway is concerned, exogenous compounds
introduced into the animal organism from outside the body can be
divided into three groups according to whether (1) they are primarily
conjugated with glucuronic acid, (2) conjugation is only moderate or
(3) although glucuronidation itself is negligible, the pathway as such
is stimulated, this being reflected by augmented release and further
metabolism of free glucuronic acid and possibly by enhanced glucuroni-

dation of another aglycone. Thus, some drugs promote conjugation, others primarily hydrolysis of UDPGA. It could be claimed that though the former drugs have not been stated to activate the enzymes of the pathway to any appreciable extent, their enhanced glucuronidation is associated with more or less increased release of free glucuronic acid, i.e., hydrolysis of UDPGA either directly, or indirectly via a glucuronide formed. Tables III and IV list some of the drugs in relation to the changes brought about by their administration. These changes are seen in enzyme activities of the pathway, and in levels of urinary bound and free glucuronate or metabolites of the L-xylulose, L-ascorbate, and glucarate routes. The tables show, however, that it is the inducer group that mainly stimulates the hepatic activities of the initial enzymes, i.e., UDPG dehydrogenase and UDPGA transferase in the glucuronate pathway and probably in consequence results in increased flow of glucuronic acid through all the three subsequent metabolic routes. That the quantitative response of individual metabolites to various aglycones is different, e.g., urinary free glucuronate and glucarate excretion are only slightly or not at all elevated by methylcholanthrene, might be due to the varying activities of diffent enzymes possibly determined by the drug-induced change in the nucleotide status or in the general metabolic pattern.

   Though the basic mechanism of action of inducer drugs is unknown, it has usually been related to the activation of the earlier enzymes of the pathway and, ultimately, to enhanced synthesis of the protein of these enzymes (cf. Conney, 1967). The latter is evidenced, for instance, by the inhibitory action of ethionine on drug-induced ascorbate production (Touster et al., 1960; Hollmann and Touster, 1962). However, an ethionine-induced decrease in protein synthesis might be secondary to a lack of ATP (Villa-Trevino et al., 1963), which would explain the marked decrease in sugar phosphates and NAD (Bartels and Hohorst, 1963). The latter factors could then inhibit the glucuronate pathway (see Section V, A and B). Inhibitors of protein synthesis, puromycin and actinomycin D, prevent the stimulation of the pathway enzymes to a certain extent but are not able to block the corresponding enhancement of ascorbate excretion (Aarts, 1966). Nor did the time-course studies favor the view that enzyme activation is the phenomenon primarily responsible for the enhanced metabolism of glucose via the glucuronate pathway during drug induction. On the contrary, within a few hours after administration of the drugs the enzyme activities are decreased, obviously owing to inhibited protein synthesis (Hollmann and Neubaur, 1967) or drug inhibition of enzyme activities, while simultaneous ascorbate and glucarate excretion is already increased (Aarts, 1966; Holl-

TABLE III
Effect of Various Compounds on Enzyme Activities of Glucuronic Acid Pathway

Compounds[a]	Enzymes[b] 1	2	3	4	5	6	7	8	9	10	11	12	13	References[c]
Borneol					+									14
Menthol					+									14, 23
Salicylamide	+	+												17
Chloretone	+0	+0			0		0	+0	0	0	-0			2, 10, 15, 21, 26, 31, 33, 36, 37
Aminopyrine	0	+												4, 15, 37
Antipyrine								+						21
Barbital	+s	+0	+0-	0	+-	0	0	+	-0	+0	0	0	0	2, 9, 15, 16, 18, 21, 26, 36, 37, 39
Cinchophen	+	+	-	-	-0	-	+	0	0	+	-	-	-	18
3,4-Benzpyrene	+0	+			+					0				3, 15, 19, 26, 28, 30, 37
3-Methylcholanthrene	0	+								0				15, 19, 26

[a] + = elevated; +s = slightly elevated; 0 = no change; - = inhibited.

[b] 1, UDPG dehydrogenase; 2, glucuronyltransferase; 3, UDPGA pyrophosphatase; 4, GA-1-P phosphatase; 5, β-glucuronidase; 6, myoinositol oxygenase; 7, glucuronate reductase; 8, uronolactone reductase; 9, aldonolactonase; 10, D-glucuronolactone dehydrogenase; 11, L-gulonolactone oxidase; 12, 3-hydroxyacid dehydrogenase; 13, glucuronolactone reductase (microsomal).

[c] References are given in footnote to Table IV.

198

TABLE IV

Effect of Various Compounds on Urinary Excretion of Metabolites Produced by the Glucuronic Acid Pathway[a]

Compounds	Metabolites excreted[a]						References[b]
	Glucu-ronides	Glucuro-nic acid	Glucaric acid	Gulonic acid	Ascorbic acid	Xylulose	
Aniline	+	+		0	0		7
Borneol	+	0		0	0	+	7, 8, 13
Menthol	+				+s	+	13, 16, 25, 29, 38
Salicylamide	+	+	+		+		17, 18, 34
Chloretone	+s	+s	+	+	+		6, 7, 15, 16, 20, 24, 26, 32, 35
Aminopyrine (amidopyrine, pyramidon)	+s		+		+	+	1, 5, 13, 24, 27
Antipyrine (phenazone)	+				+	+	5, 13, 22, 24
Barbital or phenobarbital	0	+	+	+	+		1, 6, 7, 8, 15, 20, 26, 35
Cinchophen		+	+		+		18
3,4-Benzpyrene	+?	+?	0		+		12, 13, 26
3-Methylcholanthrene		+s	0	+	+		7, 11, 15, 26, 35

[a] + = elevated; +s = slightly elevated; 0 = no change.

[b] References: 1. Aarts (1965); 2. Aarts (1966); 3. Arias et al. (1963b); 4. Arias et al. (1963c); 5. Artz and Osman (1950); 6. Burns (1957b); 7. Burns et al. (1960); 8. Burns et al. (1957); 9. Catz and Yaffe (1962); 10. Conney et al. (1961); 11. Dayton et al. (1964); 12. Elson et al. (1945); 13. Enklewitz and Lasker (1935); 14. Fishman (1940); 15. Hollmann and Touster (1962); 16. Hollmann (1964); 17. Hänninen (1966); 18. Hänninen (1968); 19. Inscoe and Axelrod (1960); 20. Kato et al. (1962); 21. Kawada et al. (1961); 22. Lawrow (1901); 23. Levvy et al. (1948); 24. Longenecker et al. (1940); 25. Longenecker et al. (1939); 26. Marsh and Reid (1963); 27. Margolis (1929); 28. Metge et al. (1964); 29. Miettinen and Leskinen (1967); 30. Pulkkinen and Hartiala (1965); 31. Salomon and Stubbs (1961); 32. Smith and Williams (1954); 33. Stubbs and Salomon (1963); 34. Taketa (1962d); 35. Touster et al. (1960); 36. Touster and Hollmann (1961b); 37. Touster and Hollmann (1961a); 38. Touster and Shaw (1962); 39. Zeidenberg et al. (1967).

199

mann and Neubaur, 1967). Enhanced activation of the pathway enzymes is a slower process and thus apparently a secondary phenomenon provoked by an unknown factor.

Recent studies have indicated that drugs induce changes of activity in the enzymes of the glucuronate pathway in the kidney also and particularly in the intestinal mucosa (Hänninen, 1966, 1968), an organ exposed directly to the stimulatory action of compounds administered orally or present in the diet. The role of this tissue in detoxication has been studied extensively by Hartiala's group (cf. Hartiala, 1961; Hänninen, 1966, 1968). It is interesting to note that drugs appear to be tissue specific. Thus, barbiturates stimulate only the pathway enzymes in the liver, while cincophene enhances many of the enzymes in the intestine and kidney as well (Hänninen, 1968). 3,4-Benzpyrene stimulates glucuronidation in the skin after local application (Dutton and Stevenson, 1962), but when given parenterally only hepatic enzymes are activated, no effect being observed on the enzymes of the intestine or kidney (Hartiala and Pulkkinen, 1964). The presence of the active glucuronic acid pathway outside of the liver indicates (see Section II, D) that, apart from ascorbic acid, the amounts of glucuronic acid metabolites in the urine are determined by enzyme activities not only of the liver but also of the kidney and intestine. These are the three organs which are metabolically very active and which, in addition, are exposed to the highest concentrations of foreign compounds introduced into the body.

Salicylamide, which is conjugated with both glucuronic acid and sulfate, has been reported to stimulate strongly enzyme activities of all three tissues and to result initially in decreased ascorbate excretion (Hänninen, 1966), probably as a result of an increased demand of UDPGA for its glucuronidation, augmented ascorbate retention by tissues, or enhanced utilization of ascorbate for hydroxylation of salicylamide. The initial fall of ascorbate excretion was followed by a secondary rise. Tables II and III show that menthol, which, analogously to borneol, increases hepatic β-glucuronidase, augments ascorbate excretion slightly; borneol, on the other hand, has no effect on urinary free glucuronate or its metabolites in the rat or the guinea pig. However, administration of borneol and menthol to pentosuric subjects is followed by marked elevation of urinary xylulose output (Enklewitz and Lasker, 1935), suggesting that even compounds of this type actually stimulate the release of glucuronic acid and its further metabolism, the detection of this phenomenon being difficult under normal circumstances.

It is apparent that it is necessary to make a serial study of all the metabolites of the glucuronate pathway after administration of compounds representative of the three groups and to relate the findings to enzyme activities of the different organs in mammals which are able as well as in those which are unable to synthesize ascorbic acid. The possibility should also be borne in mind that some drugs may, as such, be able to change renal clearance of these metabolites, for increased ascorbate excretion after administration of certain drugs has been found in man, who is unable to synthesize this vitamin (Daniels and Everson, 1936).

VI. Urinary Metabolites of the Glucuronate Pathway Related to Enzyme Activities of Various Organs

A. GENERAL FACTORS AFFECTING URINARY METABOLITES

1. *Enzyme Activities*

Under normal conditions glucuronides are major products of the glucuronic acid pathway, other metabolites, apart from ascorbate, playing a quantitatively minor role. The amount of glucuronides is determined by the type and amount of aglycone entering the body and by the capacity of the liver, intestine, and kidney to carry out conjugation. The latter depends on the availability of glucose and the activity of at least UDPG dehydrogenase and UDPGA transferase; possibly two other enzymes are involved, namely, UDPGA pyrophosphatase, which could interfere with conjugation by destroying UDPGA, and β-glucuronidase, which could either destroy UDPGA via some unknown glucuronide formed with an endogenous aglycone, or hydrolyze newly formed conjugates of exogenous aglycones *in situ*, i.e., within microsomes or later in the cytoplasm, the extracellular space, or even in the urine. The exact role of the two latter enzymes in glucuronide production *in vivo* is not known, but if anything, their reduced activity presumably favors net glucuronidation and diminishes the release of free glucuronate, even though GA-1-P phosphatase activity, which is necessary for the pyrophosphatase route, is low in the liver (see Section III, A,B).

The capacity of the two former enzymes must normally exceed the requirement of basal glucuronidation, because the latter is instantly increased three to fourfold by the presence of an acute excess of an exogenous aglycone, e.g., a menthol load (cf. Table I) even when there is no detectable increase in the enzyme activities. The enhanced flow of glucose necessary to speed up this process could have been stimulated by glucuronidation-induced initial fall of UDPGA concentration in the microsomes and subsequently in the cytoplasm. Another possibility is that under basal conditions UDPGA is formed continuously in excess, part of it being destroyed through the pyrophosphatase or transferase routes to free glucuronate. During acutely enhanced glucuronidation the excess would be used for conjugation, the release of glucuronate being reduced and the conversion of UDPG to UDPGA remaining unchanged. That at least menthol and borneol to a certain extent stimulate urinary output of glucuronic acid metabolites favors the former possibility (see Table IV). Increasing salicylamide doses proportionally augment glucuronide, free glucuronate, and ascorbate excretion (Hänninen, 1966), suggesting that UDPGA generation was actually elevated. In view of excessive UDPG dehydrogenase and transferase capacity in the normal state, the stimulation of these enzymes by certain nonglucuronogenic drugs suggests that under these conditions the flow of glucose through the pathway is enormously increased and that transferase may actually take part in the release of free glucuronate (Hollmann and Touster, 1962).

The substrate saturation of transferase is difficult to achieve in tissue homogenates and microsomal preparations, mainly owing to degradation of UDPGA by pyrophosphatase (see Section II, C). Thus, increased conversion of UDPG to UDPGA *in vivo* through enhanced UDPG-dehydrogenase activity could augment the glucuronidation capacity even without any detectable increase in transferase activity. The release of free glucuronate could similarly depend on the amount of UDPGA. Many of the inducer drugs primarily increase the formation of UDPGA, while others stimulate transferase activity, yet they augment glucuronidation of only a few glucuronogenic compounds (see Section VII, B). The possibility that the rate of entry of different compounds into the cell or smooth reticulum of the cell is rate limiting for their glucuronidation is a possible explanation for this phenomenon.

2. Enzyme Inhibition

Many of the steroid hormones have been reported to inhibit glucuronidation *in vitro* (see Section II, C, 5) but their effect on glucuronidation *in vivo* is still unknown. Administration of estradiol, testosterone, and progesterone to bitches has no effect on glucuronidation of salicylamide (Rauramo *et al.*, 1963). Many of the substrate aglycones and nonglucuronogenic compounds as well as products of the glucuronate pathway may inhibit enzymes, provided that concentrations are high enough (cf. Hänninen, 1968). The specific feedback mechanism of glucaric acid might be effective *in vivo* (see Section IV, C). Increased synthesis of this acid could favor glucuronidation through inhibition of β-glucuronidase. On the other hand, if the latter enzyme has a synthetic capacity *in vivo* (see Section II, C, 4), the effect would be the reverse. Steroid glucuronide production is actually reduced by β-glucuronidase inhibitors (Sie and Fishman, 1957). Japanese workers have repeatedly reported, however, that administration of glucuronolactone, its derivatives or glucarate derivatives, which give rise to urinary glucarate excretion (see Section IV, C), actually stimulates the glucuronidation of various glucuronogenic compounds, probably as a result of β-glucuronidase inhibition (see Section II, C, 4). Inhibition of microsomal enzymes by phenyldiallylacetate ester of ethanolamine *in vivo* has no effect on the glucuronide excretion or release of free glucuronate stimulated by inducer drugs (Nitze and Remmer, 1962).

3. *Size of Organs*

The glucuronidation activity *in vitro* and enzyme activities of the glucuronate pathway are of about the same magnitude in the liver, kidney, and intestinal mucosa per unit weight. Since the liver is the heaviest of these organs, it probably contributes at the highest rate to the production of urinary glucuronides of most aglycones and possibly also to other metabolites of glucuronic acid. Many experimental conditions, such as drug treatment (cf. Conney, 1967) and hyperthyroidism (cf. Miettinen and Leskinen, 1967) increase the size of the liver and kidneys and probably also that of the intestinal mucosa; therefore in any attempt to explain the capacity of an animal to excrete glucuronides in terms of the enzyme activities of various tissues, values for the activities should be given for the whole organ, and not per unit weight of the tissue. Inducer drugs have been reported to increase the biliary excretion of some glucuronogenic compounds without resulting in a major change of their hepatic glucuronidation per unit weight (Roberts and Plaa, 1967; Goldstein and Taurog, 1968; Klaassen and Plaa, 1968). In these cases the change in excretion is attributable to increased liver size.

4. *Aglycones*

Glucuronogenic compounds, in contrast to inducer drugs, did not stimulate the enzyme activities of the pathway in general, i.e., no adaptation took place during prolonged administration. In rats, however, continuous exposure to salicylamide results in induction of the pathway, and the increased salicylamide glucuronidation, increased free glucuronate output, and enhanced ascorbate excretion are associated with a marked increase in hepatic, renal, and intestinal UDPG dehydrogenase and UDPGA transferase, and unchanged β-glucuronidase activities (Hänninen, 1966). As in experiments *in vitro*, administration of two glucuronogenic compounds changes conjugation, probably owing to competition for the substrate or the enzyme. For instance, administration of benzoate to *p*-aminobenzoate-fed animals inhibits the glucuronidation of the latter (Koivusalo *et al.*, 1958) and in similarly treated human subjects leads to marked changes in the conjugation pattern of *p*-aminobenzoate (Koivusalo *et al.*, 1959). As already mentioned, inducer drugs may enhance glucuronidation of some, but not all (Nitze and Remmer, 1962) glucuronogenic compounds, obviously by enhancing the activity of appropriate enzymes. Since the enzyme-inducing effect of individual drugs on different tissues varies and since one aglycone may be conjugated primarily in only one of the three tissues capable of glucuronidation, the ineffectiveness of one inducer in stimulating conjugation of a given aglycone may be understandable, particularly if the uptake of the aglycone by the cell or by the intracellular compartments at a given concentration is a rate-limiting factor. If the uptake of an aglycone is the factor limiting its glucuronidation and remains unchanged during induction, the rate of glucuronidation is unaffected for that particular aglycone, even if enzyme activities are increased severalfold, but may be improved for another aglycone even in the absence of a separate transferase, provided that its uptake is not a limiting factor. Conditions characterized by transferase deficiency may resemble the latter alternative. On the other hand, it is possible that inducers known to change the structure of the transferase containing smooth reticulum of the cell (Remmer and Merker, 1963a,b) might improve the uptake and, thus, improve glucuronidation without affecting enzyme activities. Phenobarbital, which obviously increases UDPGA production in the liver, but not in the kidney and intestine (Hänninen, 1968), improves *in vivo* glucuronidation of bilirubin (cf. Crigler and Gold, 1969) which takes place primarily in the liver, (Schmid and Lester, 1966), while glucuronidation of salicylamide which is also conjugated at a high rate by the intestinal mucosa (Schachter *et al.*, 1959), is not stimulated by phenobarbitone pretreatment in the rat (Nitze and Remmer, 1962). Conflicting results have been reported in transferase deficiency states (DeLeon *et al.*, 1967). Urinary glucuronide excretion of sulfonamides (Remmer, 1964), and biliary excretion of phenolphthalein and stilbestrol are also stimulated by barbiturate pretreatment (Levine *et al.*, 1968). For phenobarbitone induction, see Section VII, B.

5. Hydrolysis of Conjugates

The amount of urinary glucuronides, free glucuronic acid, and its metabolites is also determined by hydrolysis of newly formed glucuronides which could be brought about (1) within the cell by β-glucuronidase, (2) extracellularly by β-glucuronidase or (3) by spontaneous hydrolysis of labile glucuronides. In the first case most of the free glucuronate is metabolized further and could give rise to urinary excretion of ascorbate, xylulose, and glucarate, for instance. In the two latter cases, the poor uptake of extracellular free glucuronate by cells results only in increased glucuronate output. Aniline, for instance, and some other N-derivatives increase both glucuronide and glucuronate production (Burns et al., 1960), for the labile N-glucuronides formed are apparently hydrolyzed spontaneously (Smith and Williams, 1949; Axelrod et al., 1957).

B. PHYSIOLOGICAL AND EXPERIMENTAL FACTORS

1. Age

Since Karunairatnam et al., as long ago as 1949, demonstrated low glucuronidation in neonatal liver slices, it has been generally accepted that the glucuronidation capacity of the liver during early fetal life is absent or very low and increases gradually in most species, reaching a maximum, depending on species and substrate, at birth or within a few days or weeks after birth (cf. Tenhunen, 1965; Dutton, 1966; Hirvonen, 1966). Glucuronidation of bilirubin appears to be less efficient than that of many other aglycones. Poor glucuronidation is due to low activity of UDPG dehydrogenase and particularly glucuronyltransferase. The deficiency is not limited to the liver but is also found in other organs, though in some species, such as the guinea pig, the fetal intestinal and renal transferase to some substrates may be even higher than in adults. In tissue culture the development of hepatic transferase takes place rapidly, provided the cells from young embryos are cultured in a medium containing adult and fetal serum and those from older embryos with adult serum (Ko and Dutton, 1967; Skea and Nemeth, 1969).

Since UDPGA production is low in the newborn liver, the release of free glucuronic acid and its further metabolism may also be low. The low activities of glucuronolactone dehydrogenase (Marsh and Carr, 1965) and gulonolactone oxidase (Comporti et al., 1964; Chatterjee and McKee, 1965; Chatterjee et al., 1965) in the newborn liver suggest that this might actually be the case. No systematic *in vivo* studies dealing with the urinary excretion capacity of free glucuronate and its metabolites have been carried out in newborns. In premature and full-term infants glucuronidation capacity of foreign compounds is poorly developed, the adult capacity being attained at about the age of 3 months (Vest, 1958a,b; Vest and Streif, 1959). Induction of bilirubin glucuronidation is discussed in Section VII.

Glucuronidation capacity, like the ability to metabolize drugs in general, appears to decrease again with advancing age. Thus, UDPG dehydrogenase activity or hepatic glucuronidation is higher in young rats than in old ones (Hartiala and Pulkkinen, 1964; Huttunen and Miettinen, 1965). It is well known that drug tolerance decreases in elderly people (cf. Bender, 1964), suggesting an analogous decline in glucuronidation capacity.

2. Sex

The microsomal glucuronidation of *o*-aminophenol in the liver is markedly higher in male than in female rats and is stimulated by testosterone in the latter and inhibited by estriol in the former (Inscoe and Axelrod, 1960). In Sprague-Dawley rats hepatic UDPG-dehydrogenase and transferase activities, and UDPGA concentration are higher in males than in females (Müller-Oerlingenhausen and Künzel, 1968). However, urinary excretion of glucuronides or free glucuronic acid is the same in both sexes under basal conditions, during barbiturate stimulation and after a salicylamide load, suggesting that the activity of the glucuronate pathway *in vivo* is unaffected by the sex of the rat (Nitze and Remmer, 1962). Salicylamide may be glucuronized even better by females *in vivo* (Nitze and Remmer, 1962), which might be related to high renal transferase activity (Huttunen and Miettinen, 1965), because the hepatic sex difference in transferase activity, although frequently found by many authors (see Dutton, 1966), is not an invariable observation, at least in Wistar rats (Hartiala and Pulkkinen, 1964). Testosterone glucuronidation is also unaffected by sex (Schriefers *et al.*, 1966). The same applies to thyroxine (Goldstein and Taurog, 1968). On the other hand, higher activities of aldono- and uronolactonases, glucuronate reductase, and gulonate oxidase in the liver of male than of female rats are in agreement with the better ascorbate production by the former sex (Stubbs and McKernan, 1967). That males generate glucarate at a higher rate than females is indicated by the higher activity of glucuronolactone dehydrogenase in their livers and the larger amount of glucaric acid in the urine (Marsh, 1963a; Marsh and Carr, 1965).

3. Species and Strain

As compared to other mammals, the cat is a very peculiar animal because following administration of many compounds, its urine contains only traces of their glucuronides (Robinson and Williams, 1958). Since the ability to form UDPGA appears to be normal (Dutton and Greig, 1957), the low glucuronidation capacity (Hartiala, 1955) is attributable to markedly reduced transferase activity (Dutton and Greig, 1957). Chloramphenicol, however, is instantly glucuronized by the cat (Kunin et al., 1959). Methylcholanthrene or tolerable doses of barbital or chloretone are not able to increase ascorbate excretion, suggesting that inducer drugs are unable to stimulate the pathway, probably owing to inactivity of the transferase. This inability also supports the view that the transferase route plays an important role in the release of glucuronate from UDPGA (Hollmann and Touster, 1962), generation of which according to its hepatic concentration, appears to be normal in this animal (Dutton and Greig, 1957). The extent to which glucuronate is utilized by the glucarate and L-xylulose routes in the cat is not known. The absence of the ascorbate route in man, monkeys, and guinea pig has been discussed in Section IV, B.

Drug metabolism varies between strains of the same animal species, as has been demonstrated for the half-life of antipyrine in various rat strains (Quinn et al., 1958) and for the varying effect of hexobarbital in different mouse strains (Jay, 1955). As indicated by urinary glucuronide excretion, the in vivo glucuronidation process takes place more efficiently in Sherman than in Wistar rats (Mosbach et al., 1950), and in Wistar than in Sprague-Dawley rats (Huttunen and Miettinen, 1965). The latter finding is attributable to higher UDPG dehydrogenase and transferase activities in the liver of the Wistar rats. The finding suggests that the production of UDPGA and its hydrolysis to free glucuronate is also higher in the former than in the latter strain, particularly because hepatic β-glucuronidase activity is high in the Wistar rats (Huttunen and Miettinen, 1965). This assumption also fits with the higher rate of ascorbate excretion of the latter animals (Leskinen, 1964). In the Wistar rats hypophysectomy prevents the chloretone-induced augmentation of ascorbate excretion, this inhibition being incomplete, however, in the Sprague-Dawley rats (Hollmann and Touster, 1962), suggesting that hypophyseal hormones or their induced change in the fatty acid or glucose metabolism have a different effect on the glucuronate pathway in the two strains. The action may also be mediated by lack of thyroxine but not by lack of adrenal hormones, for adrenalectomy has no effect on the drug-provoked ascorbate output (Burns et al., 1957).

Among the rat strains, Gunn rats form the most extreme variant, for they are unable to produce bilirubin glucuronides and cannot even excrete other glucuronides in large amounts, owing to low transferase activity (Schmid *et al.*, 1958). Human subjects with transferase deficiency are discussed in Section VII.

4. *Diet, Fasting, and Diabetes*

No systematic studies on the effect of a high-fat low-carbohydrate diet or low-fat high-carbohydrate diet on the glucuronate pathway have been carried out. During the former diet, as during fasting and diabetes, the energy requirement of the body is met primarily by fatty acid oxidation (cf. Krebs, 1966), which is associated with enhanced ketone body production, augmented gluconeogenesis, and depletion of liver glycogen. Accordingly, the availability of glucose for UDPG and further for the glucuronate pathway might be limited. In addition, reduced availability of insulin may as such reduce the formation of UDPGA (Müller-Oerlingenhausen *et al.*, 1967a,b). Consequently, during fat-feeding, the glucuronate pathway shows a predictably slow rate of activity, while with a high-carbohydrate diet the effect may be the opposite. As a matter of fact, though subjects on a fat-rich diet have been reported to excrete increased amounts of urinary glucuronides (Hollmann and Wille, 1952; diet probably contained exogenous aglycones), administration of glucose or of fructose stimulates glucuronide production in experimental animals, in man, and *in vitro* (Lipschitz and Bueding, 1939; Südhof, 1952, 1954a,b; Heyde and Wieland, 1960; Beck and Richter, 1962b; Taketa, 1962d). Furthermore, the feeding of carbohydrates either as a supplement or as an isocaloric substitution for fat on a 45% fat diet significantly increases ketopentose excretion in man (a finding not confirmed by others; Roe, cited in Baker *et al.*, 1960), suggesting that the release of free glucuronate and its further metabolism, at least via the xylulose route, is also enhanced by the carbohydrate-rich diet (Baker *et al.*, 1960).

Fasted animals are able to produce glucuronides *in vivo* but the capacity is reduced and they are more susceptible to drugs than the fed ones (Thierfelder, 1886; Schmid, 1936). During fasting, a markedly reduced UDPG dehydrogenase activity would limit glucuronidation via an inadequate supply of UDPGA (Miettinen and Leskinen, 1963). In addition, a reduced amount of UDPG caused by decreased availability of insulin and glucose, and an increased NADH/NAD ratio, limiting conversion of UDPG to UDPGA (Zalitis and Feingold, 1968), are factors which might be even more important than reduced UDPG dehydrogenase activity. A restoration of testosterone glucuronidation within 2 hours after injection of glucose into fasted animals (Schriefers *et al.*, 1965) also suggests that the latter factors may be involved, because enzyme protein synthesis was hardly resumed so fast. Transferase (Miettinen and Leskinen, 1963; Schriefers *et al.*, 1966) and β-glucuronidase activities remain unchanged or even increase slightly, while pyrophosphatase activity is slightly decreased (per whole organ), suggesting that the release of glucuronic acid now takes place primarily via the transferase route (Miettinen and Leskinen, 1963). If UDPGA is a limiting factor, the release of free glucuronate may be reduced, which would explain, in association with decreased gulonolactone oxidase activity (Stirpe and Comporti, 1965; Stirpe *et al.*, 1965), the diminished urinary ascorbate excretion (Musulin *et al.*, 1939; Roberts and Spiegl, 1947; Leskinen, 1964), and its reduced response to drug stimulation during starvation (Hollmann and Touster, 1962), suggesting that the ascorbate route is inhibited by fasting. On the other hand, increased gulonate dehydrogenase activity suggests that the xylulose route might even have been stimulated under this condition (Stirpe and Comporti, 1965; Stirpe *et al.*, 1965). If hydrolysis of available UDPGA is increased in relation to its utilization for the net glucuronidation and if the ascorbate route is inhibited, an increased amount of free glucuronate could be metabolized through the xylulose route during fasting as compared to the normal state and might actually contribute to the metabolism of glucose, utilization of which via both glycolysis and the pentose shunt is inhibited.

The similarity of the general metabolic profile in fat-feeding, diabetes, and fasting suggests that the activity of the glucuronic acid pathway may change analogously. As a matter of fact, the liver of diabetic rats exhibits a markedly reduced UDPG dehydrogenase activity (not found in the skin, Müller-Oerlingenhausen, 1969), decreased testosterone and *o*-aminophenol glucuronidation, and a decreased UDPGA content but unchanged UDPGA transferase activity (Müller-Oerlingenhausen *et al.*, 1967b; Schriefers *et al.*, 1966). Insulin-deficient animals excrete much less administered bilirubin into the bile as bilirubin glucuronide in experiments *in vivo* and in liver perfusions than nondiabetic ones (Müller-Oerlingenhausen *et al.*, 1967a). Thus, reduced UDPGA production through decreased UDPG dehydrogenase activity was suggested to be the factor limiting glucuronidation (Müller-Oerlingenhausen *et al.*, 1967b), particularly as the breakdown of UDPGA via pyrophosphatase is only slightly elevated in the diabetic liver (Müller-Oerlingenhausen, 1969). It may be that the action of insulin is not solely to supply more glucose to the pathway but to activate UDPG dehydrogenase directly (Müeller-Oerlingenhausen *et al.*, 1967b). Administration of insulin or tolbutamide to normal rats and mice results in enhanced biliary excretion of bilirubin glucuronide, increased hepatic bilirubin glucuronidation and UDPGA content, and unchanged UDPG dehydrogenase and transferase activities (Müller-Oerlingenhausen *et al.*, 1968). Furthermore, administration of insulin to diabetic dogs greatly increases urinary glucuronide excretion (Quick, 1926). It is apparent from these studies that the conjugation procedure of the glucuronic acid pathway is reduced by insulin deficiency in experimental animals. Increased ascorbate excretion by diabetic animals (Straumfjord and West, 1957), high serum and urinary xylulose levels and their normalization by insulin (Winegrad and Burden, 1965, 1966), and increased urinary output of glucarate by diabetic patients (Okada, 1969) indicate, however, that the hydrolysis of UDPGA to free glucuronate and the further metabolism of the latter via all three pathways is augmented in diabetes (see also Section IV, C); according to Stirpe and Comporti (1965), the xylulose route is enhanced in diabetes as in fasting, while the ascorbate route is inhibited in both conditions.

Thus there is some evidence, though not conclusive, suggesting that channeling of glucose via the glucuronate pathway and particularly via its L-xylulose route is enhanced in the three conditions. This may indicate that more glucose is shunted via the latter route to the pentose cycle, the activity of which is lessened primarily by reduction of insulin-controlled G-6-P dehydrogenase activity (cf. Weber *et al.*, 1967), and that utilization of reduced NADP is also increased in NADP-dependent reactions of the glucuronate pathway. Thus, the availability of NADPH, which is already diminished by inhibition of the pentose cycle, is still more reduced and may be a contributory factor to the reduction of fatty acid synthesis, which is known to be marked in the three conditions (cf. Numa *et al.*, 1965).

5. *Endocrinological Factors*

The effects of sex hormones and insulin have already been discussed. Of the other endocrine factors thyroid hormones have been shown to influence the glucuronic acid pathway. Thus, glucuronide excretion and the capacity to produce glucuronides after administration of an exogenous aglycone is markedly increased in rats in which hypothyroidism has been provoked by either radioiodine or thiouracil derivative (Milcu and Biener, 1958; Moudgal *et al.*, 1959; Miettinen and Leskinen, 1967). Pentose (Roe and Coover, 1950) and ascorbate excretion, and the response of the latter to menthol or glucuronolactone loads are reduced in experimental hypothyroidism (Miettinen and Leskinen, 1967). This state enhances or has no effect (Hartiala found no effect in males and a decrease in females; Hartiala and Hirvonen, 1956) on hepatic glucuronidation or transferase activity (Moudgal *et al.*, 1959; Werder and Yaffe, 1964; Miettinen and Leskinen, 1967). UDPG dehydrogenase activity also remains unchanged, suggesting that the supply of UDPGA to the pathway is adequate. Reduced activities of hepatic β-glucuronidase and pyrophosphatase may stimulate glucuronidation *in vivo* even in the presence of unchanged glucuronyl transferase activity, because breakdown of UDPGA and hydrolysis of newly formed glucuronides may have been reduced. Thus, in hypothyroidism the glucuronic acid pathway may serve primarily for detoxication purposes, the release of free glucuronic acid and particularly its further metabolism to at least pentoses and ascorbate being markedly reduced (Miettinen and Leskinen, 1967). It is conceivable that the retarded rate of general metabolism in hypothyroidism is reflected in a decreased need to metabolize glucose through the ascorbate and L-xylulose routes. Propylthiouracil fed to pregnant mice did not activate or suppress the transferase activity of the offspring (Werder and Yaffe, 1964).

Though experimental hyperthyroidism has been reported to depress the capacity for glucuronidation *in vivo* and *in vitro* (Moudgal *et al.*, 1959), other studies indicate that in rats both basal glucuronide excretion and the glucuronidation capacity after a menthol load are actually increased by administration of thyroxine to euthyroid animals or to individuals pretreated with propylthiouracil or radioiodine (Milcu and Biener, 1958; Miettinen and Leskinen, 1967). After thyroid hormone treatment, glucuronidation of *o*-aminophenol in rat liver slices was reported to be diminished by Moudgal *et al.* (1959), while Hartiala *et al.* (1956) found that thyroid treatment promoted this capacity in rats and rabbits. Total hepatic UDPG dehydrogenase activity is increased by thyroxine (Miettinen and Leskinen, 1967) even though UDPG (Reid, 1960, 1961a,b) and NAD (Glock and McLean, 1955) levels have been reported to be decreased. In addition, transferase activity is also increased by thyroxine in the liver (Reid, 1960; Miettinen and Leskinen, 1967) and intestine, suggesting that the enhanced glucuronidation *in vivo* is attributable to elevated activities of UDPG dehydrogenase and transferase, particularly because those of β-glucuronidase and pyrophosphatase remained practically unchanged (Miettinen and Leskinen, 1967). As compared to hypothyroidism, this enzyme pattern allows increased glucuronate release and may further activate the metabolism of free glucuronic acid. Elevated pentose excretion in animals and human subjects treated with thyroid hormone (Coover *et al.*, 1950; Roe and

Coover, 1950; Baker *et al.*, 1960) indicates that the xylulose route was
actually enhanced. Though thyroxine is glucuronized itself and excreted
primarily into the bile (cf. Goldstein and Taurog, 1968), it may not, even
in excessive doses, stimulate the glucuronate pathway as a drug. Stimulation is more likely to be due to general acceleration of the metabolism
or associated more specifically with activation of the NADP-dependent
reactions of the glucuronic acid pathway by an excess of NADPH. The
augmented generation of the latter nucleotide in hyperthyroidism takes
place via the enhanced pentose cycle (Glock *et al.*, 1956).

As already mentioned, lack of pituitary hormones prevents drug-
induced stimulation of the glucuronate pathway. The induction is not
apparently mediated by adrenal hormones because adrenalectomy has
no effect on it (Burns *et al.*, 1957). The formation of bilirubin glucuronide appears to be normal in hypophysectomized rats while the excretion
of bilirubin glucuronide is disturbed (Gartner, 1967). Hypophysectomy
as such reduces enzyme activities of the ascorbate route in male but not
in female rats (Stubbs *et al.*, 1967). The latter procedure markedly inhibits hepatic *o*-aminophenol glucuronidation (Gartner, 1967). Adrenalectomy prevents tolbutamide-induced stimulation of the conjugation
procedure (Müller-Oerlingenhausen *et al.*, 1969). Substitution treatment
with corticoids restored the tolbutamide effect. Though adrenalectomy
has been found to inhibit glucuronidation of borneol, *in vivo* administration of cortisone to adrenalectomized rats is not followed by improved
conjugation, while in normal rats both borneol glucuronidation and ascorbate excretion are markedly enhanced by cortisone (Tsutsumi *et al.*,
1966). Contrary to this observation, cortisol treatment has also been
found to reduce UDPG dehydrogenase activity (Marchi *et al.*, 1964).
Treatment of pregnant rats with ACTH stimulates bilirubin glucuronidation of offspring (Careddu *et al.*, 1961), although this finding is not
consistent or recorded after treatment with pituitary extract or cortisol
(Flint *et al.*, 1963). Beck studied the relation between glucuronidation
capacity of the rabbit and cortisone-induced enhancement of gluconeogenesis. During 3 hours after administration of glucocorticoids, when
gluconeogenesis was considered to be enhanced, the glucuronidation
capacity was decreased (Beck *et al.*, 1964). The latter was promoted,
however, by prolonged cortisone treatment (Beck *et al.*, 1967), which is
known to activate gluconeogenetic key enzymes during several days after
commencement of the treatment (Ashmore and Weber, 1968).

6. *Experimental Conditions*

Liver damage caused by CCl_4 results in a slight transitory decrease in
glucuronidation capacity and transferase activity, but the change appears

to be variable for different aglycones (cf. Dutton, 1966; Taketa, 1962b). Since conjugation also takes place outside the liver, glucuronidation *in vivo* may be shifted in liver damage, at least for some aglycones, to the kidney and intestine (Beck and Kiani, 1960). Thus, following hepatectomy, glucuronidation of thyroxine derivatives takes place in the kidney (Béraud and Vannotti, 1960), though normally its conjugation occurs primarily in the liver. As a result of the high capacity of UDPG dehydrogenase and transferase, and glucuronidation in extrahepatic tissues, primarily in the kidney and intestine, liver damage may sometimes be relatively extensive and yet the change in urinary glucuronide excretion may be minimal. The effect of liver injury on the xylulose and glucarate routes and the simultaneous contribution of extrahepatic tissues to these routes are not known.

Nephrectomy does not prevent drug-induced stimulation of ascorbate excretion (Burns *et al.*, 1960) because the latter is formed in the liver. Nephrosis, which in man is associated with reduced basal glucuronide excretion (Müting *et al.*, 1963), when induced in the rat with puromycin, is accompanied by a marked activation of hepatic, but not renal or intestinal transferase during the early stage of the ailment (Miettinen and Leskinen, 1968). Pyrophosphatase activity remained unaffected though it tended to decrease in the kidney. Since the primary action of puromycin is obviously exerted on the glomerular basal membrane, the increase in the activity of hepatic enzyme might be secondary to the marked metabolic alteration involving protein, carbohydrate, and lipid metabolism in consequence of the enormous loss of protein into the urine.

VII. Clinical Conditions

A. Weight Reduction on Total Fast

Total fast for long periods of time has been widely used for the treatment of intractable obesity (Duncan *et al.*, 1964). Glucuronidation capacity, as judged from the results of animal experiments and the favorable effect of glucose and fructose in man (cf. Section V, A), may be markedly reduced under such conditions. The cat in which the capacity for glucuronidation of many drugs is markedly reduced is sensitive to foreign compounds (Hollmann and Touster, 1962). In addition, studies by Dixon *et al.* (1960) suggested that oxidation, dealkylation, and hydroxylation reactions were limited for many drugs during fasting and were obvious reasons for the depressed drug tolerance of mice during caloric restriction. Accordingly, obese patients on a total fast could also be especially sensitive to foreign compounds, i.e., not only glucuronidation but also other detoxication mechanisms could have been slowed down. Bromsulfthalein, which is removed as its glutathione conjugate, is eliminated at a slower rate during fasting than during an eucaloric diet (Duncan *et al.*, 1964). Since during our clinical work we have encountered an obese woman who had been on contraceptive pills for many years without side effects, but developed severe liver damage when placed on total fast, the glucuronidation capacity of subjects on caloric restriction was studied more closely.

TABLE V

EFFECT OF TOTAL FAST ON DETOXICATION OF SALICYLAMIDE IN
OBESE HUMAN SUBJECTS

Collection period (hours)	Urinary metabolites, % of dose[a]					
	Glucuronides		Sulfates		Total	
	Fed	Fasted	Fed	Fasted	Fed	Fasted
0–1	4	0	2	0	6	0
1–2	6	3	9	3	15	6
2–4	12	26	14	16	26	42
4–8	12	14	7	9	19	23
8–24	6	8	5	7	11	15
0–24	40	50	37	36	77	86

[a] Average values from five patients fasted for 5–14 days. Salicylamide (1.5 gm) was given orally during and 2 days after the fasting period.

Basal glucuronide excretion was found to decrease during fasting by 40%, probably owing to decreased glucuronidation or, more likely, to reduced intake of aglycones in the form of proteins. The few transferase assays carried out on liver biopsy specimens of fasted subjects suggest that its activity with *p*-nitrophenol as substrate is actually decreased by 30%. Addition of EDTA to the incubation medium inhibited transferase activity similarly in specimens from both normal and fasted subjects. Pyrophosphatase activity did not show any change. Surprisingly, however, total 24-hour recovery of administered salicylamide (1.5 gm) was increased from 77 to 86% of the dose during fasting (Table V), the increment being caused by the glucuronide fraction. No free salicylamide was found, so that improved recovery was probably due to decreased formation of the hydroxylated derivative, which was not measured by the salicylate method used (see Levy and Matsuzawa, 1967). This fraction was glucuronized, suggesting that the capacity for additional sulfate conjugation was limited. Table V shows that excretion of glucuronide and sulfate conjugates was delayed during fasting, obviously owing to retarded intestinal absorption, delayed renal excretion, or more likely, reduced rate of conjugation. Delayed urinary excretion of glucuronides in loading tests has been considered to indicate impaired glucuronidation capacity (Beck and Kiani, 1960). Accordingly, glucuronidation and sulfate conjugation of salicylamide as well as its detoxication via hydroxylation appear to be limited during fasting. The availability of sulfate for salicylamide conjugation is rate limiting even in the normal state (Levy and Matsuzawa, 1967). This may be markedly aggravated during fasting, when the intake of essential sulfur-containing amino acids is limited. Since hydroxylation, and sulfate and glucuronic acid conjugation are involved with "detoxication" of steroid hormones (Jayle and Pasqualini, 1966; Schriefers, 1967), the metabolic alteration of the latter compounds could be retarded during fasting, particularly because hepatic transferase activity is reduced. The change in the latter enzyme activity suggests that the release of free glucuronate and further metabolism of this acid via the L-xylulose and glucarate routes may also be depressed, though direct evidence is lacking and animal experiments suggest that the xylulose route is actually stimulated (see Section VI, B, 4).

B. DISTURBANCES IN BILIRUBIN METABOLISM

1. Newborn Infants

Since it became apparent that bilirubin was excreted from the body as its glucuronide (Schmid, 1956; Billing and Lathe 1956; Billing et al., 1957; Talafant, 1956) and that the glucuronidation process was weakly developed in the newborn of many animal species (see Section VI, B, 1), hyperbilirubinemia of newborn infants was in most cases due to defective glucuronidation of bilirubin. In vivo and in vitro studies have shown that newborn children also have a limited capacity for glucuronidation of other compounds (Brown, 1957; Vest, 1958a,b; Vest and Streif, 1959) and that this deficiency, like that of bilirubin conjugation, is primarily due to transferase deficiency, though the supply of UDPGA via poorly developed UDPG dehydrogenase activity may also be inadequate in vivo (Brown et al., 1958). Parenteral administration of UDPG to newborn infants actually reduces serum bilirubin, probably via improved UDPGA production (Careddu and Marini, 1968). Determinations of transferase activity from liver biopsy specimens suggest that the adult level is attained at the age of about 8 weeks (Di Toro et al., 1968). Glucuronidation of N-acetyl-p-aminophenyl in vivo is fully developed at the age of 3 months (Vest, 1958a,b).

Hyperbilirubinemia due to transferase deficiency may also be an inherited disorder. This group of hyperbilirubinemias may consist of a mixture of different syndromes, of which the best known is familial non-hemolytic jaundice of the newborn, also called the Crigler-Najjar syndrome (Crigler and Najjar, 1952). This disorder is due to a recessive gene and in many respects resembles the state in Gunn rats in which bilirubin glucuronidation is practically nonexistent but in which the rats are able to conjugate many other compounds at a limited rate (Arias et al., 1958). In a few cases hyperbilirubinemia of newborn infants has been attributed to inhibition of transferase by steroid hormones transported to the infant from the mother via the milk (Arias et al., 1964).

Inducer drugs are able to stimulate production of both UDPGA transferase at least in experimental animals (see Table III) and even in the Gunn rats (Hollmann and Touster, 1962), though glucuronidation of bilirubin and the serum bilirubin level are little affected by 3,4-benzpyrene in these animals (Metge et al., 1964). Because in addition to these experiments these drugs were shown to enhance excretion of some glucuronogenic compounds (Remmer, 1964) and also to stimulate glucuronidation of bilirubin (Inscoe and Axelrod, 1960; Arias et al., 1963b), attempts have been made to induce glucuronidation also in newborn infants. Phenobarbitone treatment of children with congenital nonhemolytic jaundice was found to enhance salicylamide glucuronidation and to decrease serum bilirubin concentration by enhancing its excretion into the bile (Crigler and Gold, 1966, 1967, 1969; Yaffe et al., 1966). Studies on administration of this drug to mothers before term and to newborn infants after birth, and on newborn infants born to epileptic mothers treated with phenobarbitone and other anticonvulsants have revealed that in those babies serum bilirubin levels are significantly lower than in nontreated controls (Trolle, 1968a,b; Maurer et al., 1968). Though discordant results have also been obtained (Walker et al., 1969; Cunningham et al., 1969), it has been suggested that phenobarbitone stimulates transferase activity, thus leading to enhanced bilirubin excretion and to a decrease in its serum level. Administration of chloroquine to pregnant rats elevates glucuronyltransferase activity in the offspring but fails to diminish neonatal hyperbilirubinemia in infants of treated women (Arias et al., 1963c).

Neonatal hyperbilirubinemia has also been treated with glucuronolactone infusions (Danoff et al., 1958). But the reduction of serum bilirubin levels observed has mainly been attributed to dilution, because it was also obtained with a saline infusion (Schwob *et al.*, 1960). Though the glucarate pathway may be imperfectly developed in newborn infants (Marsh and Carr, 1965), the possibility exists that some glucarate was formed from the lactone (see Section IV, C) and that it might have had a favorable effect on glucuronidation by inhibiting a possible β-glucuron-idase-induced hydrolysis of bilirubin glucuronides. Barbiturates stimulate glucarate synthesis not only in experimental animals (Marsh and Reid, 1963) but also in adult man (Aarts, 1965; Okada, 1969) and might do so in newborn infants, too. This might offer an additional explanation for the phenobarbitone-induced reduction in bilirubin levels presented above, particularly because conjugated bilirubin introduced into the bloodstream is known to be deconjugated, probably via hydrolysis by β-glucuronidase (Acocella *et al.*, 1968; Okolicsanyi *et al.*, 1968).

Diethylnicotinamide, which stimulates bilirubin conjugation in neonatal rabbits (Careddu *et al.*, 1964), probably stimulates the glucuronate pathway in the same way as nicotinamide (see Section V) and has been reported to be effective in lowering serum bilirubin levels in newborn infants also (Sereni *et al.*, 1967).

2. *Other Conditions*

The use of inducer drugs, primarily phenobarbitone, in the treatment of jaundice has been suggested to be unsuccessful in lowering serum bilirubin levels in cases with a complete transferase deficiency for bilirubin (DeLeon *et al.*, 1967). Thus, homozygous Gunn rats and children with severe transferase deficiency (DeLeon *et al.*, 1967) fail to respond to the treatment. In patients with the Crigler-Najjar-type hyperbilirubinemia, a clear-cut effect has been reported. Thus, serum bilirubin decreased by 40–70% in two children (Yaffe *et al.*, 1966; Crigler and Gold, 1966, 1967), the decrement being caused by an enhanced rate of bilirubin excretion from the body pool and not by expansion of the latter (Crigler and Gold, 1967, 1969). Excretion of salicylamide glucuronides, but not of salicylate glucuronides, was also improved simultaneously (Yaffe *et al.*, 1966). Whelton *et al.* (1968) reported a marked reduction in the serum bilirubin level (from 8 to 2 mg %) in a hyperbilirubinemic woman by phenobarbital. Two adult patients with some variant of the Crigler-Najjar syndrome reduced their serum bilirubin in response to phenobarbitone treatment from 12.0–13.7 to 4.6 mg % and from 7.6–8.0 mg % to a minimum of 0.3 mg %, respectively (Kreek and Sleisenger, 1968). Menthol glucuronidation, which was inadequate under basal conditions, also appeared to be improved by the therapy. Since recent studies have revealed that the transferase activity for bilirubin is reduced in patients with Gilbert's syndrome (Arias, 1962; Metge *et al.*, 1964; Black and Billing, 1968), it is to be expected that those cases will respond favorably to induction therapy, provided that the reduced uptake is not a rate-limiting step. There is no information on the release of free glucuronate and its further metabolism via the L-xylulose and glucarate routes in patients with transferase deficiency or on the effect of the treatment on these phenomena. Ascorbate excretion in Gunn rats and the effect on it of inducer drugs suggest, however, that these functions may also occur at a reduced rate in the basal state but that stimulation could be obtained with phenobarbitone (Hollmann and Touster, 1962).

In patients with intrahepatic cholestasis due to postnecrotic or primary biliary cirrhosis, phenobarbitone resulted in a reduction of up to 50% in the serum bilirubin level, suggesting that transferase was activated either in the liver or outside it (for enzyme activation by different drugs see Table III), that other enzymes metabolizing bilirubin were activated or that bilirubin excretion via other routes was increased (Thompson and Williams, 1967). The serum bilirubin levels in epileptics are also lower than in normal subjects, perhaps owing to persistent induction of transferase or stimulated biliary excretion of bilirubin (Thompson *et al.*, 1969). Since these patients also exhibit elevated serum and urinary glucarate levels (Okada, 1969), obviously as a response to anticonvulsants, which usually contain barbiturates or their derivatives, the change in the bilirubin level of patients with epilepsy and obstructive jaundice might to some extent be due to the protective action of glucarate against hydrolysis of bilirubin glucuronides by β-glucuronidase. Conjugated bilirubin has been reported to be hydrolyzed in normal and bile-duct-ligated Wistar and Gunn rats (Acocella *et al.*, 1968; Okolicsanyi *et al.*, 1968). Though experimental cholestasis markedly reduces bilirubin glucuronidation *in vitro* (Metge *et al.*, 1964), the liver of patients with obstructive jaundice shows an almost normal or even increased activity (Black and Billing, 1968). Reduced or normal activity has been demonstrated in a variety of parenchymal liver diseases (Metge *et al.*, 1964; Black and Billing, 1968).

C. Endocrinological Factors

Since glucose metabolism via the phosphorylated pathways is more or less disturbed in diabetes mellitus, it has been suggested that under this condition glucose is shunted to insulin-independent pathways (Spiro, 1963). This might result in increased formation of amino sugars and glucuronic acid. The former is incorporated into glycoproteins and both these sugars are components of acid aminopolysaccharides. The association of glycoproteins with the vascular lesions of human diabetes (i.e., that polysaccharide-containing material accumulates in the capillary basement membranes) is well known. Because of the intimate relationship between the nucleotides of amino sugars and glucuronic acid, it has also been assumed that the activity of the glucuronic acid pathway may have been changed (Winegrad and Burden, 1965, 1966), particularly because the relative contribution of the glucuronic acid pathway to CO_2 production is greater in the adipose tissue of fasted and diabetic rats than of normally fed rats and because this change is restored by insulin (Winegrad and Shaw, 1964). As a matter of fact, serum xylulose levels have been found to be elevated in patients with diabetes mellitus, normalized values being obtained after administration of insulin, suggesting that the glucuronate pathway is overactive in diabetes mellitus (Winegrad and Burden, 1965, 1966). In support of this view, urinary glucarate excretion is also elevated in diabetes (Okada, 1969), indicating, in conjunction with the elevated serum L-xylulose level, that the release of free glucuronate is increased. The latter may be related to the elevated serum β-glucuronidase activity frequently found in diabetes (cf. Dohrmann, 1968) and other states associated with atheromatous manifestations (Miller et al., 1967). This enzyme might stimulate the release of glucuronate from UDPGA by splitting glucuronides within the cell, thus offering an additional explanation for the increased activity of the L-xylulose and glucarate routes in diabetes. The increased serum activity of β-glucuronidase has been assumed to reflect the atherosclerotic process when β-glucuronidase can participate in the degradation of mucopolysaccharides accumulated in vascular walls (Miller et al., 1966) primarily in the form of glycoproteins (Spiro, 1963).

Stimulation of lipolysis and subsequently increased flow of free fatty acids into the liver might be one of the primary reasons that in diabetes, as after administration of growth hormone and epinephrine to diabetic and nondiabetic subjects (Winegrad and Burden, 1965, 1966), more glucose is shunted to the glucuronate pathway. Though animal experiments suggest that lack of insulin inhibits the activity of the initial enzymes of the pathway (Müller-Oerlingenhausen et al., 1967b), patients with diabetic coma excrete large amounts of glucuronic acid into the urine (Brox, 1953; Müting, 1964) probably as a partial consequence of increased formation of endogenous aglycones via enhanced protein catabolism. Though the actual amount of free glucuronate was not determined, these findings indicate that the glucuronidation process takes place even during extreme deterioration of glucose metabolism in man. In nontreated mature-onset diabetics or most subjects with diabetic coma, basal glucuronic acid excretion is reduced, therapy with diet or tolbutamide being followed by a significant increase of the values (Müting, 1964). In agreement with enzyme activities of animal experiments (Müller-Oerlingenhausen *et al.*, 1967b), nontreated diabetic patients have been demonstrated to have a reduced glucuronidation capacity after a menthol load, a finding related to the diminished glycogen content of the liver (Vescia, 1945).

The above-mentioned findings suggest that the glucuronidation capacity is actually reduced in diabetic patients as in total fast and experimental diabetes, the release of glucuronic acid for metabolism via the glucarate and particularly L-xylulose routes being increased.

There appears to be a reciprocal relationship between thyroid hormones, drugs, and the glucuronate pathway, manifesting itself in at least three different ways. (1) Administration of drugs such as salicylates is known to replace thyroxine from thyroid-binding protein, which is followed by increased plasma-free thyroxine and its increased excretion into the bile as glucuronides, by decreased plasma PBI, and by reduced iodine uptake into the thyroid gland probably due to depressed TSH secretion in the presence of increased free plasma thyroxine (Austen *et al.*, 1958; Wolff and Austen, 1958; Osorio and Myant, 1963; Good *et al.*, 1965). In spite of these marked changes in thyroid status, the patients on long-term salicylate treatment are not hypothyroid. Thyroid hormone metabolism is similarly changed in hyperthyroid patients, but owing to the high free-thyroxine level, the decrement of PBI following salicylate administration is not associated with improved clinical status of the patients (Good *et al.*, 1966). (2) The use of inducer drugs may activate the glucuronate pathway stimulating the excretion of thyroid hormones into the bile without interfering with the binding of the hormone to plasma proteins. Thus, benzpyrene and methylcholanthrene augment the biliary excretion of thyroxine in the rat by increasing its hepatic glucuronidation (Newman and Moon, 1967; Goldstein and Taurog, 1968), while the effect of barbiturates and phenylhydantoins might partly be due to increased liver size (Goldstein and Taurog, 1968; Mendoza *et al.*, 1966). In man, barbiturates elevate PBI and increase radioactive iodine uptake by the thyroid gland (cf. Lubran, 1969). (3) Glucuronide excretion following a salicylamide load is markedly reduced in hypothyroid patients, substitution therapy with hormones or the hyperthyroid state accelerating this detoxication process (Miettinen and Leskinen, 1968). Since, in addition, the administration of thyroid hormones to euthyroid human subjects is followed by a marked increase of urinary L-xylulose excretion (Baker *et al.*, 1960), it could be concluded that in man glucuronidation and the glucuronate pathway are accelerated in the hyperthyroid state and retarded in hypothyroidism. Administration of triiodothyroxine to a child with a transferase defect did not reduce the elevated serum bilirubin level; on the contrary, the values tend to increase on a higher dose (Crigler and Gold, 1966). It is a well-known clinical finding that in the severe form of thyrotoxicosis, hepatic dysfunction, including jaundice, may occur. Furthermore, since such cases, owing to increased basal metabolic rate, are in a state of relative caloric deprivation, the glucuronidation capacity may even be decreased.

Administration of cortisone or ACTH to patients with rheumatoid arthritis, whose glucuronide excretion tends to be subnormal, results in a significant elevation of the urinary glucuronide values (Holopainen, 1958).

D. Liver Damage and Other Conditions

Glucuronic acid metabolism has been studied extensively in patients with liver injury by measuring urinary glucuronide excretion under basal conditions and after loading with glucuronogenic compounds, by assaying transferase activity from liver biopsy specimens or by determining utilization of administered glucuronolactone with the balance technique. Metabolites of the L-xylulose and glucarate routes did not appear to be studied, however.

Urinary excretion of glucuronides is either normal or even slightly increased in patients with hepatitis or cirrhosis of the liver (Nasarijanz, 1934; Südhof, 1954c; Barniville and Misk, 1959), though reduced values have also been reported (Sauer, 1930; Geller, 1952). Loading tests with some glucuronogenic compound have revealed reduced capacity to excrete glucuronides by many, but not all, cases with parenchymal liver diseases and also with obstructive jaundice, provided that the values during the first few hours after administration of the compound are compared with the respective normal ones (Nasarijanz, 1934; Snapper and Saltzman, 1947; Borgström, 1949; Barniville and Misk, 1959; Vest and Fritz, 1961; Taketa, 1962d). However, the tests appear to be less sensitive than other liver function tests and normal values can be obtained in the presence of a liver injury, probably because defective hepatic glucuronidation may be compensated by the kidneys (Beck and Kiani, 1960) and by the intestinal mucosa. Patients with increased unconjugated serum bilirubin, irrespective of etiology, frequently show abnormal glucuronide excretion in loading tests (Beck and Kiani, 1960;

Beck and Richter, 1962a), while those with anicteric cirrhosis carry out this process normally (Beck *et al.*, 1965), suggesting that elevation of free bilirubin is attributable to impaired glucuronidation. Glucurono-lactone, administered orally, is incompletely metabolized by many patients with liver damage, and consequently is excreted in abnormally large amounts into the urine (Südhof and Schellong, 1953; Südhof, 1954c).

Taketa (1962c) studied transferase activity for *p*-nitrophenol and β-glucuronidase activity as well as UDPGA concentration from liver biopsy specimens of patients with hepatitis and other liver diseases and compared these parameters with the *in vivo* glucuronidation of salicyl-amide. Glucuronyltransferase activity was reduced in hepatitis and cir-rhosis, while β-glucuronidase activity remained unchanged and the UDPGA content was decreased only in cirrhotic patients. Salicylamide glucuronidation *in vivo*, which was only slightly decreased as a result of the liver lesion, showed a positive correlation with hepatic transferase activity, suggesting that at least part of the salicylamide was conjugated in the liver. The hepatic UDPGA content was not correlated with salicylamide glucuronidation. Hepatic transferase activity with bilirubin as substrate appears to be reduced in obstructive jaundice, biliary cir-rhosis, constitutional hepatic dysfunction, hepatitis, hemochromatosis, and hemolytic anemias (Metge *et al.*, 1964). Black and Billing (1968), however, observed increased activity in cholestasis and unchanged activ-ities in disorders of the hepatic parenchyma.

Urinary glucuronic acid excretion has been measured in a great variety of other conditions (Fishman *et al.*, 1951; Holopainen, 1958; Cornillot, 1962). Though urinary excretion of large and small molecular carbo-hydrate-containing compounds increases nonspecifically in response to all kinds of experimental and clinical conditions with tissue destruction and/ or proliferation (Miettinen, 1961), augmentation of glucuronic acid ex-cretion seems to be more limited. Thus, rheumatic, cardiac, pulmonary, and infectious diseases show a normal (Cornillot, 1962) or even de-creased (Holopainen, 1958) glucuronide output, while subjects with extensive burns and skin diseases or cases with malignant neoplasms occasionally exhibit even enormously increased values (Cornillot, 1962; Cotton *et al.*, 1966). Because only the value for total glucuronides was determined, it cannot be concluded from these studies whether the in-crement was due to the free or the conjugated form. Since under these circumstances destruction of ground substance or connective tissue con-taining mucopolysaccharides might take place at an accelerated rate in the skin or malignant tissue, release of free glucuronic acid might actually have increased markedly. Increase in activeness of rheumatoid arthritis is accompanied by a tendency toward lower urinary glucuronide values, while cases with spondylarthritis ancylopoetica tend to excrete elevated amounts of glucuronides (Holopainen, 1958).

It has been suggested recently that β-glucuronidase is more active than normal in patients with etiocholanolone fever, converting nonpyrogenic conjugated etiocholanolone to the pyrogenic free form. Oral administration of glucuronic acid, its precursors, or its metabolities reduced the temperature probably via the inhibition of abnormal β-glucuronidase action (Herman et al., 1969).

E. Bladder Cancer

Carcinoma of the urinary bladder is known to be produced by chemical agents such as 2-naphthylamine (cf. Boyland, 1959). In the liver this compound is converted to its derivatives and conjugated with glucuronic acid. Hydrolysis of these conjugates by β-glucuronidase, primarily in the urinary tract, releases these carcinogenic metabolites and results in the formation of bladder cancer. Inhibition of this hydrolysis might prove to be an important preventive measure against human bladder cancer in subjects exposed to 2-naphthylamine or other industrial carcinogens. It might also have an important value in the treatment of already existing disease. Since β-glucuronidase is inhibited by saccharolactone, the latter compound has been used in clinical trials for patients with bladder cancer. No benefit was obtained, however, mainly because the compound was rapidly metabolized, so that the inhibition of β-glucuronidase remained unsatisfactory (Boyland et al., 1957). Japanese workers have studied the effectiveness of other glucarate derivatives and observed that administration of D-glucaro-1,4–6,3-dilactone and its diacetate markedly increase urinary glucarate excretion (Matsui et al., 1969a) and thus possibly also inhibition of β-glucuronidase. Accordingly, the use of these compounds in clinical trials for the purpose of preventing bladder tumor recurrences and further reducing tumor growth is in progress (Yonese et al., 1966; Yonese, 1968; Okada, 1969).

F. Clinical Use of Xylitol

Xylitol, which is formed endogenously from L-xylulose as a normal metabolite of the glucuronate xylulose cycle by mitochondrial NADP-linked dehydrogenase (see Section IV, A), is utilized rather efficiently by the animal organism after oral or parenteral administration (Mc-

Cormick and Touster, 1957; Bässler *et al.*, 1962; Schmidt *et al.*, 1964). Less than 10% of the dose is excreted as such into the urine by human subjects (Bässler *et al.*, 1962; Prellwitz and Bässler, 1963), the excretion of xylulose being only slightly increased (Bässler *et al.*, 1965). A considerable proportion of the xylitol is converted to glycogen in the liver, probably via the pentose cycle; a smaller amount is found in muscle glycogen in the normal state, but more in diabetes mellitus (Bässler and Heesen, 1963; Schmidt *et al.*, 1964). Because this pentitol did not appear to increase the blood sugar of diabetic patients (Mellinghoff, 1961; Bässler *et al.*, 1962; Prellwitz and Bässler, 1963), because it exerts an antiketogenic effect (Haydon, 1961; Bässler and Dreiss, 1963; Yamagata *et al.*, 1965), and especially because its utilization is insulin-independent (Bässler and Heesen, 1963), xylitol has been rather extensively studied by German and Japanese investigators (cf. Bässler, 1968), particularly in patients after surgery and those with diabetes mellitus. Though administration of xylitol stimulates insulin secretion in the dog (Kuzuya *et al.*, 1966; Hirata *et al.*, 1966), this effect seems to be less significant in man (Geser *et al.*, 1967; Maruhama *et al.*, 1967; Kuzuya et al., 1967).

Utilization of the pentitol is approximately the same in normal and diabetic subjects, as well as in patients with mild liver damage, while older healthy subjects and cases with severe liver injury tend to metabolize it at a lower rate, probably because the bulk of the xylitol is metabolized by the liver and only about 15% by extrahepatic tissues, muscle, and adipose tissue (Mehnert *et al.*, 1964; Müller *et al.*, 1967). Administration of xylitol is followed by a clear-cut increase in plasma lactate (Mehnert *et al.*, 1964; Yamagata *et al.*, 1965), by an unchanged pyruvate level, by a decrease in blood and urinary ketone bodies (Yamagata *et al.*, 1965), in plasma free fatty acids (FFA) (Yamagata *et al.*, 1965; Geser *et al.*, 1967)—even more than after a glucose load (Maruhama *et al.*, 1967)—and in inorganic phosphate (Geser *et al.*, 1967), and improvement in the state of comatose diabetic patients (Yamagata *et al.*, 1965). Xylitol is utilized by newborn as well as premature infants at a rate which is comparable to that found in adult subjects (Bässler *et al.*, 1966). In children with acetonemic reactions or with diabetic coma, acetonemia is markedly reduced by xylitol infusions (Toussaint *et al.*, 1966).

The mode of antiketotic action of xylitol is not known. The increased plasma FFA level caused by stimulated lipolysis of adipose tissue in diabetes, total fast, and fat-feeding results in augmented hepatic uptake of FFA followed by stimulated gluconeogenesis and enhanced conversion of FFA to ketone bodies (cf. Krebs, 1966). Xylitol, on the other hand, is known to be incorporated at a relatively high rate into the glyceride glycerol of adipose tissue (Mori *et al.*, 1967) and it also reduces FFA release from adipose tissue of diabetic rats *in vitro* (Maruhama *et al.*, 1967). Thus, a xylitol-induced decrease of plasma FFA and, further, its antiketogenic effect may primarily be due to inhibited lipolysis of adipose tissue. Improved glycolysis, changed hepatic nucleotide status, and stimulated fatty acid synthesis have also been suggested as explanations for the reduction of ketonemia by xylitol (Yamagata *et al.*, 1965). For other alternatives, see Section IV, A.

Xylitol is known also to be metabolized by human red cells *in vitro*, suggesting that the glucuronate-xylulose cycle functions normally in those cells (Asakura *et al.*, 1967). The absence of red cell microsomal enzymes, including glucuronyltransferase, indicates that the release of glucuronate for xylulose formation takes place in a special way. Since xylulose is formed from xylitol during incubation with red cells or hemolyzate, the latter pentitol is an efficient reductant of methemoglobin and results, as *in vivo* (Mehnert *et al.*, 1964; Yamagata *et al.*, 1965), in an accumulation of lactate in the red cells, obviously because of the increased NADH/NAD ratio (Asakura *et al.*, 1967). In addition, xylitol appears to be involved in many other biochemical processes in man and experimental animals. The recent progress made in unravelling its metabolism and physiology and in its clinical use have been treated in an international symposium held at Hakone, Japan (cf. Bässler, 1968).

ACKNOWLEDGMENT

A grant was received from Sigrid Jusélius Foundation during the preparation of this treatise.

REFERENCES

Aarts, E. M. (1965). *Biochem. Pharmacol.* **14**, 359.
Aarts, E. M. (1966). *Biochem. Pharmacol.* **15**, 1469.
Acocella, G., Tenconi, L. T., Armas-Merino, R., Raia, S., and Billing, B. H. (1968). *Lancet* i, 68.
Airaksinen, M. M., Miettinen, T. A., and Huttunen, J. (1965). *Biochem. Pharmacol.* **14**, 1019.
Albrecht, G. J., Bass, S. T., Seifert, L. L., and Hansen, R. G. (1966). *J. Biol. Chem.* **241**, 2968.
Anderson, L, and Coots, R. H. (1958). *Federation Proc.* **17**, 182.
Arias, I. M. (1961). *Biochem. Biophys. Res. Commun.* **6**, 81.
Arias, I. M. (1962). *J. Clin. Invest.* **41**, 2233.
Arias, I. M., Lowy, B. A., and London, I. M. (1958). *J. Clin. Invest.* **37**, 875.
Arias, I. M, Furman, M., Tapley, D. F., and Ross, J. E. (1963a). *Nature* **197**, 1109.

Arias, I. M., Gartner, L., Furman, M., and Wolfson, S. (1963b). *Proc. Soc. Exptl. Biol. Med.* **112**, 1037.

Arias, I. M., Gartner, L., Furman, M., and Wolfson, S. (1963c). *Ann. N.Y. Acad. Sci.* **111**, 274.

Arias, I. M., Gartner, L. M., Seifter, S., and Furman, M. (1964). *J. Clin. Invest.* **43**, 2037.

Arias, I. M., Gartner, L. M., and Seifter, S. (1967). *Proc. 2nd Intern. Congr. Hormonal Steroids, Milan, 1966* p. 570.

Arsenis, C., Maniatis, T., and Touster, O. (1968). *J. Biol. Chem.* **243**, 4396.

Artz, N. E., and Osman, E. M. (1950). "Biochemistry of Glucuronic Acid." Academic Press, New York.

Asakura, T., Adachi, K., Minakami, S., and Yoshikawa, H. (1967). *J. Biochem. (Tokyo)* **62**, 184.

Ashmore, J., and Weber, G. (1968). *In* "Carbohydrate Metabolism and Its Disorders" (F. Dickens, P. J. Randle, and W. J. Whelan, eds.), Vol. 1, p. 335. Academic Press, New York.

Ashwell, G., Kanfer, J., and Burns, J. J. (1959). *J. Biol. Chem.* **234**, 472.

Ashwell, G., Kanfer, J., Smiley, J. D., and Burns, J. J. (1961). *Ann. N.Y. Acad. Sci.* **92**, 105.

Austen, F. K., Rubini, M. E., Meroney, W. H., and Wolff, J. (1958). *J. Clin. Invest.* **37**, 1131.

Axelrod, J., Inscoe, J. K., and Tomkins, G. M. (1957). *Nature* **179**, 538.

Axelrod, J., Inscoe, J. K., and Tomkins, G. M. (1958). *J. Biol. Chem.* **232**, 835.

Bässler, K. H. (1968). *Klin. Wochschr.* **46**, 279

Bässler, K. H., and Dreiss, G. (1963). *Klin. Wochschr.* **41**, 593.

Bässler, K. H., and Heesen, D. (1963). *Klin. Wochschr.* **41**, 595.

Bässler, K. H., and Stein, G. (1967). *Z. Physiol. Chem.* **348**, 533.

Bässler, K. H., Prellwitz, W., Unbehaun, V., and Lang, K. (1962). *Klin. Wochschr.* **40**, 791.

Bässler, K. H., Holzmann, H., Prellwitz, W., and Stechert, J. (1965). *Klin. Wochschr.* **43**, 27.

Bässler, K. H., Toussaint, W., and Stein, G. (1966). *Klin. Wochschr.* **44**, 212.

Baker, E., Bierman, E. L., and Plough, I. C. (1960). *Metab. Clin. Exptl.* **9**, 478.

Baker, E. M., Sauberlich, H. E., Wolfskill, S. J., Wallace, W. T., and Dean, E. E. (1962). *Proc. Soc. Exptl. Biol. Med.* **109**, 737.

Barniville, H. T. F., and Misk, R. (1959). *Brit. Med. J.* i, 337.

Bartels, H., and Hohorst, H. I. (1963). *Biochim. Biophys. Acta* **71**, 214.

Beck, K., and Kiani, B. (1960). *Klin. Wochschr.* **30**, 428.

Beck, K., and Richter, E. (1962a). *Med. Klin.* **57**, 511.

Beck, K., and Richter, E. (1962b). *Klin. Wochschr.* **40**, 75.

Beck, K., Reisert, P. M., and Bayer, H. W. (1964). *Klin. Wochschr.* **42**, 524.

Beck, K., Richter, E., and Kiani, B. (1965). *Deut. Med. Wochschr.* **90**, 1005.

Beck, K., Azimi, F., and Reisert, P. M. (1967). *Klin. Wochschr.* **45**, 428.

Bender, D. (1964). *J. Am. Geriat. Soc.* **12**, 114.

Béraud, T., and Vannotti, A. (1960). *Acta Endocrinol.* **35**, 324.

Billing, B. H., and Lathe, G. H. (1956). *Biochem. J.* **63**, 6P.

Billing, B. H., Cole, P. G., and Lathe, G. H. (1957). *Biochem. J.* **65**, 774.

Black, M., and Billing, B. H. (1968). *Gut* **9**, 620, 728.

Bollet, A. J., Goodwin, J. F., and Brown, A. K. (1959). *J. Clin. Invest.* **38**, 451.

Bollet, A. J., Bonner, W. M., and Nance, J. L. (1963). *J. Biol. Chem.* **238**, 3522.
Bollman, J. L., and Mendez, F. L. (1955). *Federation Proc.* **14**, 399.
Bonsignore, A., and Ricci, C. (1957). *Sci. Med. Ital.* (*English Ed.*) **6**, 655.
Borgström, B. (1949). *Acta Med. Scand.* **133**, 7.
Borrell, S. (1958). *Biochem. J.* **70**, 727.
Boyland, E. (1959). *Ciba Found. Symp. Carcinogenesis: Mechanisms Action* p. 219.
Boyland, E., Wallace, D. M., and Williams, D. C. (1957). *Brit. J. Cancer* **11**, 578.
Breuer, H., and Dahm, K. (1967). *Symp. Deut. Ges. Endokrinol.* **12**, 280.
Bridges, J. W., and Williams, R. T. (1962). *Biochem. J.* **83**, 27p.
Bridges, J. W., Kibby, M. R., Walker, S. R., and Williams, R. T. (1968). *Biochem. J.* **109**, 851.
Brodie, B. B. (1964). *In* "Absorption and Distribution of Drugs" (T. B. Binn, ed.), p. 199. Livingstone, Edinburgh and London.
Brown, A. K. (1957). *Am. J. Diseases Children* **94**, 510.
Brown, A. K., Zuelzer, W., and Burnett, H. H. (1958). *J. Clin. Invest.* **37**, 332.
Brox, G. (1953). *Deut Z. Verdauungs- Stoffwechselkrankh.* **13**, 193, 198.
Bublitz, C., Grollman, A. P., and Lehninger, A. L. (1958). *Biochim. Biophys. Acta* **27**, 221.
Burns, J. J. (1957a). *Nature* **180**, 553.
Burns, J. J. (1957b). *J. Am. Chem. Soc.* **79**, 1257.
Burns, J. J. (1961). *Proc. Robert A. Welch Found. Conf. Chem. Res., Houston,* p. 277.
Burns, J. J., and Ashwell, G. (1960). *In* "The Enzymes" (P. D. Boyer, H. Lardy, and K. Myrbäck, eds.), 2nd ed., Vol. 3, p. 387. Academic Press, New York.
Burns, J. J., and Conney, A. H. (1966). *In* "Glucuronic Acid: Free and Combined. Chemistry, Biochemistry, Pharmacology, and Medicine" (G. J. Dutton, ed.), p. 301. Academic Press, New York.
Burns, J. J., and Evans, C. (1956). *J. Biol. Chem.* **223**, 897.
Burns, J. J., Evans, C., and Trousof, N. F. (1957). *J. Biol. Chem.* **227**, 785.
Burns, J. J., Trousof, N., Papadopoulos, N., and Evans, C. (1958). *Federation Proc.* **17**, 198.
Burns, J. J., Trousof, N., Evans, C., Papadopoulos, N., and Agranoff, B. V. (1959). *Biochim. Biophys. Acta* **33**, 215.
Burns, J. J., Conney, A. H., Dayton, P. G., Evans, C., Martin, G. R., and Taller, D. (1960). *J. Pharmacol. Exptl. Therap.* **129**, 132.
Butler, G. C., and Packham, M. A. (1955). *Arch. Biochem. Biophys.* **56**, 551.
Caputto, R., Leloir, L. F., Cardini, C. E., and Paladini, A. C. (1950). *J. Biol. Chem.* **184**, 333.
Careddu, P., and Marini, A. (1968). *Lancet* i, 982.
Careddu, P., Sereni, F., and Apollonio, T. (1961). *Minerva Pediat.* **13**, 453.
Careddu, P., Piceni-Sereni, L., Giunta, A., and Sereni, F. (1964). *Minerva Pediat.* **55**, 2559.
Castellani, A. A., De Bernard, B., and Zambotti, V. (1957). *Nature* **180**, 859.
Catz, C., and Yaffe, S. J. (1962). *Am. J. Diseases Children* **104**, 516.
Chandrasekhara, N., Rao, M. V. L., and Srinivasan, M. (1968). *Am. J. Clin. Nutr.* **21**, 736.
Charalampous, F. C., and Lyras, C. (1957). *J. Biol. Chem.* **228**, 1.
Chatterjee, I. B., and McKee, R. W. (1965). *Arch. Biochem. Biophys.* **109**, 62.

Chatterjee, I. B., Ghosh, N. C., Ghosh, J. J. and Guha, B. C. (1957). *Science* **126**, 608.
Chatterjee, I. B., Kar, N. C., Ghosh, N. C., and Guha, B. C. (1961). *Ann. N.Y. Acad. Sci.* **92**, 36.
Chatterjee, I. B., Price, Z. H., and McKee, R. W. (1965). *Nature* **207**, 1168.
Clapp, J. W. (1956). *J. Biol. Chem.* **223**, 207.
Comporti, M., Della Corte, E., and Stirpe, F. (1964). *Nature* **202**, 904.
Conney, A. H. (1967). *Pharmacol. Rev.* **19**, 317.
Conney, A. H., and Burns, J. J. (1961). *Biochim. Biophys. Acta* **54**, 369.
Conney, A. H., and Burns, J. J. (1962). *Advan. Pharmacol.* **1**, 31.
Conney, A. H., Bray, G. A., Evans, C., and Burns, J. J. (1961). *Ann. N.Y. Acad. Sci.* **92**, 115.
Coover, M. O., Feinberg, L. J., and Roe, J. H. (1950). *Proc. Soc. Exptl. Biol. Med.* **74**, 146.
Cornillot, P. (1962). *Clin. Chim. Acta* **7**, 42.
Cotton, D. W. K., Ilyas, A. W., and Mier, P. D. (1966). *Brit. J. Dermatol.* **78**, 465.
Crigler, J. F., Jr., and Gold, N. I. (1966). *J. Clin. Invest.* **45**, 998.
Crigler, J. F., Jr., and Gold, N. I. (1967). *J. Clin. Invest.* **46**, 1047.
Crigler, J. F., Jr., and Gold, N. I. (1969). *J. Clin. Invest.* **48**, 42.
Crigler, J. F., Jr., and Najjar, V. A. (1952). *Pediatrics* **10**, 169.
Cunningham, M. D., Mace, J. W., and Peters, E. R. (1969). *Lancet* **i**, 550.
Dahm, K., and Breuer, H. (1966a). *Z. Klin. Chem.* **4**, 153.
Dahm, K., and Breuer, H. (1966b). *Biochim. Biophys. Acta* **113**, 404.
Dahm, K., Breuer, H., and Lindlau, M. (1966). *Z. Physiol. Chem.* **345**, 139.
Daniels, A. L., and Everson, G. J. (1936). *Proc. Soc. Exptl. Biol. Med.* **35**, 20.
Danoff, S., Grantz, C., Boyer, A., and Holt, L. E., Jr. (1958). *Science* **127**, 759.
Dayton, P. G., Vrinten, P., and Perel, J. M. (1964). *Biochem. Pharmacol.* **13**, 143.
DeLeon, A., Gartner, L. M., and Arias, I. M. (1967). *J. Lab. Clin. Med.* **70**, 273.
Di Toro, R., Lupi, L., and Ansanelli, V. (1968). *Nature* **219**, 265.
Dixon, R. L., Shultice, R. W., and Fouts, J. R. (1960). *Proc. Soc. Exptl. Biol. Med.* **103**, 333.
Dohrmann, R. E. (1967). *Med. Klin.* (*Munich*) **62**, 537.
Dohrmann, R. E. (1968). *Arzneimittel-Forsch.* **18**, 854.
Duncan, G. G., Cristofori, F. C., Yue, J. K., and Murthy, M. S. J. (1964). *Med. Clin. N. Am.* **48**, 1359.
Dutton, G. J. (1956). *Biochem. J.* **64**, 693.
Dutton, G. J. (1959). *Biochem. J.* **71**, 141.
Dutton, G. J. (1962). *Proc. 1st Intern. Pharmacol. Meeting, Stockholm, 1961.* **6**, 39.
Dutton, G. J. (1966). In "Glucuronic Acid: Free and Combined. Chemistry, Biochemistry, Pharmacology, and Medicine" (G. J. Dutton, ed.), p. 186. Academic Press, New York.
Dutton, G. J., and Greig, C. G. (1957). *Biochem. J.* **66**, 52P.
Dutton, G. J., and Illing, H. P. A. (1969). *Biochem. J.* **112**, 16P.
Dutton, G. J., and Spencer, J. H. (1956). *Biochem. J.* **63**, 8P.
Dutton, G. J., and Stevenson, I. H. (1959). *Biochim. Biophys. Acta* **31**, 568.
Dutton, G. J., and Stevenson, I. H. (1962). *Biochim. Biophys. Acta* **58**, 633.
Dutton, G. J., and Storey, I. D. E. (1951). *Biochem. J.* **48**, xxix.
Dutton, G. J., and Storey, I. D. E. (1953). *Biochem. J.* **53**, xxxvii.
Dutton, G. J., and Storey, I. D. E. (1954). *Biochem. J.* **57**, 275.

Eisenberg, F., Jr., Field, J. B., and Stetten, D., Jr. (1955). *Arch. Biochem. Biophys.* **59**, 297.

Elson, L. A., Goulden, F., and Warren, F. L. (1945). *Biochem. J.* **39**, 301.

Enklewitz, M., and Lasker, M. (1935). *J. Biol. Chem.* **110**, 443.

Evans, C., Conney, A. H., Trousof, N., and Burns, J. J. (1959). *Federation Proc.* **18**, 223.

Evans, C., Conney, A. H., Trousof, N., and Burns, J. J. (1960). *Biochim. Biophys. Acta* **41**, 9.

Farbo, S. P., and Rinaldini, L. M. (1965). *Develop. Biol.* **11**, 468.

Fishman, W. H. (1940). *J. Biol. Chem.* **136**, 229.

Fishman, W. H. (1959). *Science* **130**, 1660.

Fishman, W. H., and Green, S. (1956). *J. Am. Chem. Soc.* **78**, 880.

Fishman, W. H., and Green, S. (1957). *J. Biol. Chem.* **225**, 435.

Fishman, W. H., Smith, M., Thompson, D. B., Bonner, C. D., Kasdon, S. C., and Homburger, F. (1951). *J. Clin. Invest.* **30**, 685.

Flint, M., Lathe, G. H., and Ricketts, T. R. (1963). *Ann. N.Y. Acad. Sci.* **111**, 295.

Florkin, M., Crismer, R., Duchateau, G., and Houet, R. (1942). *Enzymologia* **10**, 220.

Forsander, O., Räihä, N., and Suomalainen, H. (1958). *Z. Physiol. Chem.* **312**, 243.

Fouts, J. R. (1962). *Federation Proc.* **21**, 1107.

Frezza, M., DeSandre, G., Perona, G., and Corrocher, R. (1968). *Clin. Chim. Acta* **21**, 509.

Ganguli, N. C., Roy, S. C., and Guha, B. C. (1956). *Arch. Biochem. Biophys.* **61**, 211.

Gartner, L. M. (1967). *In* "Bilirubin Metabolism" (I. A. D. Bouchier and B. H. Billing, eds.), p. 175. Blackwell, Oxford.

Gaudette, L. E., and Brodie, B. B. (1959). *Biochem. Pharmacol.* **2**, 89.

Geller, W. (1952). *Schweiz. Med. Wochschr.* **82**, 599.

Geser, K. A., Förster, H., Pröls, H., and Mehnert, H. (1967). *Klin. Wochschr,* **45**, 851.

Ginsburg, V., Weissbach, A., and Maxwell, E. S. (1958). *Biochim. Biophys. Acta* **28**, 649.

Glock, G. E., and McLean, P. (1955). *Biochem. J.* **61**, 397.

Glock, G. E., McLean, P., and Whitehead, J. K. (1956). *Biochem. J.* **63**, 520.

Goldstein, J. A., and Taurog, A. (1968). *Biochem. Pharmacol.* **17**, 1049.

Good, B. F., Potter, H. A., and Hetzel, B. S. (1965). *Australian J. Exptl. Biol. Med. Sci.* **43**, 291.

Good, B. F., Hetzel, B. S., Hoffmann, M. J., Wellby, M. L., Black, M. L., Potter, H. A., and Buttfield, I. H. (1966). *Australasian Ann. Med.* **15**, 143.

Gram, T. E., Hansen, A. R., and Fouts, J. R. (1968). *Biochem. J.* **106**, 587.

Greenwood, D. T., and Stevenson, I. H. (1965). *Biochem. J.* **96**, 37P.

Grodsky, G. M., and Carbone, J. V. (1957). *J. Biol. Chem.* **226**, 449.

Hänninen, O. (1966). *Ann. Acad. Sci. Fenniae, Ser. A V* **123**.

Hnninen, O. (1968). *Ann. Acad. Sci. Fenniae, Ser. A V* **142**.

Hänninen, O., Aitio, A., and Hartiala, K. (1968). *Scand. J. Gastroenterol.* **3**, 461.

Halac, E., and Bonevardi, E. (1963). *Biochim. Biophys. Acta* **67**, 498.

Halac, E., and Frank, K. (1960). *Biochem. Biophys. Res. Commun.* **2**, 379.

Halac, E., and Weiss, P. (1967). *Pediat. Res.* **1**, 221.

Hartiala, K. (1954). *Acta Physiol. Scand. Suppl.* **31**, 114.

Hartiala, K. (1955). *Ann. Med. Exptl. Biol. Fenniae (Helsinki)* **33**, 239.

Hartiala, K. (1961). *Biochem. Pharmacol.* 6, 82.

Hartiala, K. J. V., and Hirvonen, L. (1956). *Ann. Med. Exptl. Biol. Fenniae (Helsinki)* 34, 122.

Hartiala, K. J. W., and Pulkkinen, M. (1964). *Ann. Acad. Sci. Fenniae, Ser. A V* 106, 12.

Hartiala, K. J. V., Hirvonen, L., and Kassinen, A. (1956). *Ann. Med. Exptl. Biol. Fenniae (Helsinki)* 34, 117.

Hartiala, K. J. V., Leikkola, P., and Savola, P. (1957). *Acta Physiol. Scand.* 42, 36.

Hartiala, K., Näntö, V., and Rinne, U. K. (1958). *Acta Physiol. Scand.* 43, 77.

Hartiala, K., Näntö, V., and Rinne, U. (1961). *Acta Physiol. Scand.* 53, 376.

Haydon, R. K. (1961). *Biochim. Biophys. Acta* 46, 598.

Herman, R. H., Overholt, E. L., and Hagler, L. (1969). *Am. J. Med.* 46, 142.

Hernbrook, K. R., Burch, H. B., and Lowry, O. H. (1965). *Biochem. Biophys. Res. Commun.* 18, 206.

Heyde, W., and Wieland, H. (1960). *Aerztl. Forsch.* 14, 412.

Hiatt, H. H. (1958). *Biochim. Biophys. Acta* 28, 645.

Hiatt, H. H. (1966). *In* "The Metabolic Basis of Inherited Disease" (J. B. Stanbury, J. B. Wyngaarden, and D. S. Fredrickson, eds.), p. 109. McGraw-Hill, New York.

Hirata, T., Fujisawa, M., Sato, H., Asano, T., and Katsuki, S. (1966). *Biochem. Biophys. Res. Commun.* 24, 471.

Hirvonen, T. (1966). *Ann. Univ. Turku. Ser. A II*, No. 38.

Hoffmann, W., and Breuer, H. (1968). *Z. Klin. Chem. Klin. Biochem.* 6, 85.

Hollmann, S. (1954a). *Z. Physiol. Chem.* 297, 74.

Hollmann, S. (1954b). *Z. Physiol. Chem.* 297, 83.

Hollmann, S., and Laumann, G. (1967). *Z. Physiol. Chem.* 348, 1073.

Hollmann, S., and Neubaur, J. (1966). *Klin. Wochschr.* 44, 722.

Hollmann, S., and Neubaur, J. (1967). *Z. Physiol. Chem.* 348, 877.

Hollmann, S. (1964). "Non-Glycolytic Pathways of Metabolism of Glucose" (translated and revised by O. Touster). Academic Press, New York.

Hollmann, S., and Touster, O. (1962). *Biochim. Biophys. Acta* 26, 338.

Hollmann, S., and Wille, E. (1952). *Z. Physiol. Chem.* 290, 91.

Holopainen, T. (1958). *Ann. Med. Exptl. Biol. Fenniae (Helsinki)* 36, Suppl. 8.

Hosoya, E., and Otobe, K. (1959). Cited in Fishman (1959).

Howard, C. F., Jr., and Anderson, L. (1967). *Arch. Biochem. Biophys.* 118, 332.

Hsia, D. Y. Y., Dowben, R. M., and Riabov, S. (1963a). *Ann. N. Y. Acad. Sci.* 111, 326.

Hsia, D. Y. Y., Riabov, S., and Dowben, R. M. (1963b). *Arch. Biochem. Biophys.* 103, 181.

Hübener, H. J. (1962). *Deut. Med. Wochschr.* 87, 438.

Hutton, J. J., Jr., Tappel, A. L., and Udenfriend, S. (1967). *Arch. Biochem. Biophys.* 118, 231.

Huttunen, J. K., and Miettinen, T. A. (1965). *Acta Physiol. Scand.* 63, 133.

Inscoe, J. K., and Axelrod, J. (1960). *J. Pharmacol. Exptl. Therap.* 129, 128.

Ishidate, M., Matsui, M., and Okada, M. (1965). *Anal. Biochem.* 11, 176.

Ishikawa, S. (1959). *J. Biochem. (Tokyo)* 46, 347.

Ishikawa, S., and Noguchi, K. (1957). *J. Biochem. (Tokyo)* 44, 465.

Isselbacher, K. J. (1956). *Recent Progr. Hormone Res.* 12, 134.

Isselbacher, K. J. (1961). *Biochem. Biophys. Res. Commun.* 5, 243.

Isselbacher, K. J., and Axelrod, J. (1955). *J. Am. Chem. Soc.* **77**, 1070.
Isselbacher, K. J., Charabas, M. F., and Quinn, R. C. (1962). *J. Biol. Chem.* **237**, 3033.
Jacobson, B., and Davidson, E. A. (1962a). *J. Biol. Chem.* **237**, 635.
Jacobson, B., and Davidson, E. A. (1962b). *J. Biol. Chem.* **237**, 638.
Jacobson, B., and Davidson, E. A. (1963). *Biochim. Biophys. Acta* **73**, 145.
Jay, G. E., Jr. (1955). *Proc. Soc. Exptl. Biol. Med.* **90**, 378.
Jayle, M. F., and Pasqualini, J. R. (1966). *In* "Glucuronic Acid: Free and Combined. Chemistry, Biochemistry, Pharmacology, and Medicine" (G. J. Dutton, ed.), p. 507. Academic Press, New York.
Kamil, I. A., Smith, J. N., and Williams, R. T. (1953a). *Biochem. J.* **53**, 129.
Kamil, I. A., Smith, J. N., and Williams, R. T. (1953b). *Biochem. J.* **54**, 390.
Kaplan, N. O., Goldin, A., Humphreys, S. R., Ciotti, M. M., and Stolzenbach, F. E. (1956). *J. Biol. Chem.* **219**, 287.
Karunairatnam, M. C., and Levvy, G. A. (1949). *Biochem. J.* **44**, 599.
Karunairatnam, M. C., Kerr, L. M. H., and Levvy, G. A. (1949). *Biochem. J.* **45**, 496.
Kaslander, J. (1963). *Biochim. Biophys. Acta* **71**, 730.
Kato, R., Vassanelli, P., and Frontino, G. (1962). *Experientia* **18**, 9.
Kawada, M., Yamada, K., Kagawa, Y., and Mano, Y. (1961). *J. Biochem. (Tokyo)* **50**, 74.
Kivirikko, K. I., and Prockop, D. J. (1967). *Proc. Natl. Acad. Sci. U.S.* **57**, 782.
Kiyomoto, A., Harigaya, S., Ohshima, S., and Morita, T. (1963). *Biochem. Pharmacol.* **12**, 105.
Klaassen, C. D., and Plaa, G. L. (1968). *J. Pharmacol. Exptl. Therap.* **161**, 361.
Ko, V., and Dutton, G. J. (1967). *Biochem. J.* **104**, 991.
Koivusalo, M., Luukkainen, T., Miettinen, T. A., and Pispa, J. (1958). *Acta Chem. Scand.* **12**, 1919.
Koivusalo, M., Luukkainen, T., Miettinen, T. A., and Pispa, J. (1959). *Ann. Med. Exptl. Biol. Fenniae (Helsinki)* **37**, 93.
Kornfeld, S. (1965). *Federation Proc.* **24**, 536.
Krebs, H. A. (1966). *Advan. Enzyme Regulation,* **4**, 339.
Krebs, H. A. (1967). *Advan. Enzyme Regulation* **5**, 409.
Krebs, H. A., Notton, B. M., and Hems, R. (1966). *Biochem. J.* **101**, 607.
Kreek, J. M., and Sleisenger, M. H. (1968). *Lancet* **ii**, 73.
Kumahara, Y., Feingold, D., Freedberg, I. M., and Hiatt, H. H. (1961). *J. Clin. Endocrinol. Metab.* **21**, 887.
Kunin, C. M., Glazko, A. J., and Finland, M. (1959). *J. Clin. Invest.* **38**, 1498.
Kuzuya, T., Kanazawa, Y., and Kosaka, K. (1966). *Metab. Clin. Exptl.* **15**, 1149.
Kuzuya, T., Kanazawa, Y., Shimoshige, M., and Kosaka, K. (1967). *Excerpta Med. Intern. Congr. Ser.* **140**, 77.
Landau, B. R., Bartsch, G. E., and Williams, H. R. (1966). *J. Biol. Chem.* **241**, 750.
Lathe, G. H., and Walker, M. (1958a). *Quart. J. Exptl. Physiol.* **43**, 257.
Lathe, G. H., and Walker, M. (1958b). *Biochem. J.* **68**, 6.
Lathe, G. H., and Walker, M. (1958c). *Biochem. J.* **70**, 705.
Lawrow, D. (1901). *Z. Physiol. Chem.* **32**, 111.
Lehtinen, A., Hartiala, K., and Nurmikko, V. (1958a). *Acta Chem. Scand.* **12**, 1589.
Lehtinen, A., Nurmikko, V., and Hartiala, K. (1958b). *Acta Chem. Scand.* **12**, 1585.

Lehtinen, A., Savola, P., Pulkkinen, M., and Hartiala, K. (1958c). *Acta Chem. Scand.* **12**, 1592.

Leskinen, E. (1964). Unpublished observations.

Leventer, L. L., Buchanan, J. L., Ross, J. E., and Tapley, D. F. (1965). *Biochim. Biophys. Acta* **110**, 428.

Levine, W. G., Millburn, P., Smith, R. L., and Williams, R. T. (1968). *Biochem. J.* **109**, 35P.

Levvy, G. A. (1952). *Biochem. J.* **52**, 464.

Levvy, G. A., and Conchie, J. (1966). *In* "Glucuronic Acid: Free and Combined. Chemistry, Biochemistry, Pharmacology, and Medicine" (G. J. Dutton, ed.), pp. 301–364. Academic Press, New York.

Levvy, G. A., and Marsh, C. A. (1960). *In* "The Enzymes" (P. D. Boyer, H. Lardy, and K. Myrbäck, eds.), 2nd Ed., Vol. 4, p. 397. Academic Press, New York.

Levvy, G. A., Kerr, L. M. H., and Campbell, J. G. (1948). *Biochem. J.* **42**, 462.

Levy, G., and Matsuzawa, T. (1967). *J. Pharmacol. Exptl. Therap.* **156**, 285.

Linker, A., Meyer, K., Sampson, P., and Korn, D. (1955). *J. Biol. Chem.* **213**, 237.

Lippel, K., and Olson, A. (1968). *J. Lipid Res.* **9**, 168.

Lipschitz, W. L., and Bueding, E. (1939). *J. Biol. Chem.* **129**, 335.

Longenecker, H. E., Musulin, R. R., and King, C. G. (1939). *J. Biol. Chem.* **128**, lx.

Longenecker, H. E., Fricke, H. H., and King, C. G. (1940). *J. Biol. Chem.* **135**, 497.

Lubran, K. (1969). *Med. Clin. N. Am.* **53**, 211.

McCormick, D. B., and Touster, O. (1957). *J. Biol. Chem.* **229**, 451.

Mano, Y., Suzuki, K., Yamada, K., and Shimazono, N. (1961). *J. Biochem. (Tokyo)* **49**, 618.

Marchi, S., Bertazzoni, V., and Zambotti, V. (1964). *Boll. Soc. Ital. Biol. Sper.* **40**, 1493.

Margolis, J. J. (1929). *Am. J. Med. Sci.* **177**, 348.

Marsh, C. A. (1963a). *Biochem. J.* **86**, 77.

Marsh, C. A. (1963b). *Biochem. J.* **87**, 82.

Marsh, C. A. (1963c). *Biochem. J.* **89**, 108.

Marsh, C. A. (1966). *Biochem. J.* **99**, 22.

Marsh, C. A., and Carr, A. J. (1965). *Clin. Sci.* **28**, 209.

Marsh, C. A., and Reid, L. M. (1963). *Biochim. Biophys. Acta* **78**, 726.

Maruhama, Y., Chiba, M., Anzai, M., Ohneda, A., Goto, Y., and Yamagata, S. (1967). *Excerpta Med. Intern. Congr. Ser.* **140**, 44.

Matsui, M., Okada, M., and Ishidate, M. (1965). *J. Biochem. (Tokyo)* **57**, 715.

Matsui, M., Okada, M., and Ishidate, M. (1969a). *Chem. Pharm. Bull. (Tokyo)* **17** (in press).

Matsui, M., Okada, M., and Ishidate, M. (1969b). *Chem. Pharm. Bull. (Tokyo)* **17** (in press).

Matsushiro, T. (1965). *Tohoku J. Exptl. Med.* **85**, 330.

Matthews, M. B., and Dorfman, A. (1954). *J. Biol. Chem.* **206**, 143.

Maurer, H. M., Wolff J. A., Finster, M., Poppers, P. J., Pantuck, E., Kuntzman, R., and Conney, A. H. (1968). *Lancet* **ii**, 122.

Maxwell, E. S., Kalckar, H. M., and Strominger, J. L. (1956). *Arch. Biochem. Biophys.* **65**, 2.

Mehnert, H., Summa, J. D., and Förster, H. (1964). *Klin. Wochschr.* **42**, 382.

Mellinghoff, C. H. (1961). *Klin. Wochschr.* **39**, 447.

Mendoza, D. M., Flock, E. V., Owen, C. A., Jr., and Paris, J. (1966). *Endocrinology* **79**, 106.

Metge, W. R., Owen, C. A., Jr., Foulk, W. T., and Hoffman, H. N. (1964). *J. Lab. Clin. Med.* **64**, 89.

Miettinen, T. A. (1961). *Scand. J. Clin. Lab. Invest.* **13**, Suppl. 61.

Miettinen, T. A., and Leskinen, E. (1962). *Ann. Med. Exptl. Biol. Fenniae (Helsinki)* **40**, 427.

Miettinen, T. A., and Leskinen, E. (1963). *Biochem. Pharmacol.* **12**, 565.

Miettinen, T. A., and Leskinen, E. (1967). *Ann. Med. Exptl. Biol. Fenniae (Helsinki)* **45**, 80.

Miettinen, T. A., and Leskinen, E. (1968). Unpublished observations.

Miettinen, T. A., Valtonen, E., and Nikkilä, E. A. (1968). *Scand. J. Clin. Lab. Invest.* **21**, Suppl. 101, 71.

Miettinen, T. A., Nikkilä, E. A., Taskinen, M.-R., and Pelkonen, R. (1969). *Acta Med. Scand.* **186**, 247.

Milcu, S. T., and Biener, J. (1958). *Studii Cercetari Endocrinol.* **9**, 291. Cited in *Chem. Abstr.* **53**, 15262g (1959).

Miller, B. F., Keyes, F. P., and Curreri, P. W. (1966). *J. Am. Med. Assoc.* **195**, 189.

Miller, B. F., Keyes, F. P., and Curreri, P. W. (1967). *J. Atherosclerosis Res.* **7**, 591.

Mills, G. T., Ondarza, R., and Smith, E. E. B. (1954). *Biochim. Biophys. Acta* **14**, 159.

Mills, G. T., Lochhead, A. C., and Smith, E. E. B. (1958). *Biochim. Biophys. Acta* **27**, 103.

Mori, K., Kanatsyna, T., and Kuzuya, K. (1967). *Excerpta Med. Intern. Congr. Ser.* **140**, 161.

Mosbach, E. H., Jackel, S. S., and King, C. G. (1950). *Arch. Biochem. Biophys.* **29**, 348.

Moudgal, N. R., Ragupathy, E., and Sarma, P. S. (1959). *Endocrinology* **64**, 326.

Müller, F., Strack, E., Kuhfahl, E., and Dettmer, D. (1967). *Z. Ges. Exptl. Med.* **142**, 338.

Müller-Oerlingenhausen, B. (1969). Personal communication.

Müller-Oerlingenhausen, B., and Künzel, B. (1968). *Life Sci.* **7**, 1129.

Müller-Oerlingenhausen, B., Hasselblatt, A., and Jahns, R. (1967a). *Arch. Exptl. Pathol. Pharmakol.* **257**, 314.

Müller-Oerlingenhausen, B., Hasselblatt, A., and Jahns, R. (1967b). *Life Sci.* **6**, 1529.

Müller-Oerlingenhausen, B., Hasselblatt, A., and Jahns, R. (1968). *Arch. Exptl. Pathol. Pharmakol.* **260**, 254.

Müller-Oerlingenhausen, B., Jahns, R., Künzel, B., and Hasselblatt, A. (1969). *Arch. Exptl. Pathol. Pharmakol.* **262**, 17.

Müting, D. (1964). *Lancet* **ii**, 15.

Müting, D., Reikowski, H., and Neuheisel, S. (1963). *Z. Klin. Med.* **157**, 544.

Munch-Petersen, A., Kalckar, H. M., Cutolo, E., and Smith, E. E. B. (1953). *Nature* **172**, 1036.

Musulin, R. R., Tully, R. H., Longenecker, H. E., and King, C. G. (1939). *J. Biol. Chem.* **129**, 437.

Nasarijanz, B. A. (1934). *Schweiz. Med. Wochschr.* **64**, 1090.

Neufeld, E. F., and Hall, C. W. (1965). *Biochem. Biophys. Res. Commun.* **19**, 456.

234   *Tatu A. Miettinen and Erkki Leskinen*

Neumeister, R. (1895). "Lehrbuch der Physiologische Chemie," Vol. 2, p. 346. Fischer, Jena.
Newman, W. C., and Moon, R. C. (1967). *Endocrinology* **80**, 896.
Nitze, H. R., and Remmer, H. (1962). *Arch. Exptl. Pathol. Pharmakol.* **242**, 555.
Numa, S., Bortz, W. M., and Lynen, F. (1965). *Advan. Enzyme Regulation* **3**, 407.
Ogawa, H., Sawada, M., and Kawada, M. (1966). *J. Biochem. (Tokyo)* **59**, 126.
Okada, M. (1969). Personal communication.
Okada, M., Matsui, M., Kaizu, T., and Ishidate, M. (1964). *Rept. 10th Anniv. Symp. Glucuronic Acid, Tokyo* p. 19.
Okolicsanyi, L., Magnenat, P., and Frei, J. (1968). *Lancet* **i**, 1173.
Osorio, C., and Myant, N. B. (1963). *Endocrinology* **72**, 253.
Packham, M. A., and Butler, G. C. (1954). *J. Biol. Chem.* **207**, 639.
Parke, D. V. (1952). Ph. D. Thesis, Univ. of London. Cited in Williams, R. T. (1959). "Detoxication Mechanisms," 2nd Ed., p. 492. Chapman & Hall, London.
Pitkänen, E., and Sahlström, K. (1968). *Ann. Med. Exptl. Biol. Fenniae (Helsinki)* **46**, 143.
Pogell, B. M., and Krisman, C. R. (1960). *Biochim. Biophys. Acta* **41**, 349.
Pogell, B. M., and Leloir, L. F. (1961). *J. Biol. Chem.* **236**, 293.
Prellwitz, W., and Bässler, K. H. (1963). *Klin. Wochschr.* **41**, 196.
Pulkkinen, M., and Hartiala, K. (1965). *Nature* **207**, 646.
Quick, A. J. (1926). *J. Biol. Chem.* **70**, 59.
Quinn, G. P., Axelrod, J., and Brodie, B. B. (1958). *Biochem. Pharmacol.* **1**, 152.
Rauramo, L., Pulkkinen, M., and Hartiala, K. (1963). *Ann. Med. Exptl. Biol. Fenniae (Helsinki)* **41**, 27.
Reid, E. (1959). *Biochim. Biophys. Acta* **32**, 251.
Reid, E. (1960). *Mem. Soc. Endocrinol.* **9**, 130.
Reid, E. (1961a). *Biochem. Soc. Symp. (Cambridge, Engl.)* **20**, 89.
Reid, E. (1961b). *Mem. Soc. Endocrinol.* **11**, 149.
Remmer, H. (1964). *Arch. Exptl. Pathol. Pharmakol.* **247**, 461.
Remmer, H., and Merker, H.-J. (1963a). *Klin. Wochschr.* **41**, 276.
Remmer, H., and Merker, H.-J. (1963b). *Naturwissenschaften* **50**, 670.
Richardson, K. E., and Axelrod, B. (1958). *Federation Proc.* **17**, 296.
Richardson, K. E., and Axelrod, B. (1959). *Biochim. Biophys. Acta* **32**, 265.
Roberts, E., and Spiegel, C. J. (1947). *J. Biol. Chem.* **171**, 9.
Roberts, R. J., and Plaa, G. L. (1967). *Biochem. Pharmacol.* **16**, 827.
Robinson, D., and Williams, R. T. (1958). *Biochem. J.* **68**, 23P.
Roe, J. H., and Coover, M. O. (1950). *Proc. Soc. Exptl. Biol. Med.* **75**, 818.
Rutter, W. J., and Hansen, R. G. (1953). *J. Biol. Chem.* **202**, 323.
Sadahiro, R., Hinohara, Y., Yamamoto, A., and Kawada, M. (1966). *J. Biochem. (Tokyo)* **59**, 216.
Salomon, L. L., and Stubbs, D. W. (1961). *Biochem. Biophys. Res. Commun.* **5**, 349.
Sauer, J. (1930). *Klin. Wochschr.* **9**, 2351.
Sawada, M. (1959). *Glucuronic Acid Res. Conf., 5th, Tokyo.* Cited in Fishman (1959).
Schachter, D., Kass, D. J., and Lannon, T. J. (1959). *J. Biol. Chem.* **234**, 201.
Schmid, F. (1936). *Compt. Rend. Soc. Biol.* **123**, 223.
Schmid, R. (1956). *Science* **124**, 76.
Schmid, R., and Lester, R. (1966). *In* "Glucuronic Acid: Free and Combined. Chemistry, Biochemistry, Pharmacology, and Medicine" (G. J. Dutton, ed.), p. 493. Academic Press, New York.

Schmid, R., Axelrod, J., Hammaker, L., and Swarm, R. L. (1958). *J. Clin. Invest.* **37**, 1123.

Schmidt, B., Fingerhut, M., and Lang, K. (1964). *Klin. Wochschr.* **42**, 1073.

Schriefers, H. (1967). *Vitamins and Hormones* **25**, 271.

Schriefers, H., Keck, B., and Otto, M. (1965). *Acta Endocrinol.* **50**, 25.

Schriefers, H., Ghraf, R., and Pohl, F. (1966). *Z. Physiol. Chem.* **344**, 25.

Schwob, M., Perry, R., Boyer, A., Holt, L. E., Jr., Hallman, N., Backman, A., and Hjelt, L. (1960). *Pediatrics* **25**, 686.

Sereni, F., Perletti, L., and Marini, A. (1967). *Pediatrics* **40**, 446.

Shirai, Y., and Ohkubo, T. (1954a). *J. Biochem. (Tokyo)* **41**, 341.

Shirai, Y., and Ohkubo, T. (1954b). *J. Biochem. (Tokyo)* **41**, 781.

Sie, H.-G., and Fishman, W. H. (1957). *J. Biol. Chem.* **225**, 453.

Silbert, J. E. (1966). *In* "Glucuronic Acid: Free and Combined. Chemistry, Biochemistry, Pharmacology, and Medicine" (G. J. Dutton, ed.), p. 385. Academic Press, New York.

Skea, B. R., and Nemeth, A. M. (1969). *Biochem. J.* **113**, 16P.

Smiley, J. D., and Ashwell, G. (1961). *J. Biol. Chem.* **236**, 357.

Smith, E. E. B., and Mills, G. T. (1954a). *Biochim. Biophys. Acta* **13**, 587.

Smith, E. E. B., and Mills, G. T. (1954b). *Biochim. Biophys. Acta* **13**, 386.

Smith, E. E. B., and Mills, G. T. (1955). *Biochim. Biophys. Acta* **18**, 152.

Smith, E. E. B., Munch-Petersen, A., and Mills, G. T. (1953). *Nature* **172**, 1038.

Smith, J. N., and Williams, R. T. (1949). *Biochem. J.* **44**, 242.

Smith, J. N., and Williams, R. T. (1954). *Biochem. J.* **56**, 618.

Smith, R. L., and Williams, R. T. (1966). *In* "Glucuronic Acid: Free and Combined. Chemistry, Biochemistry, Pharmacology, and Medicine" (G. J. Dutton, ed.), p. 457. Academic Press, New York.

Snapper, I., and Saltzman, A. (1947). *Am. J. Med.* **2**, 327.

Spiro, R. G. (1963). *Diabetes* **12**, 223.

Stevenson, I. H., and Dutton, G. J. (1960). *Biochem. J.* **77**, 19P.

Stevenson, I. H., and Dutton, G. J. (1962). *Biochem. J.* **82**, 330.

Stevenson, I., Greenwood, D., and McEwen, J. (1968). *Biochem. Biophys. Res. Commun.* **32**, 866.

Stirpe, F., and Comporti, M. (1965). *Biochem. J.* **95**, 354.

Stirpe, F., Comporti, M., and Della Corte, E. (1965). *Biochem. J.* **95**, 363.

Storey, I. D. E. (1950). *Biochem. J.* **47**, 212.

Storey, I. D. E. (1964a). *Biochem. J.* **90**, 15P.

Storey, I. D. E. (1964b). *Biochem. J.* **90**, 16P.

Storey, I. D. E. (1965a). *Biochem. J.* **95**, 201.

Storey, I. D. E. (1965b). *Biochem. J.* **95**, 209.

Storey, I. D. E., and Dutton, G. J. (1955). *Biochem. J.* **59**, 279.

Straumfjord, J. V., Jr., and West, E. S. (1957). *Proc. Soc. Exptl. Biol. Med.* **94**, 566.

Strömme, J. H. (1965). *Biochem. Pharmacol.* **14**, 393.

Strominger, J. L., Kalckar, H. M., Axelrod, J., and Maxwell, E. S. (1954). *J. Am. Chem. Soc.* **76**, 6411.

Strominger, J. L., Maxwell, E. S., Axelrod, J., and Kalckar, H. M. (1957). *J. Biol. Chem.* **224**, 79.

Stubbs, D. W., and McKernan, J. B. (1967). *Proc. Soc. Exptl. Biol. Med.* **125**, 1326.

Stubbs, D. W., and Salomon, L. L. (1963). *Federation Proc.* **22**, 422.

Stubbs, D. W., McKernan, J. B., and Haufrect, D. (1967). *Proc. Soc. Exptl. Biol. Med.* **126**, 464.

Südhof, H. (1952). Z. Physiol. Chem. 290, 72.

Südhoff, H. (1954a). Z. Physiol. Chem. 296, 267.

Südhof, H. (1954b). Klin. Wochschr. 32, 91.

Südhof, H. (1954c). Deut. Arch. Klin. Med. 201, 89.

Südhof, H., and Schellong, G. (1953). Klin. Wochschr. 31, 64.

Takanashi, S., Ohkubo, K., Takahashi, S., Iida, K., Okutomi, T., and Kawada, M. (1964). Rept. 10th Anniv. Symp. Glucuronic Acid, Tokyo p. 46.

Takanashi, S., Ihda, K., and Kawada, M. (1966). J. Biochem. (Tokyo) 59, 78.

Taketa, K. (1962a). Acta Med. Okayama 16, 90.

Taketa, K. (1962b). Acta Med. Okayama 16, 99.

Taketa, K. (1962c). Acta Med. Okayama 16, 115.

Taketa, K. (1962d). Acta Med. Okayama 16, 129.

Taketa, K. (1962e). Acta Med. Okayama 16, 275.

Talafant, E. (1956). Nature 178, 312.

Tenhunen, R. (1965). Ann. Med. Exptl. Biol. Fenniae (Helsinki) 43, Suppl. 6.

Thierfelder, H. (1886). Z. Physiol. Chem. 10, 163.

Thompson, R. P. H., and Williams, R. T. (1967). Lancet ii, 646.

Thompson, R. P. H., Eddleston, A. L. W. F., and Williams, R. T. (1969). Lancet i, 21.

Toussaint, W., Roggenkamp, K., and Bässler, K. H. (1966). Klin. Wochschr. 44, 663.

Touster, O. (1966). Cited in Burns and Conney (1966), p. 371.

Touster, O., and Hollmann, S. (1961a). Ann. N.Y. Acad. Sci. 92, 318.

Touster, O., and Hollmann, S. (1961b). Federation Proc. 20, 84.

Touster, O., and Reynolds, V. H. (1952). J. Biol. Chem. 197, 863.

Touster, O., and Shaw, D. R. D. (1962). Physiol. Rev. 42, 181.

Touster, O., Hutcheson, R. M., and Reynolds, V. H. (1954). J. Am. Chem. Soc. 76, 5005.

Touster, O., Hutcheson, R. M., and Rice, L. (1955). J. Biol. Chem. 215, 677.

Touster, O., Mayberry, R. H., and McCormick, D. B. (1957). Biochim. Biophys. Acta 25, 196.

Touster, O., Hester, R. W., and Siler, R. A. (1960). Biochem. Biophys. Res. Commun. 3, 248.

Trolle, D. (1968a). Lancet i, 251.

Trolle, D. (1968b). Lancet ii, 705.

Tsutsumi, S., Nakai, K., and Nakamura, H. (1966). Japan. J. Pharmacol. 16, 443.

ul-Hassan, M., and Lehninger, A. L. (1956). J. Biol. Chem. 223, 123.

Vescia, A. (1945). Boll. Soc. Ital. Biol. Sper. 20, 752.

Vest, M. F. (1958a). Arch. Disease Childhood 33, 473.

Vest, M. F. (1958b). Schweiz. Med. Wochschr. 88, 969.

Vest, M. F., and Fritz, E. (1961). J. Clin. Pathol. 14, 482.

Vest, M. F., and Streif, R. R. (1959). Am. J. Diseases Children 98, 688.

Villa-Trevino, S., Shull, K. H., and Farber, E. (1963). J. Biol. Chem. 238, 1757.

Walker, W., Hughes, M. I., and Barton, M. (1969). Lancet i, 548.

Waters, W. J., Dunham, R., and Bowen, W. R. (1958). Proc. Soc. Exptl. Biol. Med. 99, 175.

Weber, G., Lea, M. A., Hird Convery, H. J., and Stamm, N. B. (1967). Advan. Enzyme Regulation 5, 257.

Weissmann, B., Meyer, K., Sampson, B., and Linker, A. (1954). J. Biol. Chem. 208, 417.

Werder, E. A., and Yaffe, S. J. (1964). *Biol. Neonatorum* **6**, 8.

Whelton, M. J., Krustev, L. P., and Billing, B. H. (1968). *Am. J. Med.* **45**, 160.

Williams, R. T. (1967). *Ciba Found. Symp. Drug Responses in Man* p. 83.

Wilson, D. (1965). *Anal. Biochem.* **10**, 472.

Winegrad, A. I., and Burden, C. L. (1965). *Trans. Assoc. Am. Physicians* **78**, 158.

Winegrad, A. I., and Burden, C. L. (1966). *New Engl. J. Med.* **274**, 289.

Winegrad, A. I., and Shaw, W. N. (1964). *Am. J. Physiol.* **206**, 165.

Winkelman, J., and Ashwell, G. (1961). *Biochim. Biophys. Acta* **52**, 170.

Winkelman, J., and Lehninger, A. L. (1958). *J. Biol. Chem.* **233**, 794.

Wolff, J., and Austen, F. K. (1958). *J. Clin. Invest.* **37**, 1144.

Yaffe, S. J., Levy, G., and Matsuzawa, T. (1966). *New Engl. J. Med.* **275**, 1461.

Yamada, K. (1959). *J. Biochem. (Tokyo)* **46**, 529.

Yamada, K., Ishikawa, S., and Shimazono, N. S. (1959). *Biochim. Biophys. Acta* **32**, 253.

Yamagata, S., Goto, Y., Ohneda, A., Anzai, M., Kawashima, S., Chiba, M., Maru-hama, Y., and Yamauchi, Y. (1965). *Lancet* ii, 918.

Yonese, Y. (1968). *Japan. J. Urol.* **59**, 243.

Yonese, Y., Takayasu, H., Okada, M., and Ishidate, M. (1966). *Abstr. Papers 9th Intern. Cancer Congr, Tokyo* p. 700.

York, J. L., Grollman, A. P., and Bublitz, C. (1961). *Biochim. Biophys. Acta* **47**, 298.

Zalitis, J., and Feingold, D. S. (1968). *Biochem. Biophys. Res. Commun.* **31**, 693.

Zeidenberg, P., Orrenius, S., and Ernster, L. (1967). *J. Cell Biol.* **32**, 528.

SULFOCONJUGATION AND SULFOHYDROLYSIS

K. S. DODGSON and F. A. ROSE

I. Introduction

It seems certain that sulfur has played an important biological role since the beginnings of life on Earth. It is an extremely reactive element and participates in an extraordinary range of chemical linkages in a wide variety of biological compounds. In a recent review, Dodgson and Rose (1966) have stressed this fact and have pointed out that although some sulfur-containing compounds have biological functions which are well established (e.g., cysteine, methionine, thiamine, biotin, coenzyme A) the functions of many others are still ill-defined. Isothiocyanates, sulfones, sulfoxides, and sulfonic acids, all of which occur widely in the plant world, cannot yet have precise functions ascribed to them and most of the suggestions that have hitherto been offered have little or no supporting experimental evidence. A similar situation presents itself with naturally occurring sulfoconjugates, only a few of which have well-established biological functions and many of which remain as biochemical curiosities at the present time.

Sulfoconjugates constitute a very diverse and widespread group of compounds, representatives of which are to be found in microorganisms, lower and higher plants, and throughout the animal world. So diverse in their distribution and in their chemical nature are they that very few laboratories have ever considered the group as a whole. Instead, work has been concentrated on particular types of conjugate that appear to be relevant to the overall field of study of the laboratory. Thus, in the field of drug metabolism the term sulfoconjugate is generally synonymous with the sulfoconjugates of steroids and phenols, whereas the connective tissue biochemist would regard the term as synonymous with the acidic glycosaminoglycans (mucopolysaccharides). In spite of this channelling of approach, sulfoconjugates as a whole show many features in common. For example, sulfoconjugation involves a common enzymic process in which only the final sulfate transfer stage utilizes transferase enzymes specific for a particular type of conjugate. The sulfate group itself confers some general properties on all the conjugates, particularly in terms of anionic character, increased water solubility, decreased lipid solubility, ability to attract cations, water molecules, and so on.

The overwhelming majority of nature's sulfoconjugates can generally be classified as sulfuric acid esters in which the $C—O—SO_3^-$ linkage is present. However, other conjugates exist in which sulfate is covalently attached by other linkages, namely, $P—O—SO_3^-$, $N—SO_3^-$, $N—O—SO_3^-$ and $S—SO_3^-$. In this chapter all types are mentioned, although most attention is devoted to the true sulfuric acid esters. Similarly, brief mention will be made of sulfoconjugates and associated enzyme systems that may, at the present time, seem only marginally related to the theme of the book but which nevertheless illuminate the topic as a whole. From the point of view of nomenclature, it has been the common, though strictly incorrect practice, to refer to these sulfoconjugates by trivial names. For example, $C_6H_5O \cdot SO_3H$, phenyl hydrogen sulfate, is generally referred to as phenyl sulfate. The trivial nomenclature is retained in this chapter except in those cases where there is need to describe compounds more precisely.

The following list provides a general indication of the different types of sulfoconjugates that occur in nature and serves to illustrate some of the points made earlier regarding the diversity of the group as a whole.

A. Conjugates with P—O—SO$_3^-$ Linkages

These phosphosulfoconjugates take first place in the list because of their fundamental role in the biosynthesis of all other sulfoconjugates and because, as will be developed in Section II, they have other vital parts to play in the general recycling of sulfur in nature. The two principal representatives of this class are adenylyl sulfate (APS) and 3'-phosphoadenylyl sulfate (PAPS). APS is the biosynthetic precursor of PAPS and both may be regarded as "activated" forms of sulfate, rather analogous to "activated" forms of phosphate such as ATP. There are some grounds for suspecting that other compounds of this type may occur naturally. For example, Tsuyuki and Idler (1957) have reported the presence in salmon liver of a succinyladenyl phosphosulfoconjugate containing bound serine and glutamate, while Hori and Cormier (1966) suggest that the luciferin of *Renilla reniformis* contains a sulfate group bound in an "activated" form.

B. Conjugates with C—O—SO$_3^-$ Linkages

1. Arylsulfate Esters

In these sulfoconjugates a phenolic hydroxyl group is esterified with sulfuric acid. Such esters have not been detected in plants and microorganisms but are common in the animal world where many of them, including those of estrogenic steroids, are found under normal circumstances in urine. Here they constitute a major part of that fraction of urinary sulfur which has traditionally (and mistakenly) been referred to as "ethereal sulfate." Phenolic compounds or precursors not normally encountered during metabolism are also excreted in part as sulfate esters following administration to animals. In some instances arylsulfate esters have been detected in other tissue fluids and in the tissues themselves and one such compound, L-tyrosine O-sulfate, is particularly noteworthy because it occurs naturally in peptide-bound form.

2. Alkyl Sulfate Esters

a. Steroid Alkyl Sulfates. A number of different types of compound are present in this group, the distribution of which appears to be confined to the animal world. In one type, represented by such compounds as cholesteryl sulfate and dehydroisoandrosterone sulfate, sulfoconjugation involves a secondary cyclic alcohol group at C_3. In other closely related physiologically important steroids, a C_{17}, C_{20} or C_{21}-hydroxyl group is sulfoconjugated, and double sulfoconjugates, involving, for example, both C_3 and C_{17}-hydroxyls are also known. These various sulfoconjugates have been found in urine and plasma and in some instances in tissues, and it has become apparent in recent years that some of them probably constitute important participants in the biotransformations of steroids. The sterol vitamins, D_2 and D_3 can also be sulfated at C_3 by mammalian tissue preparations, but the biological significance of this is not yet clear.

A further type of steroid alkyl sulfate is represented by bile salts such as scymnol sulfate and ranol sulfate which occur in the biles of many primitive vertebrates. Primary or secondary alcohol groups situated in various positions of the C_{17} side chains are sulfated (sometimes an additional sulfate group is present at C_3), producing conjugates with pronounced surface-active properties.

Finally, there is some evidence, not yet adequately confirmed, that steroid alkyl sulfate diesters containing glycerol and fatty acids may occur in mammalian plasmas.

b. Simple Alkyl Sulfates. It will be seen in Section III,B that mammalian and toad liver preparations are able to sulfoconjugate simple alcohols such as ethanol, propanol, and butanol. However, with one exception (isopropyl sulfate), such sulfoconjugates have not been detected as natural products in the animal world.

In plants and microorganisms, choline sulfate is an alkyl sulfoconjugate of considerable importance. It appears to act as a nonionized store of sulfur that can be made available as inorganic sulfate in times of sulfur deficiency. In contrast, the biological role of 1,14-docosyl disulfate, which is present in certain microorganisms, is probably concerned with membrane structure (Mayers and Haines, 1967).

3. Carbohydrate Sulfate Esters

These constitute an extremely complex group of compounds of widely varying chemical structure and physiological function. Most of the representatives of the group are polysulfated polymers in which many hexose and/or hexosamine and/or uronic acid units are sulfated. Primary or secondary hydroxyl groups are sulfoconjugated and in one or two instances sulfate groups are present on hexosamine amino groups (N—SO$_3^-$ linkage). In all cases, molecules of strong anionic character are produced with interesting physicochemical and biological properties. Examples of the simplest types of these compounds are to be found in the polyhexose polysulfates of seaweeds and of relatively primitive organisms such as sea urchins and mollusks. The group reaches maximum complexity in the acidic proteoglycans (protein–acidic glycosaminoglycan complexes) and sulfated glycoproteins of animal tissues.

Other important carbohydrate sulfoconjugates occur as glycolipid sulfate esters (sulfatides) in nervous tissue and in at least one microorganism. Biochemical curiosities are represented by holothurin A, the neurotoxin of the sea cucumber (*Actinopyga agassizi*) and by the lactose sulfate, neuramin lactose sulfate and uridine diphospho-N-acetyl galactosamine 4-sulfate of animal tissues.

C. Conjugates with N—SO$_3^-$ Linkages

The presence of nitrogen-linked sulfate groups in certain complex carbohydrate sulfoconjugates has already been mentioned in Section I, B, 3. Heparin is the best-known example of compounds containing this type of linkage. However, it also contains many C–O–S linked sulfate groups and will therefore be considered as a carbohydrate sulfoconjugate.

The only other known examples of nitrogen-linked sulfoconjugates are to be found in the arylamine sulfates (strictly speaking, arylsulfamates) that can be detected in urine following the administration of compounds such as aniline and 2-naphthylamine. However, it is possible that the amino nitrogen of such physiologically important compounds as tyramine, serotonin, and noradrenaline may also undergo sulfoconjugation *in vivo*.

D. Conjugates with N—O—SO$_3^-$ Linkages

So far as is known, representatives of this group are confined to the plant world, where they exist as the so-called mustard-oil glycosides in the seeds and tissue of members of the Cruciferae. They can be regarded as sulfate esters of substituted hydroxylamines. These conjugates will not be discussed further.

E. Conjugates with S—SO$_3^-$ Linkages

Both cysteine and glutathione can give rise to S-sulfoconjugates when incubated with scrapings of rat intestinal mucosa in the presence of inorganic sulfate (Robinson and Pasternak, 1964; Robinson, 1965). PAPS is necessary for this sulfoconjugation to occur, but it is not yet established whether sulfate is transferred directly from PAPS to the preformed thiol or whether PAPS must first be reduced to yield inorganic sulfite. The biological importance of these conjugates as far as animals are concerned is obscure, although S-sulfoglutathione is known to be a component of the lens of the eye (Waley, 1959). The compounds will not be considered further in this chapter.

The attention of the reader is drawn to a number of earlier reviews on sulfate esters or on enzymes involved in their synthesis and degradation. These include reviews by Dodgson (1956, 1959, 1966), Dodgson and Lloyd (1968), Dodgson and Rose (1966), Dodgson and Spencer (1957 a,b), Gregory and Robbins (1960), Lloyd (1966), Ney (1959) and Roy (1960a).

II. Conjugates with P—O—SO$_3^-$ Linkages

A. Occurrence

The discovery of the two principal activated forms of sulfate, APS (I) and PAPS (II), and their relationship to each other announced by Hilz and Lipmann (1955), followed a series of pioneering studies by DeMeio and various other workers (see DeMeio and Arnolt, 1944; Bernstein and McGilvery, 1952a,b; DeMeio et al., 1953; Segal, 1955) on the sulfoconjugation of phenols in rat liver.

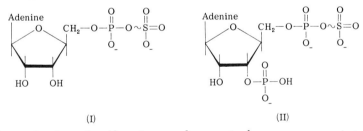

The activation of sulfate is now known to be necessary not only in sulfoconjugation but also in vital processes such as the biosynthesis of reduced sulfur compounds in plants and microorganisms, and it is therefore not surprising to find that APS and PAPS are formed by a wide variety of animal and plant tissues and by microorganisms of many kinds. Thus, apart from mammalian tissues, the synthesis has been demonstrated in lower animals, e.g., toad and tench (*Tinca tinca*) (unpublished results, these laboratories), the clam *Spisula solidissima*, and the marine snail

Busycon (Goldberg and Delbruck, 1959), plants (Ashai, 1964), marine algae (Goldberg and Delbruck, 1959), yeasts and fungi (Wilson and Bandurski, 1956), and other microorganisms (see Peck, 1962; Wilson, 1962). All these observations refer to the formation of APS and PAPS by cell preparations *in vitro,* but at least one worker (Pasternak, 1960) has demonstrated the formation of PAPS *in vivo* following the intraperitoneal injection of ^{35}S-labeled sulfate into mice. The wide natural distribution of these phosphosulfoconjugates suggests that they have important roles to fulfill in living organisms and further comment on this will be made later in this section.

Compared with other group donor substances, not a great deal is known about the energy of hydrolysis of the sulfate group in APS and PAPS. From enzymic studies, however, Gregory and Lipmann (1957) have concluded that for PAPS it is probably higher than that of the pyrophosphate linkages of ATP (cf. Banerjee and Roy, 1968). It is therefore not surprising that APS and PAPS are labile compounds and readily undergo hydrolysis in acid media, even at relatively low temperatures. In alkali they are much more stable and are claimed to withstand elevated temperatures.

B. Biosynthesis and Chemical Preparation

The biosynthetic production of APS and PAPS and also their preparation by chemical means has been adequately reviewed by a number of workers (see Gregory and Robbins, 1960; Robbins, 1962, 1963).

Briefly, APS is the immediate precursor of PAPS, which is formed from sulfate and 2 moles of ATP in a two-stage process. APS is formed in the first stage by the interaction of ATP and inorganic sulfate under the influence of the enzyme ATP-sulfate adenylyl transferase [Eq. (1)].

$$ATP + SO_4{}^{2-} \rightleftharpoons APS + PP_i \tag{1}$$

The formation of APS is a thermodynamically unfavorable process since $\triangle F'$ equals $+ 11$ kcal, but the removal of PP_i by inorganic pyrophosphatase and the removal of APS by subsequent transformations effectively displaces the equilibrium. APS is not utilized directly in sulfation in the mammalian system but interacts with a second molecule of ATP to produce PAPS in an irreversible reaction catalyzed by ATP-adenylyl sulfate 3'-phosphotransferase [Eq. (2)].

$$APS + ATP \rightarrow PAPS + ADP \tag{2}$$

The sulfate activating system has been purified from baker's yeast (Robbins and Lipmann, 1958a,b; Wilson and Bandurski, 1958) and, together with the mammalian liver system (Spencer, 1960), has been examined in some detail. Studies on the system from other sources, although limited, have revealed many similarities but also some striking species differences, particularly with respect to the optimum concentrations of the reactants. For instance, the rat liver and yeast systems have similar broad optimal ATP concentration requirements (Mattock, 1967), whereas those of toad liver and mouse mastocytoma tissue (unpublished results, these laboratories; Balasubramanian et al., 1967) have very critical requirements for a particular concentration of this substrate. Attention to these factors is therefore important when attempting to detect the presence of the sulfate activating system in other species.

Both APS and PAPS can be prepared by chemical synthesis (Baddiley, et al., 1957, 1959; Reichard and Ringerts, 1957; Cherniak and Davidson, 1964). When required in substrate amounts, APS is most conveniently obtained in this way but PAPS is more readily prepared biosynthetically, employing liver or yeast preparations (Brunngraber, 1958; Banerjee and Roy, 1966). Both compounds will keep for several days without appreciable deterioration if stored in the frozen state.

C. Enzymic Hydrolysis

Both APS and PAPS are broken down by rat liver and other tissue preparations to yield a variety of products (Brunngraber, 1958; Lewis and Spencer, 1962; Fujiwara and Spencer, 1962) including adenosine, inosine, hypoxanthine, sulfate, and phosphate. These observations have demonstrated the role of known 3'- and 5'-nucleotidases as well as various deaminases in the breakdown of APS and PAPS, but also indicated the possible participation of sulfohydrolase enzymes.

The presence of PAPS sulfohydrolase activity in extracts of animal tissues has been confirmed by several workers (e.g., Suzuki and Strominger, 1960b; Adams, 1964; Abraham and Bachhawat, 1964) and recent studies with rat liver (Koizumi et al., 1969) show that activity is localized in both the lysosomal and supernatant fractions of the cell. Inorganic sulfate and 3'-phosphoadenosine 5'-phosphate (PAP) are the reaction products of the sulfohydrolase, which is strongly activated by cobalt ions. A similar enzyme has been found in sheep brain (Balasubramanian and Bachhawat, 1962) and recent studies in our own laboratories indicate that this type of activity may be widely distributed throughout the various phyla of the animal world.

A second and quite distinct sulfohydrolase which hydrolyzes APS to inorganic sulfate and AMP has recently been identified in rat liver

(Bailey-Wood *et al.*, 1969). Enzyme activity is localized in both the lysosomal and soluble fractions of the liver cell and a purified preparation (150-fold) of the enzyme, free from the PAPS sulfohydrolase and other APS degrading enzymes, has been obtained from the latter fraction. The purified enzyme shows no activity toward the structural analog ATP, which, in common with certain other nucleotides and ADP, has a pronounced inhibitory effect. In contrast, AMP, the product of the desulfation of APS, has no effect on the enzyme which, unlike the PAPS sulfohydrolase, is inhibited by cobalt ions.

Other pronounced differences between the APS and PAPS sulfohydrolases have been demonstrated by inhibition and activation studies and the existence of two separate enzymes in the rat liver cell is clearly established. The liver cell can therefore hydrolyze as well as synthesize APS and PAPS. Both of these facilities are features of the soluble fraction of the cell and it may be wondered whether an integrated interplay of the activities of the enzymes constitutes a mechanism for controlling the cellular levels of APS and PAPS.

D. Biological Significance of APS and PAPS

Following the discovery of these compounds it soon became apparent that they played vital roles, not only in sulfoconjugation processes, but also in dissimilatory and assimilatory pathways of sulfur metabolism in plants and microorganisms and possibly in sulfate-transport processes.

1. Utilization in Sulfate Reduction

Dissimilatory and assimilatory sulfate-reduction processes are important features of the constant recycling of sulfur in Nature and these and other related points have been considered in detail by Dodgson and Rose (1966). Sulfate activation is an essential prerequisite for these reductive processes, and this may be deduced from the fact that the initial stage of the reduction (i.e., conversion of sulfate to sulfite) is thermodynamically unfavorable with a standard free-energy change of +14 kcal.

Certain relatively primitive bacteria such as *Desulfovibrio desulfuricans* are able to produce active sulfate in the form of APS and to utilize this in a dissimilatory reduction process leading to the formation of hydrogen sulfide with the concomitant liberation of energy. In this process, sulfate is serving as a terminal electron acceptor for energy-yielding electron transfers, a role that may have featured prominently on the primitive Earth when the atmosphere was largely anaerobic.

During the course of the evolution of microorganisms, the requirement for reduced organic forms of sulfur presumably led to the early development of pathways for the conversion of inorganic sulfate into such compounds as cysteine and methionine. PAPS is produced as the active sulfate intermediate in such assimilatory reduction processes. Similar processes leading to the formation of reduced organic sulfur compounds also occur in plants and, through the intervention of rumen microorganisms, in certain herbivorous animals.

In the animal world, reduction of inorganic sulfate has been claimed to occur in chick embryonic tissue (Machlin et al., 1955; Chapeville and Fromageot, 1957) and in bovine corneal epithelium (Wortman, 1963), while preparations of rat intestinal mucosa are apparently able to produce PAPS and subsequently reduce it to inorganic sulfite (Robinson, 1965). However, it seems probable that further reduction of sulfite cannot occur in higher animals (cf. Huovinen and Gustafsson, 1967) and the appearance in tissues of ^{35}S-labeled sulfur-containing amino acids following the injection of labeled inorganic sulfate (Boström and Åqvist, 1952; Dziewiatkowski and Di Ferrante, 1957; Rambaut and Miller, 1965) is almost certainly the result of the activity of intestinal microorganisms. Generally speaking therefore, higher animals appear to have lost the facility to utilize active forms of sulfate for reductive assimilatory (or dissimilatory) purposes and the role of APS and PAPS appears to be concerned only with the process of sulfoconjugation. Such animals are, however, ultimately dependent on the APS- and PAPS-producing systems of plants and microorganisms for their supplies of reduced organic sulfur compounds.

2. Utilization in Transport

There are some indications that PAPS may be involved in sulfate transport mechanisms. Although a nonenergy dependent active transport of sulfate, involving a specific sulfate-binding protein, is known to be a feature of at least one microorganism (Pardee et al., 1966), plants and animals probably utilize energy-dependent mechanisms. This has been established for simple plants and an "active" form of sulfate has been implicated (Wedding and Black, 1960; Abraham and Bachhawat, 1964; Bidwell and Gosh, 1963). In mammals, several workers have established the energy-dependent nature of the process (Astudillo et al., 1962; Deyrup, 1963; Omachi, 1964; Anast et al., 1965) and sulfate transport through the intestinal wall of the rabbit is believed to involve PAPS (Astudillo et al., 1964). However, much more work is needed before meaningful conclusions about the role of PAPS in transport phenomena can be made.

3. Utilization in Sulfate Ester Biosynthesis

From the foregoing discussion it will be apparent that, during the course of evolution, changes in the relative functional significance of activated forms of sulfate have taken place. Bacteria employ these compounds mainly in sulfate reduction processes, although APS is used as an intermediate in certain organisms which can oxidize thiosulfate to sulfate (see Dodgson and Rose, 1966) and at least one sulfoconjugate has been detected in a bacterium (Kates *et al.*, 1967). In fungi, algae, and plants sulfoconjugation becomes more common, but the ability to reduce PAPS for assimilatory purposes persists. In higher organisms reductive processes involving PAPS virtually disappear but sulfoconjugation becomes much more versatile and elaborate, reaching its maximum complexity in mammals.

In Section I attention was drawn to the great variety of naturally occurring sulfate esters of widely differing chemical structures. Formation of all these compounds involves PAPS and appropriate transferring enzymes (sulfotransferase) which transfer the activated sulfate group to appropriate acceptors. The present view is that many such sulfotransferases exist, some of them displaying a remarkable degree of acceptor specificity. Many of the known sulfotransferases are discussed in later sections, but it is probable that more enzymes still remain to be discovered.

E. THE ROLE OF VITAMIN A IN PAPS BIOSYNTHESIS

There have been numerous reports that the metabolism of sulfate in mammals is markedly influenced by vitamin A (e.g., Fell *et al.*, 1956; Dziewiatkowski, 1954; Balakhovski and Kuznetsova, 1958). However, the assignment of a *specific* role of the vitamin has been the subject of controversy. Thus, low activity of ATP-sulfate adenylyltransferase is claimed to be a significant feature of the vitamin A-deficient animal (Varandani *et al.*, 1960; Subba Rao *et al.*, 1963; Sundaresan, 1966). Experiments on enzyme isolated from the livers of vitamin A-deficient rats established that its low activity *in vitro* could be markedly enhanced by addition of retinol or retinoic acid (Subba Rao and Ganguly, 1964, 1966), or by an unknown metabolite of vitamin A which could be extracted from normal liver (Sundaresan *et al.*, 1964). Recently, 5,6-monoepoxyretinoic acid has been shown to have activity comparable with that of the unknown liver metabolite. The work of Carroll and Spencer (1965) on the enhancement of the PAPS-synthesizing activity of fetal liver by vitamin A derivatives supports the idea of a specific cofactor role for the vitamin or a derivative. However, other workers have completely failed to detect any significant disturbance of PAPS synthesis in a variety of tissues from vitamin A-deficient rabbits and rats (Pasternak *et al.*, 1963; Pasternak and Pirie, 1964; Mukherji and Bachhawat, 1967).

Some of the present confusion can probably be traced to the lack of standardization of experimental conditions, particularly with respect to the degree of avitaminosis of the animals used. Much more attention has been given to this point in the more recent studies but, even so, other factors such as sex and species differences and the nature of the tissue under investigation have often been overlooked. Moreover, the general nutritional state of the animals must be taken into account in such experiments, since protein deprivation alone results in a considerable reduction of the tissue levels of ATP-sulfate adenylyltransferase (Levi *et al.*, 1968; Geison *et al.*, 1968). For the present therefore, the precise relationship between PAPS biosynthesis and vitamin A cannot be said to have been unequivocally established.

F. ALTERNATIVE SULFOCONJUGATION MECHANISMS

Although PAPS is well established as the sulfate donor for sulfoconjugation reactions, the possibility that alternative sulfate donors might exist requires brief mention. The early claims (Suzuki *et al.*, 1957) that sulfation of polysaccharides in mollusk preparations might occur via an unknown arylsulfate donor of high sulfate group potential, have never received adequate experimental support and can probably be dismissed.

More recently it has been proposed that ascorbic acid may be involved in sulfation (Ford and Ruoff, 1965). Ascorbic acid sulfated at position 3 has been prepared synthetically and shown to be capable of nonenzymically sulfating certain alcohols (Mumma, 1968) and steroids (Chu and Slaunwhite, 1968) in the presence of suitable oxidizing agents. It is not known if ascorbic acid sulfate is formed *in vivo* or whether it would fulfill the role of a biological sulfating agent, but Chu and Slaunwhite speculate that their observations would be in accord with the high concentration of ascorbic acid found in adrenal tissue (Szent-Györgyi, 1928; Sayers *et al.*, 1944) and with the oxidation of ascorbic acid that occurs during adrenal steroidogenesis (Salomon, 1957). Another novel nonenzymic process has been described by Adams (1962) who observed the sulfation of hexosamines and N-acetylhexosamines by PAPS in the presence of charcoal. Similarly, sulfur analogs of 1-phosphoimidazole can participate in sulfate transfer reactions (Mayers and Kaiser, 1968). Reactions of this kind presumably have no biological significance but could be important factors in chemical manipulations involving PAPS.

III. Conjugates with C—O—SO$_3$$^-$ Linkages

A. ARYLSULFATE ESTERS

1. *Occurrence*

The existence in mammalian urines of more than one form of sulfate has been known since the early part of the nineteenth century but the isolation (Baumann, 1876) of potassium phenyl sulfate (III) from the urine of dogs which had been fed with phenol was the first clear evidence that phenolic compounds could be esterified with sulfuric acid *in vivo*. It has since become clear that sulfoconjugation is one of the principal ways in which many phenols are metabolized and subsequently excreted.

(III)

A large part of the total sulfoconjugate ("ethereal sulfate") fraction of normal urines (total in humans is equivalent to between 0.1 and 0.3 gm SO$_4$$^{--}$/day) is composed of phenolic sulfoconjugates (arylsulfates) formed from phenolic materials normally present in the diet or which arise as a result of metabolism within the tissues or in the intestinal tract. The phenols that may form arylsulfates under natural circumstances are many and varied (see Williams, 1959) and include simple phenols and phenolic acids, heterocyclic phenols such as indoxyl and skatoxyl and phenolic hormones, including phenolic steroids such as estrone. However, the evidence for this is frequently indirect and comparatively few of the sulfated products have actually been isolated from urine and positively identified. To some extent this is due to the difficulties of isolating arylsulfates from urine (see Dodgson *et al.*, 1955a), particularly as they are relatively labile, extremely soluble and present, individually, in small amounts only. Among the more biologically significant ones that have been positively identified in normal human urine have been estrone sulfate (Schacter and Marrian, 1938; McKenna *et al.*, 1961), estriol 3-sulfate (Troen *et al.*, 1961), indoxyl sulfate, and skatoxyl 6-sulfate (Acheson and Hands, 1961; Heacock and Mahon, 1964), serotonin *O*-sulfate (Davis et al., 1966), L-tyrosine O-sulfate (Tallan et al., 1955; John et al., 1966), and 2-amino-3-hydroxyacetophenone O-sulfate (Dalgliesh, 1955).

The occurrence of arylsulfates in mammalian tissues and in tissue fluids other than urine is much less clearly defined, probably because systematic searches for such compounds have rarely been made. However, various estrogen sulfates (including an interesting double conjugate, estriol-3-sulfate-16-glucuronosiduronate; Touchstone et al., 1963; Levitz and Katz, 1968) have been recognized in bile and other tissue fluids and tissues (e.g., Menini and Diczfalusy, 1961; Purdy et al., 1961). Other non-estrogenic arylsulfates have also been detected, including 3:5-3'-triiodo-L-thyronine O-sulfate (Roche et al., 1959) and 3-methoxyphenylglycol-4-sulfate (Schanberg et al., 1968). Bilirubin disulfate has been detected in bile but the precise nature of the mode of attachment of the sulfate groups is not yet clear (Noir et al., 1966).

Finally, and perhaps one of the most significant findings yet made, peptide-bound L-tyrosine O-sulfate residues are present in mammalian and other fibrinogens (e.g., Bettelheim, 1954; Doolittle and Blombäck, 1964), in the polypeptide hormone of the stomach, gastrin II (H. Gregory et al., 1964) and, among lower organisms, in phyllokinin and caerulein, two biologically active polypeptides isolated respectively from the skin of a South American tree frog (Anastasi et al., 1966) and from the skins of various Australian amphibians of the genus Hyla (Anastasi et al., 1968). Free L-tyrosine O-sulfate has also been detected in the tissues of a number of relatively primitive organisms such as starfishes, sea urchins, and shrimps (Kittredge et al., 1962).

Probably many arylsulfate esters that still remain to be identified are normally present in living organisms. Thus, following injection of Na_2 $^{35}SO_4$ into experimental animals, a large number of radioactive sulfate esters can be detected on paper chromatograms of the urines. In the rat the chromatographic pattern alters considerably if the animals are bred and maintained under germ-free conditions (Boström et al., 1963) and sulfoconjugates of phenol and indoxyl, which are present in "normal" rat urines cannot be detected. This illustrates the probable importance of the gut flora in determining the overall pattern of urinary arylsulfate excretion. The use of more indirect methods (Boström and Vestermark, 1960) has revealed that human urine also contains a complex pattern of sulfate esters. The relative movements of many of these esters on two-dimensional paper chromatograms suggest that they are arylsulfates. The biles of animals receiving $Na_2^{35}SO_4$ also give complex sulfoconjugate chromatographic patterns (e.g., Boström and Vestermark, 1959; Spencer, 1960) probably reflecting, at least in part, the presence of arylsulfates.

Extracts of various rat tissues contain endogenous sulfate acceptors, several of which are phenolic in character (Spencer, 1960) and cell sap preparations from liver and other tissues are certainly able to sulfate many important phenols, including L-adrenaline, L-noradrenaline, tyramine, thyroxine, pyridoxine, and pyridoxal (Spencer, 1960; Boström and Wengle, 1964; Sato et al., 1959). By inference it may be supposed that sulfoconjugates of these compounds can occur naturally and may ultimately be recognized.

The view might be held that interest in naturally occurring arylsulfates has tended to become obscured by the many studies that have been made on arylsulfates produced as a result of externally contrived or abnormal circumstances. For example, in the field of drug metabolism, numerous studies have been made on arylsulfates that appear in urine following administration of phenols or phenolic precursors to intact animals. Other analogous studies have involved a similar type of approach in vitro, where phenols have been added to suitably fortified liver slices or cell sap preparations in the presence of $Na_2{}^{35}SO_4$ (e.g., Sato et al., 1956). From these various approaches information has accumulated regarding the ability of mammalian organisms to form arylsulfates as a means of "detoxicating" phenols, but surprisingly little information has accrued to our knowledge and understanding of naturally occurring arylsulfates.

2. General Properties

Sulfoconjugation of phenols results in marked changes in physical, chemical, and biological properties. The sulfoconjugates are usually strongly acidic and must be stored in the laboratory as alkali salts at low temperatures in order to minimize autocatalytic degradation. The salts are usually extremely soluble in aqueous media and, in solution, they tend to undergo spontaneous and autocatalytic hydrolysis which, in at least one instance (p-nitrophenyl sulfate, a much-used substrate for arylsulfohydrolases), is photochemically accelerated (Havinga et al., 1956). They are readily hydrolyzed by acid but are relatively stable to alkali. Acid hydrolysis is facilitated when the sulfate group is present at a position of low electron-availability in the aromatic nucleus, and by the presence of electrophilic substituent groups (Burkhardt et al., 1936). Scission of the O-S bond of the C-O-S linkage occurs (Spencer, 1958), probably via attack by the electrophilic hydroxonium ion. In contrast, alkaline hydrolysis requires drastic experimental conditions and is believed to involve an attack on the C-O bond, although this could not be confirmed by Spencer (1958).

Sulfoconjugation eliminates or greatly modifies a number of important properties characteristic of the phenolic hydroxyl group. For example, the marked shifts in position of λ_{max} and increases in intensity of visible or ultraviolet absorption associated with the conversion of the phenolic hydroxyl to its anionic form are not shown by the corresponding sulfoconjugate. The absorption of the latter in either acid or alkali tends to approximate to that of the cation of the parent phenol. Advantage has been taken of these facts in developing spectroscopic methods for determining the rates of enzymic hydrolysis of arylsulfates (Dodgson and Spencer, 1953a, 1957b). Marked changes also occur in the infrared spectrum following sulfoconjugation.

Finally, it may be reasoned that sulfoconjugation may reduce or abolish the physiological properties of those compounds in which such activities are at least partly due to the free phenolic grouping. Unfortunately, direct evidence for this concept is extremely scanty and represents a serious gap in our knowledge when attempting to suggest biological roles for arylsulfates and for the enzymes that utilize them as substrates. Moreover, in at least one case (the polypeptide, phyllokinin), physiological activity is enhanced by the presence of the sulfate group. Further comments on these points are made in Section III,A,6.

3. Biosynthesis

Both *in vitro* and *in vivo* studies have provided information concerning the biological sulfation of phenols and, on the whole, the results of the different approaches have been complementary and consistent with the view that sulfate is transferred from PAPS to phenols under the influence of specific sulfotransferases. Almost the whole of the work that led to the discovery of PAPS and to the elucidation of its biosynthesis was based on experiments involving the formation of arylsulfates. During some of the early studies (see Nose and Lipmann, 1958) it became apparent that the crude extracts of tissues such as liver contained more than one sulfotransferase enzyme and it has since become obvious that a whole family of such enzymes exists, some members apparently showing a high degree of specificity. However, few of these enzymes have been purified sufficiently for their specificity to be precisely defined.

So far as phenols and phenolic steroids are concerned, the indications are that at least three distinct sulfotransferases exist in mammalian tissues, namely, phenolsulfotransferase, estrogen sulfotransferase, and L-tyrosine methyl ester sulfotransferase. The use of the word "indications" stresses the tentative nature of this statement and reflects a degree of confusion in the field at the present time. The reasons for this are not hard to find. First, investigators have used different animal species and tissues for their studies. Second, the sulfotransferases tend to be extremely unstable enzymes that are difficult to purify, and third, mixtures of different sulfotransferases are not easily resolved, so that the distinction between enzymes often has to be based on kinetic and similar evidence rather than on clean separation by enzyme fractionation techniques. These difficulties materialize in apparently contradictory findings; for example, Adams and Poulos (1967) claim to have separated an estrogen sulfotransferase from bovine adrenal glands that shows no other sulfotransferase activity except a feeble one toward the synthetic estrogens, stilbestrol and hexestrol. In contrast, Roy and Banerjee (1966) claim that guinea pig estrogen sulfotransferase can also transfer sulfate to simple phenols and arylamines.

a. Phenol Sulfotransferase. The ability to sulfoconjugate simple phenols such as phenol and *p*-nitrophenol is a feature of many animal organisms, including birds (Layton and Frankel, 1951), amphibians (DeMeio, 1945), mollusks (Goldberg and Delbruck, 1959), and insects (Smith, 1955). In mammals this ability is present in very many tissues (see, e.g., Holcenberg and Rosen, 1965) and, so far as the collective results can be interpreted, it seems clear that the bulk of the sulfotransferase activity is present in the soluble fraction of cell preparations—that is, in the same fraction as the system responsible for PAPS biosynthesis. Although many microorganisms and plants are able to produce PAPS, they do not appear to possess phenolsulfotransferases.

Mammalian livers are probably the most convenient source of phenol sulfotransferase, although not necessarily the richest source. Unfortunately, no really sound quantitative distribution studies of the enzyme have been made. Such studies as have been made have frequently measured the total synthetic process, including PAPS formation, and have usually disregarded the fact that sulfohydrolases and phosphatases capable of degrading APS and PAPS are present in crude tissue preparations. Similar problems are posed if PAPS is first prepared and used directly as sulfate donor. A novel system for phenolsulfotransferase assay was developed for rabbit tissues by Gregory and Lipmann (1957) who took advantage of the reversibility of the enzyme under circumstances where the arylsulfate ester involved has a high sulfate group potential approaching that of PAPS itself. *p*-Nitrophenyl sulfate is such a compound and, if present in sufficiently high concentrations, can donate sulfate to PAP in the presence of phenol sulfotransferase [Eq. (3)].

$$p\text{-Nitrophenyl sulfate} + PAP \xrightarrow{\text{phenolsulfotransferase}} p\text{-nitrophenol} + PAPS \quad (3)$$

This reaction is particularly effective if a sulfate acceptor such as phenol or m-aminophenol is also present. The sulfoconjugates of these compounds have very low sulfate group potentials and, under these circumstances, sulfate can be transferred from p-nitrophenylsulfate via PAPS to phenol [Eq. (4)].

$$p\text{-Nitrophenyl sulfate} + phenol \xrightarrow[\text{PAP}]{\text{phenolsulfotransferase}}$$
$$p\text{-nitrophenol} + phenyl\ sulfate \quad (4)$$

The p-nitrophenol released may then be measured spectrophotometrically.

Ingenious though this assay system is, its general validity must be in doubt following the recent failure of Banerjee and Roy (1968) to achieve a similar transfer with a purified phenolsulfotransferase from guinea pig liver. Whether this reflects species differences or some other factor is not yet clear.

In spite of the problems inherent in assaying phenolsulfokinase activity in crude tissue preparations, attempts have been made to investigate enzyme levels under different physiological circumstances. Thus, Wengle (1963), employing $PAP^{35}S$ as sulfate donor and phenol as acceptor, found low activity in fetal and neonatal rats and concluded that these animals were poorly equipped for the "detoxication" of phenols as sulfoconjugates. Similar findings were made in rats (Carroll and Spencer, 1965) and guinea pigs (Spencer and Raftery, 1966), and in these instances the low levels of enzyme were attributed to fetal deficiency of an essential cofactor related to vitamin A, reflecting the low maternal transfer of the vitamin to the fetus. However, much more work is required before this explanation can be accepted unequivocally. Carroll and Spencer (1965) also noted transfer of sulfate from PAPS to endogenous acidic glycosaminoglycans in fetal skin and this type of transfer would, of course, be a basic necessity for biosynthesis of all connective tissue glycosaminoglycans in the developing fetus. With this in mind, the low phenolsulfotransferase activity of fetal rats and guinea pigs is seen as reflecting the low "detoxicating" needs of the fetus. Some indirect support for this view comes from the observation (Spencer and Raftery, 1966) that the situation is quite different in the enclosed system of the developing hen's egg, where phenolsulfotransferase levels are extremely high.

Early work on phenolsulfotransferase showed a singular lack of agreement between investigators as to the general and kinetic properties of the enzyme, and these discrepancies have been adequately summarized in a review by Roy (1960a). The most reliable published data are those of Banerjee and Roy (1966, 1968) for the guinea pig liver enzyme which was partially purified from cell sap preparations and clearly separated from sulfotransferase activity toward estrone, arylamines, and steroids such as dehydroisoandrosterone, cholesterol, and deoxycorticosterone. However, when tested in our own laboratories, Roy's preparation was active toward L-tyrosine methyl ester and further comment on this is made in Section III,A,3,c.

When p-nitrophenol was used as the acceptor for assaying "purified" phenolsulfotransferase, maximum activity was obtained at pH 5.6 and, in contrast to some other sulfotransferases, there was no requirement for Mg^{++} ions. The enzyme is very unstable and, under test conditions at $37°C$, will give zero-order kinetics for about 10 minutes only. Kinetic studies of the forward reaction and the type of inhibition produced by the products show that there are independent binding sites for the phenol and PAPS and that the reaction catalyzed is a rapid equilibrium random bi bi reaction with one dead-end complex of p-nitrophenol and 3'-phosphoadenylic acid. The reaction can be represented by Eq. (5)

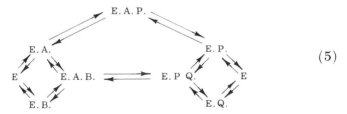

$$\tag{5}$$

where E is the enzyme, A is p-nitrophenol, Q is p-nitrophenyl sulfate, B is 3'-phosphoadenylyl sulfate, and P is 3'-phosphoadenylic acid.

In terms of the maximum velocity for the forward reaction (V_1), the enzyme showed the following order of specificity toward a variety of phenolic acceptors: 1-naphthol > p-nitrophenol > 2-naphthol > 5,6,7,8-tetrahydro-2-naphthol > 2-phenanthrol > 4-nitro-1-naphthol > phenol. Slight activity toward equilin and equilenin was observed but estrone was not an acceptor, nor did it inhibit transfer to p-nitrophenol. The molecular weight of the enzyme appeared to be of the order of 70,000.

It may probably be assumed that this type of sulfotransferase is normally responsible for the sulfation of simple phenols by mammalian tissues, although it is a pity that information is not yet available regarding the ability of the enzyme to transfer sulfate to biologically important phenols such as serotonin, 3:5:3'-triiodothyronine, and adrenaline. It will be interesting, when a really pure enzyme is obtained, to investigate the relative specificity of the enzyme more thoroughly in order to see what correlation (if any) exists between specificity and the ability of intact animals to deal with exogenous phenolic materials. The one investigation which appears to have been made on intact animals is that by Williams (1938) who studied the ability of rabbits to sulfoconjugate orally administered phenols containing different substituent groups in the benzene ring. The results generally indicated that electrophilic *ortho*-substituent groups tended to depress sulfoconjugation, whereas nucleophilic *ortho*-substituent groups had the opposite effect. A similar, but less pronounced effect was obtained with *meta*-substituent groups, while different *para*-substituent groups had relatively little effect. *o*-Hydroxybenzoic acid was not conjugated at all by the rabbit, and this was attributed to "*ortho*-effect." Of course, feeding experiments of this type have inherent weaknesses on which one need not elaborate; some of them have been considered by Bray *et al.* (1952a,b); nevertheless it would certainly be of interest to see whether similar relationships existed at the enzyme level.

A further interesting point about the specificity of phenolsulfotransferases emerges when one considers the situation where more than one phenolic hydroxyl group is available for sulfoconjugation. Vestermark and Boström (1960) have indicated that polyhydric phenols such as catechol and phloroglucinol may become polysulfated by suitably fortified liver preparations. However, a similar polysulfation may not occur to a significant extent *in vivo*. Thus, orally administered catechol and quinol (Garton and Williams, 1948, 1949) appear to give rise to monosulfate esters *in vivo*. In the case of the substituted dihydric phenol, 4-chloro-2-hydroxyphenol (4-chlorocatechol), Dodgson *et al.* (1955a) were able to isolate a monosulfate only from the urines of rabbits receiving the phenol. This ester was characterized as 4-chloro-2-hydroxyphenyl sulfate, thus establishing the preferential biological sulfation of the phenolic hydroxyl furthest removed from the substituent chlorine atom. Moreover, the monosulfate ester obtained by chemically sulfating 4-chlorocatechol with sufficient reagent to sulfate one hydroxyl only, was identical with that isolated from the rabbit urine. This provides another pointer to the probability that the nature of substituent groups may profoundly influence the sulfoconjugation process and suggests an interesting field of study in relation to the phenolsulfotransferases.

It will be seen in Section III, A, 5 that conjugation of a phenol with sulfate does not necessarily mean that the resultant conjugate is destined for excretion without further metabolic change.

b. Estrogen Sulfotransferase. As with phenolsulfotransferase, the natural distribution of estrogen sulfotransferase appears to be restricted to the animal world. Included in the tissue distribution studies that have been made are those on the ox (Holcenberg and Rosen, 1965), adult (Boström and Wengle, 1967) and fetal (Wengle, 1964; Diczfalusy *et al.*, 1961) humans, the hen (Raud and Hobkirk, 1968), and sea urchin (Creange and Szego, 1967). The collective findings of these and other studies show the enzyme to be localized in the cell sap and that its tissue distribution generally, but not always, follows that of phenolsulfotransferase. Because of the hormonal nature of the acceptor molecules, particular attention has been paid to the placenta and to fetal tissues as a source of the enzyme, but the results obtained by different workers have not always added up to form a coherent pattern. Thus bovine and guinea pig placentas possess high enzyme activity whereas activity in human placenta appears to be quite low (Holcenberg and Rosen, 1965; Levitz *et al.*, 1961; Bolte *et al.*, 1964). Again, human fetal liver, kidney, and lung are all able to sulfoconjugate estrogens at a relatively high rate (approx. 40% of that of the adult liver, Wengle, 1964), whereas fetal and neonatal rat livers have little or no estrogen sulfotransferase activity compared to that of adults (Carroll and Spencer, 1965; Raud and Hobkirk, 1966).

Several workers (e.g., Wengle and Boström, 1963; Payne and Mason, 1963) have shown that estradiol and estriol can form disulfated conjugates but in these instances it must be presumed that the estrogen sulfotransferase is responsible for the sulfation of the phenolic hydroxyl only (cf. Adams and Poulos, 1967).

The two most detailed studies on estrogen sulfotransferases have involved the guinea pig liver enzyme (Banerjee and Roy, 1966; Roy and Banerjee, 1966) and that from bovine adrenal glands (see Adams and Poulos, 1967; Adams and Chulavatnatol, 1967). No clean separation of the guinea pig enzyme from sulfotransferases other than phenolsulfotransferase could be achieved, but it was possible to conclude that the enzyme was distinct from others that transferred sulfate to steroid alkyl hydroxyl groups. The optimum pH of the enzyme is in the region of 6.0 when estrone is the acceptor molecule and, in contrast to phenolsulfotransferase, full enzyme activity depends on the presence of Mg^{++} ions. The rates of sulfation of estrogens by the impure enzyme decreased in the order estrone > 17-deoxyestrone > estradiol > 2-methoxyestrone > estriol.

The bovine adrenal enzyme has been separated from other sulfotransferases (N.B. L-tyrosine methyl ester has not been tested with the final enzyme preparation) and a pH optimum of 8.0 was noted when estrone was used as the acceptor. Enzyme activity was considerably enhanced by

Mg^{++} ions. The rate of sulfoconjugation of the three classical estrogens decreased in the reverse order to that described for the guinea pig enzyme. A free thiol group (or groups) appears to be important for enzyme activity and kinetic studies suggest that the mechanism of action of the enzyme is of the sequential type.

DEAE-cellulose chromatography and electrophoresis on Geon resin have established that two forms (A and B) of the bovine enzyme exist. Form B is converted to A on standing and only B is obtained if the enzyme is isolated in the presence of mercaptoethanol. The relationship of B to A appears to be that of a trimer to a monomer, the latter having a molecular weight of about 67,000 (i.e., close to that for phenolsulfotransferase). The effect of mercaptoethanol is taken to indicate that association requires a conformation that can be maintained only when a thiol group(s) is in the reduced state. Both A and B forms on gel electrophoresis give multiple bands representing individual isoenzymes. Kinetic studies on both forms of enzyme suggest that the fully associated enzyme exhibits allosteric properties and point to the possibility that estrone is the true substrate for the enzyme. Further support for this latter possibility has come from an elegant study (Adams, 1967) that revealed that the isolated sulfotransferase preparation actually contained estrone which had remained attached to the enzyme throughout the purification procedure. The estrone was released as estrone sulfate simply by incubating the enzyme with PAPS. The finding promoted an examination of the attachment of other estrogens to the enzyme and led Adams to conclude that the enzyme interacted with the rear or α-face of the estrogen.

Further work is now required to establish the biological significance of the allosteric and other properties of the enzyme in relation to the general problem of estrogen transformations and their biological control.

c. L-*Tyrosine Methyl Ester Sulfotransferase.* Mention has been made in an earlier section (III,A,1) that L-tyrosine O-sulfate is a unique arylsulfate ester in that it occurs in nature in a peptide-bound form, as well as in urine in free form. A more detailed account of this ester is provided in section III,A,6, but some attention is given here to the unresolved problem of its biosynthesis. Briefly, following the discovery of naturally occurring L-tyrosine O-sulfate, a number of laboratories attempted to sulfoconjugate free L-tyrosine employing suitably fortified cell sap preparations from liver and other sources under conditions in which phenolsulfoconjugation would normally operate (e.g., Grimes, 1959; Nose and Lipmann, 1958; Suzuki and Strominger, 1960a). Segal and Mologne (1959) also studied this problem briefly and concluded, on the basis of indirect evidence, that only tyrosine derivatives in which the carboxyl group was absent or substituted and the amino group was free would undergo sulfoconjugation by the PAPS-phenolsulfotransferase system. This problem has been investigated in some detail in these laboratories (Jones and Dodgson, 1965; Jones *et al.*, 1966; Basford *et al.*, 1966; Dodgson *et al.*, 1967) and it has been firmly established by direct evidence that L-tyrosine can be sulfoconjugated only when the carboxyl group is blocked. The methyl and ethyl esters of L-tyrosine, L-tyrosylglycine, and L-tyrosylalanine will all act as acceptors. Moreover, the products can subsequently lose the carboxyl-blocking group *in vivo* and *in vitro* to yield free L-tyrosine *O*-sulfate. Tyramine is also a substrate for the enzyme. The responsible enzyme has been partially purified some 450-fold from rat liver but a clean separation from all other sulfotransferases has not yet been achieved. The preparation is not active toward estrone and mixed substrate and other evidence shows that the enzyme is distinct from that responsible for the sulfoconjugation of dehydroisoandrosterone. The important question is whether the enzyme is identical with the ordinary liver phenolsulfotransferase and the collective evidence suggests that it is not. A partial separation of the activity toward L-tyrosine methyl ester and that toward *p*-nitrophenol has been achieved by column chromatography; the two activities have quite different stabilities and their behavior in the presence of varying ratios of oxidized to reduced glutathione are virtually diametrically opposed. The ratios of activities in different tissues vary considerably, as do those shown at varying stages during purification. The collective findings indicate that transfer of sulfate from PAPS to *p*-nitrophenol on the one hand, and to L-tyrosine methyl ester, L-tyrosylglycine, and tyramine on the other, is accomplished by different sulfotransferases. For the present, the enzyme responsible for the latter activities is referred to as L-tyrosine methyl ester sulfotransferase. Work now reaching completion also establishes that the reaction catalyzed by the enzyme is of the rapid-equilibrium random bi bi type, the sites for PAPS and L-tyrosine methyl ester binding therefore being independent. The enzyme is specific for a nonprotonated amino group and probably a protonated phenolic hydroxyl group so far as L-tyrosine methyl ester and tyramine are concerned. There is a group in the PAPS-binding site with a pK value of about 8.5 which may be a thiol group in view of the fact that PAPS, but not tyrosine methyl ester, protects the enzyme from inactivation by iodoacetamide. There is evidence with this enzyme that time-dependent conformational changes take place which affect its molecular weight and chromatographic behavior, and this is also true of the rat-liver phenolsulfotransferase (unpublished results). The present belief is that these changes, known to be a feature of at least four coexisting sulfotransferases, are responsible for the conflicting results regarding the specificity of the enzymes and the apparent overlapping of specificity in "purified" enzyme preparations.

4. *Enzymic Hydrolysis*

The first demonstration of an enzyme that could hydrolyze a sulfate ester came from Derrien (1911) who noted that preparations of the snail, *Murex trunculus,* could hydrolyze indoxyl sulfate. A similar sulfo-hydrolase (sulfatase) was later observed in fungi (Neuberg and Kurono, 1923) and mammals (Neuberg and Simon, 1932). For many years this type of enzyme was referred to as "phenolsulfatase," a term later modified to "arylsulfatase" and, more recently, to "arylsulfohydrolase."

It is now known that arylsulfohydrolases are extremely widely distributed in nature and that considerable differences exist in relative specificity and general properties of enzymes from different sources. Failure to recognize this fact led to much confusion in work published prior to 1955, particularly with mammalian sources of the enzyme. Successive workers who assumed that they were studying a single enzyme were, as a result of using different substrates and mammalian tissue preparations, actually measuring one or more of at least three distinct arylsulfohydrolases. Recognition of this emerged as a result of two independent studies of the intracellular localization of enzyme activity in the livers of rat (Dodgson *et al.*, 1953a) and mouse (Roy, 1954) respectively. In our own laboratories potassium *p*-acetylphenylsulfate (IV) was chosen as assay substrate whereas Roy selected dipotassium 2-hydroxy-5-nitrophenyl sulfate (V) (nitrocatechol sulfate) for this purpose.

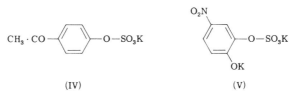

$$CH_3 \cdot CO-\langle\bigcirc\rangle-O-SO_3K$$

(IV)

(V)

With the former substrate, enzyme activity was localized in the so-called "microsomal" fraction whereas with the other substrate, activity was concentrated in the "mitochondrial" fraction. Moreover, in the latter case there was evidence that two distinct enzymes were coexisting in the fraction.

Dodgson *et al.*, (1954a, 1955b, 1956a) subsequently reconciled these conflicting findings by demonstrating that mammalian livers possessed three distinct enzymes, two of which were associated with the "mitochondrial" fraction (later shown to be in lysosomes) and the other associated with the "microsomal" fraction. By general agreement the microsomal enzyme was called C and the other two A and B respectively. C has a high optimum pH, is particularly active toward simple arylsulfate esters such as p-acetylphenyl sulfate and p-nitrophenyl sulfate, and is inhibited by cyanide but not phosphate. In contrast, A and B have low pH optima, are much more active toward nitrocatechol sulfate than toward the simpler esters, and are inhibited by phosphate but not cyanide. It later became apparent that the behavior of arylsulfohydrolases from sources other than mammals tended to fall into one or the other of these two main categories in terms of substrate specificity and behavior toward inhibitors. It therefore became convenient to refer to the C-type enzymes as Type I enzymes and the others as Type II (Dodgson and Spencer, 1957b), although there are some grounds for supposing that this may be an oversimplification of the situation.

Methods for assaying these enzymes have been reviewed by Dodgson and Spencer (1957b).

a. Mammalian Arylsulfohydrolase A. Enzymes A and B were originally reported as "mitochondrial" in origin when their cellular distribution was studied by the relatively crude techniques available in the early 1950's, but it was soon made clear that both were lysosomal in origin (Viala and Gianetto, 1955). Enzyme A is of considerable interest from the kinetic point of view and many features of its unusual behavior are still not adequately explained. Roy (1953) first noted that ox-A preparations acting on nitrocatechol sulfate gave anomalous enzyme concentration–activity curves and interpreted the phenomenon as a function of enzyme polymerization. However, other studies (Dodgson and Spencer, 1956a) on human-A showed this explanation to be untenable, the basic anomaly being that the reaction was not of zero-order. Attempts to ascribe the anomalies to substrate impurities (Roy, 1956a) were subsequently disproved (Dodgson and Spencer, 1956b) and it has become clear that the anomalies are a feature of the enzyme per se. Incubation of the enzyme with substrate appears to result in the slow conversion of the enzyme into an inactive or less active form that is markedly affected by various ions. The inactive enzyme can be reconverted into its original active form by incubation with sodium acetate at pH 8.0 (Baum *et al.*, 1958; Baum and Dodgson, 1958a). Some support for these conclusions has come from the studies of Anderson (1959a,b,c).

More recently, in a series of elegant studies on pure ox-A (Nichol and Roy, 1964, 1965, 1966) it has been shown that the enzyme exists as a monomer (mol. wt. 107,000) at pH 7.5 and as a tetramer (mol. wt. 411,-

000) at pH 5.0 under certain conditions of protein concentration and ionic strength. At very low protein concentrations and pH 5.0, the tetramer dissociates, while at pH 6.3 and relatively high protein concentrations, a dimer coexists in rapid equilibrium with monomer, trimer, and tetramer. These observations have not, as yet, served to clarify and account for the anomalous kinetic behavior of the enzyme, but it is obvious that any future kinetic studies must be made under circumstances in which the polymerization problem can be eliminated. Various physical studies have provided some indication of the intramolecular forces involved in determining the structure of the tetramer but a detailed description of these studies is beyond the scope of this chapter.

Recent work from our own laboratories has shown that both human and rat arylsulfohydrolase A exhibit behavior which is very similar to that of the ox enzyme in that they exist as monomers or tetramers, depending on the pH or the ionic strength.

The latest study on the ox enzyme by Roy and his colleagues (Jerfy and Roy, 1969) has been concerned with the chemical nature of the active site of the enzyme and points to tyrosyl and histidyl residues as important features of the active site. These authors suggest that a tyrosyl residue at the active site is sulfated and its subsequent desulfation is aided by an adjacent histidyl residue. However, much further work will be required in order to test this hypothesis adequately.

Very little work has been done on the specificity of enzyme A, although the human enzyme is known to be quite active toward other nitrocatechol sulfate isomers. Its activity and affinity toward p-nitrophenyl sulfate is relatively low but the enzyme still shows anomalous kinetics toward this substrate under appropriate experimental conditions (Baum *et al.*, 1958). The enzyme shows only feeble activity toward L-tyrosine O-sulfate and phenyl sulfate (Dodgson *et al.*, 1959). A surprising development in connection with specificity has come from the observations of Mehl and Jatzkewitz (1963, 1964, 1965, 1968) who claim that A is part of an enzyme system that is responsible for the desulfation of cerebroside sulfates. The complete system requires A and a heat-stable factor, the nature of which is unknown. This work throws an entirely new light on the possible functions of A and this point is discussed at greater length in Section III,C,2.

b. Mammalian Arylsulfohydrolase B. This enzyme coexists with A in the lysosomes of the cell although the precise relationship between the two enzymes within the lysosome is uncertain. Roy (1960c) has demonstrated that ox-A can be more readily extracted from liver lysosomal particles than B and concluded that the two enzymes were localized in different parts of the lysosome or in different lysosomes. An alternative explanation, that the two were present in different types of cell, was not considered. This last possibility seems unlikely following the recent finding that isolated rat liver parenchymatous cells contain both enzymes (unpublished results, these laboratories).

Early observations on partially purified human arylsulfohydrolase B indicated that this enzyme exhibited unusual kinetic anomalies under certain circumstances. The enzyme was affected by the concentration of buffer used in incubation mixtures and the pH-activity and substrate concentration—activity curves show anomalies which vary from preparation to preparation (Dodgson and Wynn, 1958).

It was assumed at the time that these anomalies arose from the varying tendency of B to combine with inert protein under different experimental conditions, because further purification yielded a product with apparently normal kinetics. Similar anomalies were also apparent with the ox enzyme, but some interesting differences exist between this enzyme and that from humans, particularly with regard to the effects of chloride ions, which activate the former but inhibit the latter. More striking still is the fact that both human- and ox-B, under normal circumstances, exhibit little activity toward p-nitrophenyl sulfate but become increasingly active toward this substrate in the presence of increasing amounts of chloride ions (see, e.g., Webb and Morrow, 1959, 1960).

More recent studies on arylsulfohydrolases B from ox liver (Allen and Roy, 1968), ox brain (Bleszynski, 1967), ox and rabbit cornea (Wortman, 1962), and rat liver (unpublished results, these laboratories) reveal that the enzyme can be resolved into further fractions by chromatographic or electrophoretic techniques. Allen and Roy (1968) have purified and separated two of these forms (B_α and B_β) from ox liver. Both have molecular weights of about 25,000 and are kinetically indistinguishable. Under certain conditions of low ionic strength, aggregates of these enzymes can occur which yield mixtures of polymers with molecular weights of up to 300,000. Electrostatic interaction is probably the main factor involved in this aggregation process.

Comparatively little work on the specificity of B has been attempted. The studies of Dodgson and Wynn (1958) on partially purified human liver enzyme indicate that, in the absence of Cl^-ions, the activity of the enzyme is greatly enhanced by the presence of a free phenolic grouping in the benzene ring of the substrate, in addition to the presence of a second substituent. There is some evidence that maximum activity depends on the free phenolic grouping being present in an unionized form. Human B is not active toward phenyl sulfate or L-tyrosine O-sulfate.

c. *Mammalian Arylsulfohydrolase C.* This microsomal enzyme is normally assayed spectrophotometrically with either *p*-acetylphenyl sulfate or *p*-nitrophenyl sulfate as substrates, although the latter is unsuitable for use with whole-liver preparations which are able to metabolize the liberated *p*-nitrophenol (Dodgson and Spencer, 1953a). Most detailed studies have been made with the rat liver enzyme but some information is also available for the ox and human enzymes. In all cases, the enzyme is localized in the so-called "microsomal" fraction of liver cell homogenates where it exists in a tightly bound insoluble form (Dodgson *et al.*, 1954b, 1957a,b). Most extraction procedures that are known to rupture particle membranes fail to release the enzyme into solution, but a pseudo-solubilization of rat-C can be achieved by treating washed microsomes with cationic and nonionic surface-active agents. Both solubilization and activation of the enzyme are associated with the formation of detergent micelles and removal of detergent results in reversion of the enzyme to its insoluble state. Anionic detergents also "solubilize" the enzyme but act as strong competetive inhibitors. True solubilization has been achieved by treatment of the microsomal fraction with crude pancreatic enzyme preparations in the presence of the nonionic detergent, Lissapol-N, but there is appreciable loss of activity during the process. However, the enzyme remains soluble after detergent removal.

Recent work (Milsom *et al.*, 1968) has shown that rat-C is localized in the smooth endoplasmic reticulum and a method has been described for assaying the rat enzyme under conditions where no contaminating lysosomal arylsulfohydrolases can interfere. The cell membrane and membranes of mitochondria, nuclei, and lysosomes do not contain the enzyme. C is therefore a very useful "marker" enzyme for the membrane component of the endoplasmic reticulum of the rat liver cell, but it is not yet clear how far it can be used for other tissues and species.

Human- and ox-liver C have not been obtained in a truly soluble form and do not appear to respond to treatment of microsomes with pancreatic enzyme preparations. Other differences exist between the enzymes of different species, including the fact that those of human and rat are not affected markedly by K^+ and Na^+ ions, whereas ox-C is appreciably inhibited. Guinea pig liver does not appear to contain C (Roy, 1958) but the enzyme is present in guinea pig testes (unpublished work, these laboratories).

All C enzymes, in contrast to A and B, operate at pH values on the alkaline side of neutrality (optimum conditions for rat liver: pH 8.0 at a substrate concentration of 40 mM potassium *p*-acetylphenylsulfate in the presence of 0.1 M phosphate buffer). Very few studies on their specificity have been attempted, but ox-, rat-, and human-C have only feeble activity toward phenyl sulfate and none toward L-tyrosine O-sulfate.

d. Mammalian Estrogen Sulfohydrolase. Some years ago Hanahan *et al.* (1949) observed that subcutaneous injection of estrone sulfate-^{35}S into pregnant and nonpregnant rats resulted in the appearance of labeled inorganic sulfate in urine. It was further claimed (without detailed supporting evidence) that rat liver contained a sulfohydrolase capable of hydrolyzing the conjugated estrogen. Subsequently, Pulkinnen and Paunio (1963) noted that liver, placental, and kidney preparations were able to desulfate estrone sulfate and that the responsible enzyme was localized in the "microsomal" fraction of the cell and was probably identical with arylsulfohydrolase C. It is not without interest that several workers in our own laboratories have unsuccessfully attempted to detect activity toward estrone sulfate with washed rat liver microsomal preparations. The reason for this discrepancy is not yet apparent but may reside in the use of different methodological approaches to the detection of enzyme activity.

Meanwhile, French and Warren (1967) have examined the ability of "microsomal" preparations of human placenta to hydrolyze *p*-nitrophenyl sulfate and estrone sulfate. Neither activities could be obtained in a truly soluble form, but kinetic and other studies provided strong evidence that two quite different enzymes were involved. Unfortunately, attempts to resolve the two enzymes by various procedures and treatments were unsuccessful. It will perhaps not pass unnoticed that this situation is rather analogous to that of the sulfotransferases, where different enzymes exist for transfer of sulfate from PAPS to estrogens and to *p*-nitrophenol.

e. Distribution of Arylsulfohydrolases in Mammalian Tissues. Many studies of the distribution pattern of these enzymes have been made, both by conventional and by histochemical (light and electron microscope) techniques. However, many of the results are difficult, if not impossible, to interpret adequately because of the failure of investigators to appreciate the complexity of the enzymes. Three enzymes similar to (and presumably identical with) liver A, B, and C are present in human pancreas, kidney, lung, brain, heart, intestine, and bronchial and tracheal epithelium (Dodgson *et al.*, 1956a; Spencer, 1959a), and the collective evidence from a large number of laboratories suggests that most mammalian tissues possess the enzymes. A and B activity has also been detected in human serum and urine (Dodgson and Spencer, 1957c,d). Some C may also be found in urine but is associated with cell debris and can be separated by briefly centrifuging the urine.

Examination of the relative activities of these three enzymes in mammalian tissues under normal and abnormal physiological conditions could yield valuable information about their possible biological role, but the difficulties inherent in the assay of the individual enzymes have not been fully overcome. These difficulties were considered at length by Dodgson and Spencer (1957b) and include such problems as competition for assay substrate when more than one enzyme is present in the same tissue, the presence of endogenous inhibitors such as phosphate and sulfate, and the anomalous kinetics of A and B. Similar problems regarding the histochemical localization of the enzymes have been discussed by Roy (1961a). The assay of rat C can now be done simply and without interference from A or B but the problem of the independent assay of the latter enzymes is a thorny one. Baum et al. (1959) showed how this could be achieved in human urine and use of this method subsequently led to the exciting discovery that human A virtually disappears from urine and tissues of patients with metachromatic leukodystrophy (Austin et al., 1964), a disease which is featured by an increase in cerebroside sulfate esters in the kidney and the myelin of the nervous system (see Section III, C, 2). Unfortunately, the same method cannot be applied to rat tissues because of the different properties exhibited by the human and rat enzymes (unpublished results, these laboratories) and numerous attempts to devise a method for the precise and independent assay of rat A and B have been singularly unsuccessful. These various difficulties have largely been ignored by most other workers, with the result that the literature is full of cytochemical and histochemical papers concerning the enzymes, most of which are largely meaningless from the quantitative point of view. In many instances much time and labor would have been saved by a careful scrutiny of the results of previous work on the enzymes.

f. Distribution of Arylsulfohydrolases in Other Organisms. This type of enzyme is very widely distributed in the animal kingdom and has been noted in birds, amphibians, fishes, insects, and in a wide range of relatively primitive marine and terrestial organisms. In fact, it seems likely that arylsulfohydrolases are a feature of all animal life (e.g., Ney and Ammon, 1959). In addition, arylsulfohydrolase activity has been detected in some plants (Baum and Dodgson, 1957; Poux, 1966), fungi, (e.g., Harada and Spencer, 1962), and bacteria (e.g., Dodgson et al., 1954c; Harada, 1954).

It is beyond the scope of the present chapter to detail the many studies that have been made, but key papers to some of the more detailed ones discuss enzymes from lower vertebrates (Roy, 1963), mollusks (Takahashi,

1960; Dodgson *et al.*, 1953b; Dodgson and Spencer, 1953b; Leon *et al.*, 1960; Dodgson and Powell, 1959a,b; Jarrige, 1962), fungi (Harada and Spencer, 1962), and bacteria (Harada, 1959; Rammler *et al.*, 1964; Fowler and Rammler, 1964; Milazzo and Fitzgerald, 1967).

The collective results of these various studies indicate that the distinction between Types I and II arylsulfohydrolases is not always clear cut and one is left with an impression that a range of enzymes exists between the two extremes. As far as Type I enzymes are concerned, they appear to be more sparsely distributed than those of Type II. Although common features of higher mammals, they appear to have a restricted distribution in lower mammals and lower vertebrates. Many of the bacterial and fungal enzymes that have been described also seem to possess Type I features. Type II enzymes, on the other hand, seem to be present in most animal organisms but those with anomalous kinetics analogous to arylsulfohydrolase A appear to be restricted to higher mammals. In other animal organisms (see Roy, 1963), the enzymes more closely resemble arylsulfohydrolase B, but there are a number of instances where the properties are sufficiently different from those of mammalian liver B to support the view that they constitute intermediary types.

Insufficient work has been done on the mechanism of action of arylsulfohydrolases in spite of the possibility that such studies might provide further indications of the essential differences between different enzymes. Spencer (1958, 1959b) in a study of three examples of Type I enzymes and one Type II enzyme has established unequivocally that the position of cleavage of the sulfate ester linkage of the appropriate substrate is at the O-S bond of the C-O-S structure (cf. Section III,A,2). In this respect therefore, the overall action of the different types is identical, probably involving a displacement reaction resulting from attack by hydroxonium ions. With regard to the detailed mechanism of action, only in the case of the Type I enzyme of *Alcaligenes metalcaligenes* has a close study been made (Dodgson *et al.*, 1955c, 1956b). The affinity of this enzyme for substrate and the maximum velocity of hydrolysis are enhanced by the introduction of electrophilic substituent groups into the benzene ring and are decreased by nucleophilic groups. Generally speaking, the respective effects can be approximately related in degree to the Hammet value (substitution constant, σ) of the substituent group. On the basis of these and other results a mechanism has been proposed for enzyme action.

Clearly, much more remains to be done before the typing of arylsulfohydrolases can be precisely defined but, for the moment, the present system does serve a useful purpose, providing its limitations are recognized. Certainly it has drawn attention to the complexity of the enzymes and points to the often disregarded conclusion that searches for and studies of these enzymes must be conducted with due regard for the wide variations in substrate specificity and other properties that exist.

5. *Relationship between Biosynthesis and Hydrolysis of Arylsulfate Esters*

Nature has provided a wide variety of living organisms with enzymic mechanisms whereby arylsulfates can be synthesized and degraded. Unfortunately, the reason why such provision should have been made cannot be stated with certainty. Until comparatively recently the view has always been held that all arylsulfate esters are end products of the metabolism of phenols and phenolic steroids. Concurrent with this early view was the idea that arylsulfohydrolases participated in some way in the biosynthesis of such esters. Arylsulfoconjugation could therefore be regarded as a protective mechanism whereby toxic phenols were rendered less toxic and arylsulfohydrolases neatly fitted into place as a factor involved in this process.

Of course, these views must now be modified; the existence of proteins containing L-tyrosine *O*-sulfate residues alone make them untenable. Furthermore, arylsulfohydrolases are not directly involved in arylsulfate biosynthesis (Baum and Dodgson, 1958b) and, indeed, certain fungi produce considerable amounts of arylsulfohydrolase but are unable to synthesize arylsulfate esters. It is also difficult to reconcile the complex pattern of arylsulfohydrolases of differing specificities in the particulate matter of mammalian cells with the view that all arylsulfates are end products of metabolism. Why should a mammalian organism go to great lengths to protect itself by sulfoconjugating phenols and possess, at the same time, enzymes that are theoretically able to release the phenol from the conjugate—and why four enzymes in the same cell? These factors, together with others to be considered presently, indicate the need for rethinking and raise the possibility that in some instances controlled synthesis and degradation of arylsulfates in different specific regions of the cell might be of some biological significance. This is not to say that a purely "detoxicating" role for the sulfotransferases is not of considerable importance to living organisms so far as some phenols are concerned. Apart from probable decrease in toxicity arising from sulfoconjugation of phenols, the mechanism achieves a basic "detoxication" aim, namely, the production of a strongly acidic, water-soluble compound that can rapidly be excreted via the kidneys. Curtis (1966) has shown that the sulfoconjugates of phenol, 2-naphthol, and 4-nitrocatechol all bind extensively to plasma protein and are secreted by the renal tubules at a rate which suggests that an active process is at least partly involved (cf. Sperber, 1948). However, this is not true for all arylsulfates and, in sharp contrast, L-tyrosine O-sulfate does not bind to plasma proteins and is subject to reabsorption by the renal tubules.

It is probably safe to conclude that many of the arylsulfate esters present in urine do indeed represent metabolic end products, in which event it should follow that they should be rapidly excreted in an unchanged form after being reinjected into the living animal. This postulate is supported by some experimental evidence. For example, Hawkins and Young (1954) found that injected phenyl sulfate-^{35}S and 2-naphthyl sulfate-^{35}S were rapidly excreted by rats without undergoing significant desulfation. Similar findings have been made (Dodgson et al., 1961b) with p-hydroxyphenylacetic acid sulfate-^{35}S. None of these esters is a good substrate for mammalian arylsulfohydrolases, so that the results are not unexpected.

In contrast, injection of p-nitrophenyl sulfate-^{35}S (the substrate for mammalian arylsulfohydrolase C) into rats is followed by the appearance in the urine of up to 30% of the radioactivity of the dose as inorganic sulfate-^{35}S (Dodgson and Tudball, 1960). Moreover, estrone sulfate-^{35}S (substrate for estrone sulfohydrolase) is almost completely desulfated in vivo in the rat (Hanahan et al., 1949) and human (Twombly and Levitz, 1960) and nitrocatechol sulfate-^{35}S (substrate for arylsulfohydrolases A and B) is partly desulfated in vivo, particularly if steps are taken to prevent its rapid excretion (Flynn et al., 1967). The sulfate ester of 3,5,3'-triiodothyronine is also rapidly desulfated in the intact rat (Roche et al., 1960). In other instances quite different transformations can take place in vivo; for example, in the rat, injected nitrocatechol sulfate-^{35}S is excreted in part as a double sulfo- and glucuronoconjugate. Injected cyclohexylphenyl 4-sulfate-^{35}S and 2-sulfate-^{35}S are both rapidly eliminated in bile as double sulfoglucuronoconjugates (Hearse et al., 1969a) and injected L-tyrosine O-sulfate-^{35}S is rapidly and completely deaminated without loss of the ester sulfate group (Dodgson et al., 1961b). Similar types of transformation have been noted with the estrogens; for example, conversion of estrone sulfate to 15α-hydroxyestrone sulfate has been observed in humans (Jirku et al., 1967). In connection with metabolic transformations of this type, attention might be drawn to a most useful autoradiographic method whereby the tissue site of the metabolism may be pinpointed (Powell et al., 1967; Hearse et al., 1969b).

These collective observations lead one to speculate whether there may be two types of arylsulfate ester from the point of view of physiological importance: first, those that are metabolically inert *in vivo* and are presumably end products and, second, those that can undergo metabolic transformations *in vivo* and which may therefore be of some value to the living organism. From speculation to confirmation is a long step however, and it is at this point where the gaps in our knowledge become all too apparent. This arises partly from the natural tendency to study enzymes with those substrates that can be most readily obtained and that have potential value in terms of ease of assay of the enzyme action in which they are involved. Thus with both the sulfotransferases and the sulfohydrolases, *p*-nitrophenol and its sulfate ester have, respectively, been useful substrates for following enzyme action. A great deal of valuable information has accrued from the use of such "synthetic" substrates but has provided little guidance as to the natural substrates for the enzymes.

However, accepting the fact that much more work needs to be done, there are good indications that in one case, at least a close relation may exist between sulfoconjugation and sulfohydrolysis. This case concerns the estrogens and there is a growing suspicion that the carefully regulated sulfation and desulfation of estrogens, particularly estrone, is an important feature of estrogen function *in vivo*. Thus, the work of Purdy *et al.* (1961) suggests that estrone sulfate is an important circulating transport form of estrogen in human plasma, while that of Levitz *et al.* (1960) has indicated that sulfation of estrone is involved in the transport of estrogens across the placental barrier. The work of Diczfalusy's group has led him to suggest that sulfoconjugation of estrogens may be a means whereby the fetus is protected from the high concentrations of maternal estrogens that circulate during pregnancy. Other work (e.g., Pulkinnen, 1957; Pulkinnen and Hakarainen, 1965; Bolte *et al.*, 1964; French and Warren, 1965) also supports the view that there may be a delicate balance operating during pregnancy whereby the levels of active estrogens are controlled by a balance of sulfoconjugation and sulfohydrolysis. This does not exclude the possibility that estrogen sulfate esters do not possess biological activity in their own rights, for example, as regulators of amino acid metabolism through inhibitory activity toward aminotransferases (Mason and Gullekson, 1960; Riggs and Walker, 1964; Scardi *et al.*, 1962).

Collectively then, there are good indications that the interplay of estrogen sulfotransferase and estrogen sulfohydrolase is biologically important and the time is clearly ripe for a determined attack on the problem from the point of view of providing more positive proof.

In one other case, namely the arylsulfohydrolases A of mammals, evidence is accumulating which allows one to suggest a function for the enzyme. Thus the work of Austin and his colleagues and Mehl and Jatzkewitz, which has already been briefly discussed and which is elaborated in Section III,C,2, provides a strong indication that this enzyme is a typical lysosomal acid hydrolase whose role in the lysosome is to participate in the degradation of cerebroside sulfate esters. In this way it could then play a major part in maintaining normal cerebroside sulfate levels during growth and adult life. It will be appreciated that cerebroside sulfates really bear no structural resemblances whatsoever toward nitrocatechol sulfate, the "synthetic" assay substrate. Nevertheless, the fact that a heat-stable factor must apparently also be present before A can act on cerebroside sulfate esters is a significant clue which needs to be investigated more closely before firm conclusions about specificity can be drawn. It is not inconceivable that a complete enzyme system which normally acts on cerebroside sulfates could, when dissociated into enzyme and cofactor, have a profoundly different specificity toward substrates, including attacking those that it would not normally meet in the course of metabolism.

This still leaves the unsatisfactory position of mammalian arylsulfohydrolases B and C, the function of which, and their relationship to arylsulfotransferases, is still quite obscure. In the case of B it seems important to remember its cellular localization in lysosomes and to recall De Duve's (1959) view that lysosomal hydrolases "hardly ever synthesize, rarely transfer, occasionally arise artificially from nonhydrolyzing precursors and most of the time are just what they appear to be, hydrolases." Certainly B (and many other arylsulfohydrolases that have been tested in these laboratories) cannot directly transfer sulfate to other acceptors and we must accept De Duve's view. What substrate does it then hydrolyze? The obvious candidate would be fibrinogen with its sulfated tyrosine residues. Unfortunately, B has no activity toward L-tyrosine O-sulfate and injection of di- and polypeptides containing sulfated tyrosine residues is followed by liberation of free L-tyrosine O-sulfate without loss of ester sulfate (Jones et al., 1963; Basford et al., 1966). Is there then some other macromolecule, as yet undiscovered, containing bound or associated arylsulfate residues, the degradation of which may involve arylsulfohydrolase B? Certainly, it is difficult to visualize the enzyme as involved in the hydrolysis of simple free arylsulfate esters in the lysosomes, as there seems to be no rationale behind such a possibility.

In the case of mammalian arylsulfohydrolase C the situation is also obscure if it is accepted that the enzyme is different from estrogen sulfohydrolase. The question may indeed be asked whether the enzyme can be of vital importance when it is not present in most guinea pig tissues and in those of many lower organisms. However, this would not be in accord with evolutionary principles, and the fact that arylsulfohydrolases become more complex as one ascends the evolutionary scale suggests that the enzymes must serve some useful purpose.

The biological role of C therefore remains a mystery and will continue to do so until more information is obtained about its specificity. Could it be concerned with the transport and the regulation of cellular levels of hormones such as serotonin, adrenaline, and triiodothyronine, all of which readily form arylsulfoconjugates which, there is some grounds to suppose, may have greatly reduced biological activity? Here again is a worthwhile field of investigation.

It might be imagined that useful information about the biological roles of enzymes involved in sulfoconjugation and hydrolysis would be forthcoming from studies on the variation of enzyme levels in tissues and in urine and blood under different circumstances. With the exception of the studies on the relationship between arylsulfohydrolase A and metachromatic leukodystrophy, little of real value has emerged from this type of approach. Such investigations as have been made have usually been confined to the sulfohydrolases, frequently without regard to the complex problem of individual assay of these enzymes. Various fluctuations (sometimes considerable in extent) in tissue and urine levels of these enzymes have been reported in such conditions as surgical trauma, cancer, kwashiorkor, fatty liver, vitamin-A deficiency and so on and many of these studies have been documented by Dzialoszynski and Gniot-Szulzycka (1967). It is impossible to deduce anything concerning the physiological role of the enzymes from these studies.

In nonmammalian organisms the relationship between sulfoconjugation and hydrolysis is even more obscure so far as arylsulfate esters are concerned and it seems pointless to speculate further about this at present. In fungi and bacteria, arylsulfohydrolase activity is frequently present or can be induced and it seems likely that enzyme production is associated with the need to acquire sulfur for the purpose of synthesizing cysteine and other important sulfur-containing compounds. No microorganisms have yet been found that possess arylsulfotransferase activity and there is no record in the literature of any arylsulfate ester of microbial or fungal origin.

6. L-*Tyrosine O-Sulfate*

This arylsulfoconjugate has already been mentioned several times during the earlier parts of this section. However, the unique nature of the compound merits a further brief discussion.

Four biologically important polypeptides have now been recognized as containing one or more sulfated tyrosine residues. Three of these are similar in that they are relatively small molecules possessing potent physiological activity and each containing a single sulfated tyrosine residue. The structures of caerulein (VI), gastrin II (VII) and phyllokinin (VIII) (see Section III, A, 1) are compared below with that of brady-kinin (IX).

```
1   2    3    4           5    6    7    8    9      10
Pyr-Gln-Asp-Tyr(SO₃H)-Thr-Gly-Try-Met-Asp-Phe-NH₂                (VI)
```

```
1   2    3    4    5    6    7
Pyr-Gly-Pro-Try-Met-Glu-Glu-                                      (VII)
```

```
8   9    10   11   12          13  14  15  16  17
Glu-Glu-Glu-Ala-Tyr(SO₃H)-Gly-Try-Met-Asp-Phe-NH₂
```

```
1  2   3   4   5   6   7   8   9    10   11
H-Arg-Pro-Pro-Gly-Phe-Ser-Pro-Phe-Arg-Ileu-Tyr(SO₃H)             (VIII)
```

```
1  2   3   4   5   6   7   8   9
H-Arg-Pro-Pro-Gly-Phe-Ser-Pro-Phe-Arg-OH                         (IX)
```

The similarity between caerulein and gastrin II in the C-terminal portion of the chain will be apparent, as will the extremely close resemblance of phyllokinin to bradykinin. Details of the physiological (vasoactive and kininlike) properties of these compounds may be found, for example, in papers by Bertaccini et al. (1968a,b), Gregory (1968), Kenner and Sheppard (1968), Morley (1968), and Anastasi et al. (1966). No obvious function can yet be ascribed to the sulfate group in the case of gastrin II but phyllokinin and caerulein lose appreciable physiological activity when the sulfate group is removed.

The fourth polypeptide containing L-tyrosine O-sulfate residues is the protein fibrinogen. Here the situation is somewhat confused in that appreciable differences exist in the distribution of the residues in fibrinogens from different species. Briefly, during the conversion of fibrinogen to fibrin by the action of thrombin, two main peptides (fibrinopeptides A and B) are sequentially released immediately prior to polymerization to form the insoluble fibrin gel. Fibrinopeptides B of most mammalian species contain a single L-tyrosine O-sulfate residue and, occasionally (dog and cat), two such residues lying adjacent to each other (see Krajewski and Blombäck, 1968). In contrast, the analogous peptide released from the fibrinogens of man, certain other primates, and rat and guinea pig, contain no such residues. In addition, present knowledge sug-

gests that, with the possible exception of the rabbit, the fibrinogens of all
species contain one or more L-tyrosine O-sulfate residues in addition to
any that might be present in fibrinopeptide B (Jevons, 1963; unpublished
results from these laboratories). The evolutionary significance of these
variations has been discussed by Doolittle and Blombäck (1964) and
Krajewski and Blombäck (1968) and further elaboration on the importance of the residues will not be made here, except to comment that their
physiological importance (if any) is still quite obscure. Rather, attention
will be drawn to one or two important factors concerned with the biosynthesis of the residues and their ultimate metabolic fate.

   a. Biosynthesis. This problem has briefly been discussed in terms of
L-tyrosine methyl ester sulfotransferase (Section III, A,3,c) where it was
pointed out that L-tyrosine cannot undergo sulfoconjugation unless the
carboxyl group is first blocked with a further amino acid or other
suitable blocking group. How then does the biosynthesis of the peptide-
bound tyrosine residues of fibrinogen occur? Two possibilities may be
envisaged; first, that sulfation occurs at an early stage in the assembly of
fibrinogen immediately following the formation of an amino acyl AMP-
enzyme complex or the formation of the amino acyl-transfer RNA complex, in which case a specific code for incorporation of sulfated tyrosine
residues must be envisaged. Second, that sulfation takes place subsequent
to the incorporation of the appropriate tyrosine residues into the polypeptide chain. The first possibility is attractive in terms of specific sulfation
of a very few tyrosine residues in a protein of a molecular weight in the
region of 340,000 but will hardly appeal to the coding experts. The
second seems more rational and analogous to other situations such as the
phosphorylation of proteins or the hydroxylation of collagen proline
residues, but is perhaps rather less satisfactory in terms of the subtle
species differences that exist in the location and immediate amino acid
environment of the sulfated tyrosine residues.

   This is a fascinating problem that is being tackled in our own laboratories but without, as yet, any clear indication as to the favored route. It
is, however, of some immediate interest that the formation of tyrosyladenyl
complexes, prior to insertion of the amino acid into a polypeptide chain,
would satisfy the necessity for the amino group of tyrosine to be blocked
before sulfation via tyrosyl methyl ester sulfotransferase could proceed.
Furthermore, synthetic tyrosyl adenylate can undergo biological sulfation
with the transferase enzyme (unpublished results). The use of this
particular route would also help to explain the important fact that in
phyllokinin the tyrosine residue which carries the sulfate group is in the

C-terminal position of the polypeptide chain (i.e., unblocked carboxyl group).

b. Metabolic Fate of L-*Tyrosine O-Sulfate.* Mention has been made in earlier parts of this section that the mammalian arylsulfohydrolases exhibit little or no activity toward L-tyrosine O-sulfate, although the arylsulfohydrolases of *Aspergillus oryzae* and *Alcaligenes metalcaligenes* are quite active toward the ester (Dodgson *et al.*, 1959). It is perhaps not surprising to find that neither free nor peptide-bound L-tyrosine O-sulfate undergoes desulfation when injected into experimental animals (Dodgson *et al.* 1961b; Jones *et al.*, 1963; Powell *et al.*, 1964; Basford *et al.*, 1966). However, in the rat and rabbit at least, the injected free ester is extensively metabolized, without loss of sulfate, to yield the sulfate esters of *p*-hydroxyphenylpyruvic acid and *p*-hydroxyphenylacetic acid (Dodgson *et al.*, 1961b; Powell *et al.*, 1963). In the case of the rat, Rose *et al.* (1966) have established that a specific aminotransferase enzyme is responsible for the initial loss of amino group. The enzyme is pyridoxal phosphate-dependent and uses 2-oxoglutarate as amino group acceptor. It is localized in the mitochondria of the cells and is quite distinct from either L-tyrosine-2-oxoglutarate aminotransferase or L-phenylalanine-pyruvate aminotransferase. The enzyme is widely distributed in rat tissues and is also known to be present in a number of other animal species, including humans. This interesting aminotransferase is now being studied further.

c. Relationship of Biosynthesis to Metabolic Fate. To summarize the various findings described in the previous paragraphs, first, free L-tyrosine cannot undergo biological sulfation and, second, mammalian tissues possess an enzyme that can deaminate L-tyrosine O-sulfate. This enzyme certainly operates *in vivo* when the ester, either in free or peptide-bound form, is injected into rats or rabbits. These facts may now be considered in relation to the presence of L-tyrosine O-sulfate in mammalian urines. First noted as a constituent of human urine (Tallan *et al.*, 1955), the ester is now known to be present in other mammalian urines and can be routinely assayed by an autoanalytical procedure (John *et al.*, 1966). The normal urinary concentrations, expressed in terms of milligrams L-tyrosine O-sulfate excreted in 24 hours per kilogram of body weight, increase in the order: rat, 0.11; sheep, 0.30; human, 0.32; calf, 0.36; rabbit, 0.64; mouse, 0.82. In the human this quantity is equivalent to an excretion of the order of 20–25 mg per day.

If free L-tyrosine cannot undergo biological sulfation, it must be presumed that the urinary L-tyrosine O-sulfate must come from bound or carboxyl-blocked ester. The obvious source in that case would be the daily turnover of fibrinogen and gastrin II. However, simple calculations, based on fibrinogen turnover rates quoted in the literature, show that such turnover could not possibly account for the large amounts of urinary L-tyrosine O-sulfate. Moreover, when considered in relation to the presence in tissues of an enzyme that deaminates L-tyrosine O-sulfate, it may well be wondered whether the amount of the ester in urine represents perhaps only a fraction of the amount actually produced each day. The interesting implication of these findings is either that other proteins that are being degraded also possess sulfated tyrosine residues or, alternatively, that some carboxyl-blocked tyrosine residues (tyrosyladenylate complexes?) are being rapidly sulfated and degraded to yield free sulfated tyrosine.

L-Tyrosine O-sulfate thus emerges as one of the most exciting arylsulfate esters yet discovered and it may well be supposed that much more will be heard about this compound in the future.

B. ALKYLSULFATE ESTERS

1. Steroid Alkyl Sulfates

a. Occurrence. The occurrence of conjugates of adrenal steroids as normal constituents of urine has been recognized for many years. This localization in association with many other conjugates regarded as "detoxication products," together with their apparent lack of physiological activity was undoubtedly responsible for the long-held opinion that they represented nothing more than end products of steroid metabolism.

The so-called 17-ketosteroid fraction of urine is claimed to be composed mainly of glucuronic acid conjugates but the remainder consists of sulfoconjugates. Principally three sulfated ketosteroids are represented in this fraction, viz., dehydroisoandrosterone sulfate, androsterone sulfate, and etiocholanolone sulfate, the most prominent being dehydroisoandrosterone sulfate (3β-sulfatoxyandrost-5-en-17-one) (X).

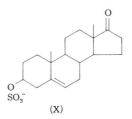

(X)

The characteristic feature of this and certain other compounds is the sulfated secondary cyclic alcohol group at C_3 but, in addition, sulfoconjugates of steroid C_{17}, C_{20}, and C_{21}-hydroxyl groups are known to occur in human urine and bile (see Drayer and Giroud, 1964; Laatikainen *et al.*, 1968). The excretion of several C_{21} steroid sulfoconjugates also follows the administration of cortisol and corticosterone to humans (Pasqualini and Jayle, 1961; Pasqualini, 1964).

Apart from monosulfate conjugates, a number of double conjugates in which a second hydroxyl group is sulfated have been recognized. Thus Laatikainen *et al.* (1968) detected the presence in human bile of disulfates involving the C_3, C_{17}, and C_{20} hydroxyl groups of a variety of steroids. The production of similar double conjugates has also been reported to occur in humans receiving various precursor steroids (Pasqualini and Jayle, 1962; Baulieu and Corpéchot, 1965), while recently Arcos and Lieberman (1967) have discovered a novel double conjugate involving both sulfate and N-acetylhexosamine. Studies with *in vitro* systems (Wengle and Boström, 1963) have also established that a considerable number of steroids can undergo disulfation but not all of the products have been shown to occur naturally.

The presence in plasma of an interesting sulfate diester following the administration of ACTH to humans has been reported (Oertel, 1961). The compound contains glycerol, fatty acid, sulfate and 17-ketosteroid and is considered to be a sulfatidyl 17-ketosteroid complex in which sulfate acts as a diester bridge between the 17-ketosteroid moiety and glycerol. Further complexes of this type involving dehydroisoandrosterone, androsterone, and etiocholanolone were subsequently recognized (Oertel and Kaiser, 1962) and it is claimed that 80% of the plasma steroid sulfoconjugates exist in this form mainly in association with the plasma α-lipoprotein fraction (Oertel *et al.*, 1962). In spite of the fact that the synthesis and metabolism of these steroid sulfatides have been studied in some detail (Oertel, 1963; Oertel and Treiber, 1966), their significance is still largely a matter for conjecture.

The occurrence of steroid alkyl sulfates elsewhere in nature seems to be restricted to the biles of many primitive vertebrates. The general characteristic of these compounds is the presence of a sulfated primary or secondary alcohol grouping in the C_{17} side chain. Examples of compounds of this type are provided by scymnol sulfate (3α, 7α, 12α, 24, 26, 27-hexahydroxy-5β-cholestane 26- or 27-sulfate) which is present in the bile of certain fishes of the Elasmobranchii group and ranol sulfate (3α, 7α, 12α, 24, 26-pentahydroxy-27-nor-5α- and 5β-cholestane 24-sulfate) the bile salt of the ranidae family of frogs (see Haslewood, 1967). More recently a disulfated compound, myxinol disulfate, which in addition to sulfate in the C_{17} side chain is sulfated in the C_3 position, has been found in the bile of the hagfish (Anderson and Haslewood, 1967). The occurrence of these compounds is of interest in relation to the problem of bile salt evolution, particularly in the light of the recent discovery of numerous steroid sulfates in human bile (Laatikainen *et al.*, 1968).

The finding that steroid conjugation is not necessarily followed by excretion (Roberts *et al.*, 1961), together with the demonstration of steroid conjugates in tissues and plasma, has led to a dramatic change in attitude toward these compounds and current opinion holds that they are important intermediates in steroid biotransformation processes. Early support for this concept came from the work of Roberts *et al.* (1964) who, stimulated by earlier observations from their laboratories on the role of steroid conjugates as biosynthetic intermediates, examined the role of cholesteryl sulfate in such processes. The results demonstrated that the compound could serve as a precursor for the synthesis of dehydroisoandrosterone sulfate and a variety of other $\Delta^5 - 3\beta$ steroid sulfates in a patient with an adrenocortical carcinoma. Subsequently, cholesteryl sulfate was isolated from a variety of mammalian tissues (Drayer *et al.*, 1964; Drayer and Lieberman, 1965; Moser *et al.*, 1966) and its biosynthesis by tissue preparations was demonstrated (Banerjee and Roy, 1967; Rice *et al.*, 1968). Nevertheless, the current opinion seems to be that cholesteryl sulfate must be of minor importance only in normal metabolic processes (Gurpide *et al.*, 1966; Oertel *et al.*, 1968) since it does not appear to undergo conversion to other steroids in normal subjects.

However, biosynthetic pathways involving other steroid sulfoconjugates as intermediates have been clearly substantiated and a detailed account of these and their significance is presented elsewhere in this book. This aspect of the subject need not therefore be considered further here but the sulfotransferases and sulfohydrolases which have come to light during studies on steroid sulfoconjugates are worthy of some attention.

b. Sulfoconjugation. Soon after the discovery of the sulfate activating system, alkyl steroid sulfotransferase activity was noted in crude preparations of the liver of rat, rabbit, and ox (Roy, 1956b; Schneider and Lewbart, 1956; DeMeio *et al.*, 1958). Activity toward a wide range of steroids, including 3α- and 3β-hydroxy steroids of the 5α, 5β, and Δ^5 series as well as the 17α-, 17β- and 21-hydroxy steroids, could be demonstrated. The possibility that these activities were due to more than one enzyme was first shown by Nose and Lipmann (1958) and further support for this has come from Holcenberg and Rosen (1965) who noted that many ox tissues could sulfoconjugate dehydroisoandrosterone whereas relatively few were able to sulfoconjugate testosterone to any significant extent.

Detailed enzymological studies have been relatively few but the work of Banerjee and Roy (1966) on the sulfotransferases of guinea pig liver has clearly distinguished that a minimum of two types of enzyme exist. One of these will transfer sulfate to dehydroisoandrosterone, androsterone, and cholesterol (cf. Banerjee and Roy, 1967), i.e., to the secondary cyclic alcohol group at C_3, and all the information available at present suggests that a single enzyme is responsible for this activity. The other type transfers sulfate to testosterone and deoxycorticosterone, i.e., to either a C_{17}-secondary or C_{21}-primary alcohol grouping. Partial separation of the two types of activity can be achieved on DEAE-Sephadex but complete resolution has not proved possible. It is also impossible to say whether the activities toward testosterone and deoxycorticosterone are the result of the activity of a single enzyme or whether more than one such enzyme exists. In this connection it will be recalled that other work has established that C_{17}, C_{20} and C_{21} sulfoconjugates occur naturally and it is possible that more than one sulfotransferase is necessary to produce these conjugates. Both of the types of alkyl steroid sulfotransferases mentioned above depend on magnesium ions for maximum activity and, as stated earlier, both can be distinguished from estrogen and phenol sulfotransferases.

Other enzymological studies on the bovine adrenal gland alkyl steroid sulfotransferases have been made by Adams and Edwards (1968) who noted unusual kinetic properties when dehydroisoandrosterone was used as the acceptor molecule. Gel filtration and sucrose density gradient centrifugation experiments led them to suggest that monomeric and polymeric forms of the enzyme could exist in association with each other. This work, however, should be treated with caution as no attempt was apparently made to separate the relatively crude enzyme preparation into fractions corresponding to those of Banerjee and Roy (1966) and the possibility that two or more sulfotransferases were participating in this association phenomenon seems to exist.

Presumably, when double sulfoconjugates are found, a minimum of two sulfotransferases are involved.

Both Wengle and Boström (1963) and Carrol and Spencer (1965) have shown that alkyl steroid sulfotransferases are virtually absent from fetal liver preparations but that full activity quickly develops after birth. This finding may be of significance in relation to the possible role of alkyl steroid sulfoconjugates as important metabolic intermediates.

Recently, the biosynthesis of sulfoconjugates of vitamin D_2 (XI) and D_3 has been shown to occur in rat liver and a variety of other tissues

(Higaki *et al.*, 1965). The sulfotransferase involved has not been studied closely but it may well be identical with the enzyme responsible for the sulfoconjugation of dehydroisoandrosterone.

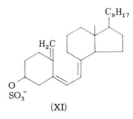

(XI)

c. Enzymic Hydrolysis. The interest in enzymes that are able to hydrolyze sulfate esters of steroid alcohols arose from their usefulness as tools in the analysis of urinary steroid conjugates and enzymes of molluskan origin have frequently been used for this purpose. Enzyme activity in mammalian tissues was first reported by Gibian and Bratfisch (1956) who observed that rat and ox liver preparations were capable of hydrolyzing dehydroisoandrosterone sulfate. The ox enzyme was further studied by Roy (1957) who showed that its specificity was restricted to the 3β-sulfates and the 5α and Δ^5 series of steroids. Other isomeric 3-sulfates, two 17-sulfates of the androstane series and two 20-sulfates of the pregnane series were not substrates for the enzyme. Subsequently Roberts *et al.* (1961) obtained *in vivo* evidence of the presence of a dehydroisoandrosterone sulfohydrolase in human tissues and Pulkinnen (1961) and Warren and Timberlake (1962) demonstrated that placental tissue also possessed similar activity. Immature and full-term placenta were equally active, but fetal and maternal blood, both of which contain appreciable quantities of dehydroisoandrosterone sulfate (Migeon and Plager, 1955), were inactive, as are fetal tissues in general (Warren and French, 1965).

The observations that 3β-steroid sulfate esters are involved in estrogen biosynthesis in the human female has focused even more attention on placental steroid sulfohydrolase activity. Recently, Pasqualini *et al.* (1967) have reinvestigated the situation by perfusing human term placenta with a variety of steroid sulfate esters and have confirmed that 3-sulfates of the 3β and Δ^5 series are the only ones appreciably hydrolyzed by this tissue (see also, French and Warren, 1966). Steroid sulfohydrolase activity of placental tissue has been studied in detail by French and Warren (1967) who claim that it is localized in the microsome fraction of the tissue cell but is distinct from other sulfohydrolase enzymes present in this fraction.

At the present time therefore, only one mammalian steroid sulfohydrolase enzyme can be recognized and this is specific for the sulfate group on the secondary cyclic alcohol group at C_3. However, others may yet await discovery; for example, the administration of cortisone 21-[^{35}S]sulfate to rats is followed by the appearance of considerable amounts of labeled inorganic sulfate in the urine (Dodgson *et al.*, 1965) although a sulfohydrolase active toward cortisone 21-sulfate cannot be demonstrated in the tissues. In this connection it is interesting to note that sulfohydrolase activity toward dehydroisoandrosterone sulfate and cortisone 21-sulfate has been detected in the digestive juices of the mollusk *Patella vulgata* (Roy, 1956c) and the edible snail *Helix pomatia* (Dodgson, 1961). Attempts to demonstrate the existence of a specific cortisone 21-sulfohydrolase in these preparations have, however, been only partially successful (Jarrige, 1962; Dodgson *et al.*, 1965).

2. Simple Alkyl Sulfates

The sulfate esters of simple alcohols, with one exception, have not been found to occur naturally. The exception is isopropyl sulfate which occurs in the hen's egg in appreciable quantities (Yagi, 1966) although the significance of this is quite obscure.

In spite of this, several workers have demonstrated the sulfoconjugation of simple aliphatic alcohols by mammalian and other tissue preparations. Thus, Vestermark and Boström (1959) and Spencer (1960) have studied the formation of conjugates of methanol, ethanol, propanol, and butanol by mammalian liver preparations while a comparable sulfoconjugating activity is also known to be a feature of the livers of lower animals e.g. frog, toad (unpublished results, these laboratories). The sulfotransferases involved in these processes have so far received relatively little detailed attention.

Sulfohydrolase enzymes active toward simple alkyl sulfates have not yet been detected in mammalian tissues. A widely distributed enzyme which will liberate sulfate from L-serine O-sulfate was discovered some years ago in these laboratories (Dodgson *et al.*, 1961a) but it is now clear that the enzymic reaction is not a true sulfohydrolysis but an αβ-elimination resulting in the liberation of pyruvic acid, sulfate, and ammonia (Tudball and Thomas, 1965). Similarly, rat tissues can degrade the long-chain alkyl conjugate, sodium dodecyl sulfate, but the process involves chain shortening without loss of sulfate grouping (Denner *et al.*, 1969).

In the higher fungi the alkyl sulfate ester of choline is found in appreciable quantities (e.g., Wooley and Peterson, 1937; Harada and Spencer, 1960; Itahashi, 1961). The role of this ester in fungi has been studied in detail by Spencer and his colleagues (Orsi and Spencer, 1964; Hussey *et al.*, 1965; Scott and Spencer, 1965) and it appears that the ester is utilized as a convenient store of sulfate. In times of sulfur sufficiency, sulfur is converted into choline sulfate through the action of a choline sulfotransferase and, in times of sulfur deprivation, choline sulfohydrolase activity releases inorganic sulfate which can then be used for cysteine production. The ability to form choline sulfate also occurs in plants, possibly for a similar purpose, but its formation in the mammalian system has not been demonstrated.

C. Carbohydrate Sulfate Esters

Sulfoconjugates of carbohydrates have an extremely wide natural distribution and occur for the most part as polymers but occasionally as relatively simple molecules. Many of the polymers play important roles as structural components of living organisms; roles for which they are particularly suited by virtue of their unique physicochemical properties. The story of many of these compounds is only now beginning to emerge in a coherent pattern. Progress has been hindered by experimental difficulties of isolation, purification, and chemical structure, and throughout much of the work there has been a tendency to oversimplify. This tendency, which is still apparent to some extent, has resulted in a profuse but frequently confused literature and it would far exceed the scope of this chapter to consider the group in complete detail. Many compounds or groups of compounds will therefore be mentioned only in passing and attention will be focused on those esters that are to be found in mammals. In these cases also much detail will be omitted, the major emphasis being placed on sulfate group biosynthesis and hydrolysis.

1. *Polysaccharide Sulfate Esters*

Sulfated polysaccharides may well be present in most if not all animal organisms. In the plant world they occur in high concentrations in seaweeds where they probably act as important structural components in helping to provide the necessary strength and flexibility which such organisms need for survival. In relatively primitive animal organisms polysaccharide sulfates frequently appear as important components of mucus secretions, usually in the form of polyhexose polysulfates. In compounds such as charonin sulfate (Egami and Takahashi, 1962) and *Buccinum* polysaccharide sulfate (Hunt and Jevons, 1966) the structure is that of a $\beta 1 \rightarrow 4$ linked polysulfated polyglucose. In others, polyfucose or polygalactose chains are sulfated, for example, in the jelly coats of sea urchins where it is supposed that such compounds may be important factors in relation to the fertilization process (Immers, 1962; Ishihara, 1968). In at least one case (*Buccinum* polysaccharide sulfate) the sulfated polysaccharide exists as a proteoglycan complex (Hunt and Jevons, 1965). More complex sulfated polysaccharides also occur in relatively primitive organisms, for example, the horatin sulfates of *Charonia lampas*, which contain a multiplicity of sugars as well as sialic acid (Inoue, 1965); the mactins of *Spisula solidissima* which are also extremely complex and probably associated with protein (Burson *et al.* 1956; Cifonelli and Mathews, 1968) and the Lorenzan sulfates of elasmobranch fish (Doyle, 1967). In addition, a great number of nonmammalian organisms contain glycosaminoglycan sulfates resembling those of mammalian connective tissue, as well as other complex sulfoconjugates.

Interesting as these various compounds are, they are only marginally relevant to this chapter and will not therefore be discussed further.

In mammals the polysaccharide sulfates constitute a group of complex polymers, the mucopolysaccharides or (more correctly) acid glycosaminoglycans. Most of them are important components of the amorphous extracellular "ground substance" of connective tissues where they exist as covalently bound complexes with proteins to form the so-called proteoglycans. The roles of the proteoglycans in these tissues are principally concerned with stabilizing and supporting (and possibly orientating) the collagen and elastin fibers and the cellular elements of the tissue, in maintaining a controlled ionic and aqueous environment within the tissue, and in contributing to the characteristic flexible and load-bearing properties of anatomical surfaces (see Dodgson and Lloyd, 1968). Other possible functions have been suggested from time to time, including an involvement in the calcification process (e.g., Matukas and Krikos, 1968), the concentration of urine (e.g., Pinter, 1967), and the aggregation of cells (e.g., Rappaport, 1966). Heparin proteoglycan is probably an exception to these generalizations since it is mainly intracellular in its localization, being present in the granular cytoplasmic inclusions of mast cells where its function is probably concerned with the binding and release of histamine.

Other polymers of even greater complexity are to be found in tissues and in mucins; for example, the sulfated, sialic acid-containing polymers from corneal stroma (Robert and Dische, 1963), colonic mucosa (Kent and Marsden, 1963; Kent *et al.*, 1967), and gastric mucosa (Schrager and Oates, 1968). Comparatively little is known about the precise structure of these compounds and they will not be referred to again.

a. Chemical Structure of Polysaccharide Sulfate Chains. The complete structures of the various proteoglycans are still unresolved but it is possible to define the structure of the sulfated polysaccharide components of most of them with a reasonable degree of accuracy even though many of the fine details are obscure. The major results of structural studies have been considered in a number of reviews to which the reader is referred (Jeanloz, 1963a,b; Muir, 1964; Brimacombe and Webber, 1964; Cifonelli, 1968; Dodgson and Lloyd, 1968; Guthrie et al., 1968).

Each of the sulfated polysaccharides is characterized by substituted hexosamine residues alternating in a regular fashion with either hexuronic acid or hexose residues. The hexosamine moiety is invariably either D-glucosamine (2-deoxy-2-amino-D-glucose) or D-galactosamine (2-deoxy-2-amino-D-galactose), the uronic acid moiety is either D-glucuronic acid or L-iduronic acid and the hexose is galactose. The hexosamine residues are generally substituted by N-acetyl groups but in heparin N-sulfate (sulfamate) groups are found and in heparan sulfate both groups occur in about equal amounts. In all the polymers O-sulfate ester groups are located on the hexosamine residues (except in the immediate region where linkage to protein occurs) but some polymers carry additional groups on the hexose or uronic acid moieties. The linkage between uronic acid (or hexose) and hexosamine is of the β-type, except in heparin and heparan sulfate where it is α-type.

Broadly speaking, the polymers may be regarded as linear structures displaying a considerable molecular weight dispersity and exhibiting strong polyanionic characteristics resulting from hexuronic acid carboxyl groups or from sulfate groups, or both. The main basic features of the chemical composition of the compounds are summarized in Table I; while, with the exception of heparan sulfate, the presently accepted structures of the *main* disaccharide repeating periods are given in Fig. 1. In each case the compounds have been subdivided into three groups in terms of their chemical components.

In the case of heparan sulfate, no satisfactory repeating period can yet be assigned. In the first instance the term heparan sulfate embraces a family of related polymers of fairly low molecular weight. When isolated from any source it apears to possess both N-acetyl and N-sulfate groups in approximately equal amounts. Moreover, two fractions can usually be resolved, one with high N-acetyl and low N-sulfate contents, and the other having the reverse (Knecht et al., 1967).

Other structural deviations from the normal straight-chain, regular repeating structure exist in the various polymers and these deviations establish the polymers as being more complicated than was at first thought. Thus both heparin and heparan sulfates may well have some

TABLE I
Major Components of the Sulfated Glycosaminoglycans[a]

Glycosaminoglycan	Hexosamine	Other major monomer	N-Acetyl groups	O-Sulfate groups	N-Sulfate groups
Group I					
Chondroitin 4-sulfate	D-Galactosamine	D-Glucuronic acid	+	+	—
Chondroitin 6-sulfate	D-Galactosamine	D-Glucuronic acid	+	+	—
Dermatan sulfate	D-Galactosamine	L-Iduronic acid	+	+	—
Group II					
Keratan sulfate	D-Glucosamine	D-Galactose	+	+	—
Group III					
Heparin	D-Glucosamine	D-Glucuronic acid	—	+	+
Heparan sulfates	D-Glucosamine	D-Glucuronic acid	+	+	+

[a] Excluding polysaccharide–protein linkage sequences.

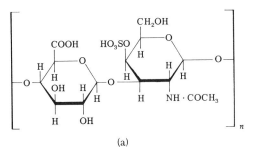

(a)

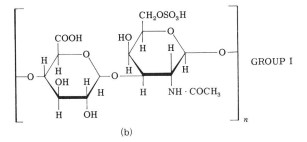

GROUP I

(b)

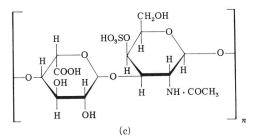

(c)

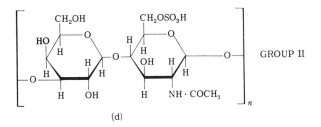

GROUP II

(d)

FIG. 1.   Repeating periods assigned to the acidic glycosaminoglycans. (a) Chondroitin 4-sulfate. (b) Chondroitin 6-sulfate. (c) Dermatan sulfate. (d) Keratan sulfate. (e) Heparin.

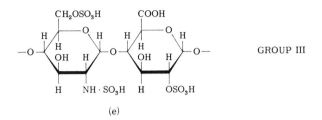

(e)

small degree of branching (e.g., Cifonelli and Dorfman, 1961; Linker and Hovingh, 1968). Both compounds also contain L-iduronic acid as a minor component (e.g., Cifonelli and Dorfman, 1962; Radhakrishnamurthy *et al.*, 1968). Dermatan sulfate is also a hybrid molecule containing appreciable quantities of D-glucuronic acid in addition to L-iduronic acid. The sections of the chains containing the latter appear to be sulfated at the 4-position on the hexosamine, while sections containing the former are sulfated at position 6 (Fransson and Rodén, 1967a,b; Fransson, 1968 a). Variation in the degree of sulfation is a fairly common phenomenon among these polymers and there is evidence that "undersulfated" and "oversulfated" repeating periods of the chains occur (e.g., Mathews and Decker, 1968).

Keratan sulfates, some of which may have some degree of branched structure (e.g., Bhavanandan and Meyer, 1968), appear to contain significant amounts of sialic acid and L-fucose (Seno *et al.*, 1965; Bray *et al.*, 1967). Frequently, polysulfation can be recognized in keratan sulfate and sulfate groups can be detected at position 6 of the hexose residues as well as at position 6 of the hexosamines.

In summary, although the major features of the sulfated glycosaminoglycans have been established, there still remains much work to be done on the fine structure. Absence of positive information about these fine details adds further to the inherent difficulties of studying the biosynthesis and degradation of these polymers and of their protein complexes.

b. Chemical Structure of Proteoglycan Complexes. As mentioned earlier, the sulfated glycosaminoglycans do not exist as such within the tissues but are always present in firm covalent association with protein. It is not without interest that early workers were preoccupied with the need to remove protein from the polysaccharide preparations and, in so doing, failed to recognize the intimate nature of the relationship between the two. Following the pioneering work of Shatton and Schubert (1954) and Malawista and Schubert (1958), who used very mild procedures for extracting proteoglycans from cartilage, more and more attention has been paid to the intact proteoglycans.

Most of the initial studies were made on a chondroitin 4-sulfate proteoglycan of mammalian cartilage which could be resolved into a heavy fraction (PP-H) and a light fraction (PP-L) by centrifuging. These proved to be useful systems for study but it has since been established that each fraction can be further resolved into multiple fractions (e.g., Rosenberg et al., 1967) of varying molecular weight. Other proteoglycans, with the possible exception of heparin proteoglycan (Serafini-Fracassini and Durward, 1968), probably exhibit a similar polydispersity. A further complication is that the composition of any one proteoglycan may vary considerably from tissue to tissue; for example, the keratan sulfate proteoglycan of cornea is quite similar to that of cartilage in terms of polysaccharide structure but very different in terms of protein structure (e.g., Meyer and Anderson, 1965). The composition may also vary with age (e.g., Hoffman et al., 1967a). Moreover, it now seems reasonably well established that two different sulfated glycosaminoglycans can sometimes be present in the same proteoglycan. Thus a number of workers (e.g., Heinegard and Gardell, 1967; Hoffman et al., 1967b; Franek and Dunstone, 1967) have provided good evidence which indicates that both chondroitin 4-sulfate and keratan sulfate are present in the same proteoglycan in both nasal cartilage and nucleus pulposus. The ratio of the two sulfated glycosaminoglycans to each other is not always constant. Similarly, Buddecke et al. (1963) claim that the human costal cartilage chondroitin sulfate proteoglycan contains both chondroitin 4-sulfate and chondroitin 6-sulfate.

Many attempts are now in progress to examine proteoglycans in detail in order to bring about a better understanding of the structure and conformation of the complexes as a whole. Different workers at the present time have their own views on the basic unit structure of the chondroitin 4-sulfate–keratan sulfate–proteoglycan. One of the most recent ones is that of Serafini-Fracassini (1968) for the bovine nasal cartilage complex. He considers the *basic* molecular unit as a core (mol. wt. approx. 89,000) of three polypeptide chains in lateral alignment. An N-terminal valine chain and an N-terminal aspartate chain each carry two laterally extending chondroitin 4-sulfate chains of about 15,000–18,000 molecular weight. The third chain is variable in that its N-terminal amino acid is leucine, in which case it also carries two laterally extending chondroitin 4-sulfate chains, or it is isoleucine, in which case it carries an unknown number of keratan sulfate chains. Multiples of this basic unit are linked end to end to give aggregates which have very high molecular weight in the tissues and contain variable amounts of keratan sulfate, presumably depending on age and other factors.

c. *Linkage of Polysaccharide to Protein.* Some years ago, Muir (1958) examined the amino acid composition of pig tracheal chondroitin 4-sulfate proteoglycan before and after treatment with papain and observed an enrichment of serine in the undegradable chondroitin 4-sulfate peptide which remained. Quantitative considerations suggested that this amino acid was probably involved in forming the link between polysaccharide and protein. Further evidence for this slowly accumulated (e.g., Anderson *et al.*, 1963, 1965) and culminated in a series of excellent papers by Rodén and his colleagues (J. D. Gregory *et al.*, 1964; Lindahl and Rodén, 1965, 1966; Rodén and Armand, 1966; Lindahl *et al.*, 1965; Lindahl, 1966, 1967; Rodén and Smith, 1966; Fransson, 1968b) who established that the linkage region in the cases of the proteoglycans of chondroitin 4-sulfate, heparin and dermatan sulfate, were structurally identical. The immediate link is a glycosidic one between a xylose residue in the polysaccharide chain and a serine hydroxyl group in the protein chain. Two galactose residues lie immediately adjacent to the xylose and then follows the first uronic acid residue of the main polysaccharide chain sequence (XII). β-Glycosidic linkages are involved in all cases and the uronic acid is always D-glucuronic acid even though this may not be the principal uronic acid of the polysaccharide chain.

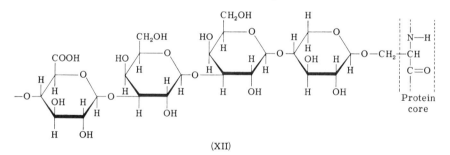

(XII)

Other studies on heparan sulfate (Knecht *et al.*, 1967) suggest that a similar linkage to protein is involved but it seems probable that keratan sulfate is an exception in that both serine and threonine are involved as linking amino acids, possibly through O-glycosidic bonds to N-acetylgalactosamine residues (Bray *et al.*, 1967). The possibility exists that linkages other than those through xylose and serine occur to a minor extent in the other proteoglycans also, following the work of Katsura and Davidson (1966a,b).

At present the regions in the polysaccharide chain which are immediately adjacent to the linkage regions are being closely studied (e.g., Lindahl, 1968) as there are indications that they may be different from the remainder of the polysaccharide main chain.

  d. Biosynthesis. i. General comments. It will be clear from the foregoing that mammalian proteoglycans are among the most complex molecules that nature has yet devised. It will also be obvious that their biosynthesis must be an extremely intricate process. There are many inherent difficulties in studying biosynthesis, not the least of which is that of the relative metabolic inertness of connective tissues and the slow turnover of their polymeric components, particularly in the mature animal. The investigator is faced, not only with the difficulty of finding a suitable system with which to study the problem, but also with the difficulty of separating the relatively sparse cellular elements responsible for the synthesis from the mass of extracellular material in which they are distributed. Further complications arise from the fact that in many instances several proteoglycans coexist in the same tissue and the study of the tissue distribution is a difficult problem in itself. Furthermore, three-dimensional variations in the distribution in any one tissue are common (e.g., Antonopolous et al., 1965; Solheim, 1966) and probably significant, and age is only one of the factors involved in these variations. Other no less formidable difficulties arise as the result of the lack of fine detail on the chemical structures of the polymers themselves, particularly in the cases of heparin and heparan sulfate.

It is not therefore surprising that the progress in elucidating the mechanisms whereby the proteoglycans are formed has been less than spectacular. However, there has been no shortage of effort, and in spite of the difficulties, it is now possible to understand some of the mechanisms that are involved and to make an intelligent assessment of some of the others. The main efforts have been concentrated on a relatively few tissue systems, particularly calf and chick embryonic cartilage and hen's oviduct for the study of chondroitin sulfate synthesis, the cornea and nucleus pulposus for keratan sulfate, skin for dermatan sulfate and mast cells and mast cell tumors for heparin and heparan sulfate. Cultured fibroblasts have also been used and, more recently, leukocytes have emerged as a useful system for chondroitin sulfate synthesis (e.g., Olsson et al., 1968).

The elucidation of biosynthesis requires a knowledge of the mechanisms whereby the polysaccharide sulfate chains and protein cores are assembled and the manner in which these two components are finally united and transported to their ultimate cellular or extracellular destination. Briefly, the accumulated evidence from several laboratories suggests the following principal sequence of events in the biosynthesis of the uronic acid-containing proteoglycans. Synthesis of the protein core occurs on the ribosomes of the rough endoplasmic reticulum of the cell and, as expected, can be inhibited by puromycin which also inhibits glycosaminoglycan synthesis (e.g., de la Haba and Holtzer, 1965; Kleine et al.,

1968). This supports the idea that correct production of core protein is an essential prerequisite for correct production of glycosaminoglycan chains. Probably the linkage sugars, xylose and galactose, are also added primarily at the rough endoplasmic reticulum and enzyme systems can be detected in this cell fraction which are capable of incorporating UDP-xylose and UDP-galactose into the linkage region (e.g., Horwitz and Dorfman, 1968; Telser et al., 1965; Robinson et al., 1966; Grebner et al., 1966; Helting and Rodén, 1968). After formation of the linkage region, uronic acid and N-acetylhexosamine residues are alternatively added via their appropriate UDP derivatives (Telser et al., 1966). This alternate addition appears to occur in both the rough and (particularly) the smooth endoplasmic reticulum. Probably, each glycosaminoglycan requires its own polymerizing enzyme system and, of course, the mechanism of polymerization must take into account the inversion of linkage for β-linked polymers such as dermatan sulfate and the chondroitin sulfates, but retention of linkage for the α-linked heparin and heparan sulfate. Final completion of chains probably occurs in or close to the Golgi complex of the cell from whence the final assembled proteoglycans are exported to their extracellular (or intracellular, in the case of heparin) destinations. The problem of sulfation has not been mentioned in this brief summary but will be dealt with separately in some detail.

Of course, the broad picture presented here still requires many details to be filled in and adequate explanations for certain structural features have yet to be offered. For example, how is polysaccharide chain formation terminated, how does chain branching occur in some of the polymers, how is the switch from glucuronic acid to iduronic acid made in the synthesis of dermatan sulfate chains? How is keratan sulfate synthesized? Presumably an analogous mechanism is involved with UDP-galactose substituting for UDP-uronic acid, and presumably specific enzymes exist for synthesizing the novel (and, as yet, inadequately characterized) linkage sequence. How is the insertion of keratan sulfate into chondroitin 4-sulfate proteoglycan achieved?

Other questions remain to be solved, including the vital ones of the overall control of the biosynthesis. This problem has already begun to receive attention (e.g., Kleine et al., 1968; Kornfeld et al., 1964) and is likely to be a focal object of future studies.

ii. Sulfation of polysaccharide chains.   Present evidence suggests that a number of different sulfotransferase enzymes of high specificity are involved in the sulfoconjugation of the polysaccharide moieties of proteoglycans. Thus Suzuki et al. (1961), working with hen's oviduct preparations produced evidence which indicated that different sulfotransferases were responsible for incorporation of sulfate into heparan sulfate, chondroitin 4-sulfate, and chondroitin 6-sulfate. Similarly, Hasegawa *et al.* (1961) have distinguished between sulfotransferases for dermatan sulfate and for the chondroitin sulfates (see also Adams and Meaney, 1961; Davidson and Riley, 1960; Adams, 1964). The collective evidence, although scanty and certainly not unequivocal, tends to support the view of a multiplicity of sulfotransferases, including a distinct one for the sulfoconjugation of amino groups (Eisenman *et al.*, 1967). Considering the variations in the steric disposition of sulfate groups and the differences in chemical constitutions and configurations of the carbohydrate moieties, it might be suspected on theoretical grounds alone that specific enzymes would be involved. Moreover, there is certainly a precedent for this multiplicity when one considers the analogous situation of the different sulfotransferases which are active toward phenolic hydroxyl groups (Section III, A, 3).

Unequivocal resolution of the question will probably be difficult. Previous reference has been made to the problems associated with the clean separation of the soluble arylsulfotransferases but such problems may be multiplied with the glycosaminoglycan sulfotransferases because they all appear to be particle bound. This interesting fact was first brought to light as a result of studies on the sulfoconjugation of heparin by Silbert (1967a,b,c), Eisenman *et al.* (1967), and Rice *et al.* (1967). Sulfotransferase activity for both N- and O-sulfoconjugation of heparin and heparan sulfate were shown to sediment from mast cell tumor tissue principally as a microsomal fraction (probably smooth endoplasmic reticulum), while the PAPS-producing enzymes remained in solution in the supernatant. This has since been confirmed for an entirely different system, namely, the chick embryo cartilage enzyme responsible for the transfer of sulfate to chondroitin 4-sulfate proteoglycan (DeLuca and Silbert, 1968). If this turns out to be true for all glycosaminoglycan sulfotransferase enzymes, then the interesting situation exists that all the enzymes responsible for the assembly of proteoglycans from their individual units are bound in the membranous system constituted by rough and smooth endoplasmic reticulum and by the Golgi apparatus. The complete system for the production of a proteoglycan molecule could thus be visualized as a sort of assembly line of enzymes arranged along a channel which culminates in the Golgi apparatus. This, in turn, may have interesting genetic implications in terms of the location of cistrons governing the manufacture of the enzymes. This view of the total biosynthetic process probably also explains why glycosaminoglycan sulfotransferase systems have generally worked best when transferring sulfate to endogenous acceptor rather than to chemically prepared material added to the system.

One important source of controversy which is still not completely resolved concerns the precise timing of sulfation in the production of the complete proteoglycan. As early as 1954, Davidson and Meyer (1954), following the discovery of chondroitin (chondroitin sulfate without sulfate groups), suggested that the compound was the immediate precursor of chondroitin sulfate. In other words, they envisaged that polymer formation preceded sulfation. However, this view subsequently received a setback following the isolation of UDP-N-acetyl-D-galactosamine 4-sulfate from hen oviduct (e.g., Suzuki and Strominger, 1960b,c,d). Oviduct preparations were capable of transferring sulfate from PAPS to N-acetyl-D-galactosamine and could also transfer sulfate from PAPS to N-acetyl-D-galactosamine 4-sulfate to give the corresponding 4-6-disulfate (see also, Harada et al., 1967). These and other findings made by Suzuki and his colleagues (see Dodgson and Lloyd, 1968) were not incompatible with the idea that sulfation of proteoglycan was stepwise. That is to say that uronic acid and hexosamine residues would be added alternately to the growing glycosaminoglycan chain and a sulfate group would be introduced into each hexosamine residue immediately following its addition to the nonreducing end of the growing chain.

Subsequently, Telser et al. (1966), working with a cell-free chick cartilage system and using as acceptors even- and odd-numbered nonsulfated oligosaccharides prepared from chondroitin sulfate, found that a sulfated terminal N-acetyl-D-galactosamine residue blocked the addition of a further uronic acid residue. This may have pronounced significance in terms of chain termination but it is difficult to see how any chain could form if polymerization and sulfation proceeded at precisely the same rate, as suggested by Suzuki and his colleagues.

Other workers have not been slow to join in the controversy and the reader is referred to the work of Rice et al. (1967), Silbert (1967b,c), and Kleine et al. (1968) for the various views that have been expressed. The weight of evidence indicates that sulfation probably occurs both during and after synthesis of the polysaccharide chain but final proof of this is still awaited.

Further points of interest concerning sulfation are concerned with dermatan sulfate and with heparin and heparan sulfates. It will be recalled that dermatan sulfate is a hybrid polymer containing both iduronic acid residues linked to N-acetyl-D-galactosamine 6-sulfate, and glucuronic acid residues linked to N-acetyl-D-galactosamine 4-sulfate. Fransson

(1968a) has isolated a hybrid octasaccharide containing both features after enzymic degradation of dermatan sulfate and has pointed to the intriguing possibility that two sulfotransferases may be involved in constructing adjacent areas of the same molecule. The type of uronic acid associated with the hexosamine would then be determining which sulfotransferase operated. Alternatively, he suggests that if only one sulfotransferase is involved, the type of uronic acid would direct sulfoconjugation to either position 6 or 4 on the hexosamine. In this connection, it is interesting to note that Meezan and Davidson (1967) have suggested that the conformation of the protein moiety might be a factor which helps to determine the position of sulfoconjugation in the case of the isomeric chondroitin 4- and 6-sulfates.

In the case of heparin, Silbert (1967c) has provided good evidence that N-acetyl-D-glucosamine residues are first incorporated into the growing polysaccharide chains but that the acetyl groups are subsequently exchanged for N-sulfate groups when sulfoconjugation begins. Thus, mouse mast cell tumor preparations, in the absence of PAPS, were able to catalyze the formation of an N-acetylated nonsulfated glycosaminoglycan related to heparin. When PAPS was subsequently added, a loss of N-acetyl occurred and N-sulfate appeared. The fact that N-acetyl-D-glucosamine residues persist in the immediate vicinity of the linkage region of heparin, would perhaps imply that this exchange is sterically hindered in this region. It is possible that a similar exchange mechanism perhaps operates during the biosynthesis of heparan sulfate proteoglycan, since both N-acetyl and N-sulfate groups are present in the polysaccharide chain.

e. Enzymic Hydrolysis. i. Catabolism of protein cores. The dramatic effects of injected papain on cartilaginous tissues of animals was first demonstrated by the noteworthy rabbit "ear-flop" experiments of Thomas (1956). Injection is followed by a rapid destruction of cartilage matrix and the liberation of acid glycosaminoglycans into the blood and urine (Bryant *et al.*, 1958; McCluskey and Thomas, 1959; Tsaltas, 1966). Similar effects on cartilage (either in tissue culture or in the intact animal) during states of hypervitaminosis A can also be attributed to proteolytic activity, featured by an increased release and enhanced synthesis of an acid cathepsin localized within the lysosomes of cells (e.g., Dingle, 1961; Fell and Dingle, 1963; Fell, 1965). The study of Saunders and Silverman (1967) has shown that, under the electron microscope, chondroitin sulfate proteoglycan is revealed as a linear stranded structure and this linearity is retained to some extent when polysaccharide chains are degraded by hyaluronidase, but vanishes completely following

treatment with papain. The importance of the protein core in maintaining the conformation of the proteoglycan as a whole is thus clearly
revealed. A number of other workers have confirmed and extended the
general concept that connective tissue and other cells contain one or more
proteolytic enzymes that can participate in the degradation of the protein core of proteoglycans (see Dziewiatkowski *et al.*, 1968; Ali *et al.*,
1967). Such enzymes are almost certainly lysosomal in origin and coexist
with other lysosomal enzymes that are known to be capable of attacking
the glycosaminoglycan chains.

ii. Hydrolysis of linkage region. Glycosaminoglycan chains of varying
size and free or relatively free from protein, can be detected in normal
urine (e.g., Knecht *et al.*, 1967; Manley *et al.*, 1968; Orii, 1968) and O-xy-
losyl-serine has also been shown to be present (Tominga *et al.*, 1965).
These findings suggest that proteoglycan degradation may involve the
linkage region between protein and glycosaminoglycan. Support for this
comes from the observation that β-xylosidase activity occurs in many
mammalian tissues, including cartilage, spleen, kidney, liver, and skin
(Fisher *et al.*, 1966; 1967; Robinson and Abrahams, 1967; Esterley *et al.*,
1968). The lysosomal origin of the rat liver and pig kidney enzymes has
been demonstrated and there is some evidence that this type of enzyme
is active toward the xylose-serine linkages of chondroitin sulfate-peptide
preparations, although not necessarily toward intact proteoglycan.
Lysosomal β-galactosidases are also widely distributed in animal tissues
(e.g., Conchie and Hay, 1963; Öckerman, 1968) and the liver levels of
the enzyme depart markedly from normal in certain inherited diseases of
proteoglycan metabolism (e.g., Van Hoof and Hers, 1968). The present
inference to be drawn from these various observations is that one or both
of these glycosidases can perhaps participate in the hydrolytic fission of
either intact or partially degraded proteoglycan, but positive proof is not
yet available.

iii. Degradation of glycosaminoglycan chains. The chains of chondroitin 4-sulfate and chondroitin 6-sulfate proteoglycans can be degraded
by mammalian and other hyaluronidases. This type of degradation has
previously been considered at length by Dodgson and Lloyd (1968) and
will not be discussed in detail here. Briefly, the mammalian enzyme is an
endo-poly-β-hexosaminidase hydrolyzing β1 $\longrightarrow$ 4 hexosaminide bonds
and yielding even-numbered oligosaccharides, particularly tetrasaccharides. Although the most widely studied enzyme has been that from
mammalian testes, the enzymes has been detected in rat liver and bone
(Aronson and Davidson, 1965; Hutterer, 1966; Vaes, 1967), human
serum (Bollet *et al.*, 1963), canine submandibular gland (Tan and Bow-

ness, 1968), and bovine aorta (Buddecke and Platt, 1965). The liver enzyme is lysosomal in origin and has been purified and shown to have properties which are rather different from those of the testicular enzyme (Aronson and Davidson, 1967). The enzyme, like other mammalian hyaluronidases, is not active toward heparin, heparan sulfates, keratan sulfate, and dermatan sulfate (some degradation of dermatan sulfate occurs in those regions where D-glucuronic acid is present), and no mammalian enzyme capable of depolymerizing these other glycosamino-glycans has yet been recognized.

Two other lysosomal glycosidases can also participate in the degradation of chondroitin 4- and 6-sulfate proteoglycans. β-Glucuronidase can remove single terminal uronic acid residues from the nonreducing ends of even-numbered oligosaccharides resulting from hyaluronidase action and, subsequently, β-N-acetylhexosaminidase can remove single terminal N-acetylhexosamine residues from the nonreducing ends of odd-numbered oligosaccharides thus formed, provided that sulfate groups are first removed (see Dodgson and Lloyd, 1968).

iv. Sulfohydrolysis. A number of enzymes of bacterial or molluskan origin are known which can remove sulfate from sulfated glycosaminogly-cans, and the reader is referred to the reviews of Dodgson (1966) and Dodgson and Lloyd (1968) for details. Mammalian sulfohydrolases attacking these polymers have been more difficult to find but three such enzymes have been noted.

The first of these appears to be a genuine chondrosulfohydrolase in that it utilizes polymeric chondroitin sulfate as its substrate. The enzyme has been partially purified from bovine aorta (Held and Buddecke, 1967) and its low optimum pH (4.4) perhaps indicates its lysosomal origin. The preparation was free from hyaluronidase and its action was not enhanced by addition of that enzyme to incubation mixtures. It was possible to separate the chondrosulfohydrolase from associated arylsulfohydrolase activity. The enzyme does not appear to be very active and long incubation periods were necessary for its assay.

The second enzyme cannot be regarded as a true chondrosulfohydrolase because it shows no activity toward polymer chondroitin 4-sulfate (cf. Dodgson and Lloyd, 1957). The enzyme has been partially purified from rat liver lysosomes (Tudball and Davidson, 1969) and appears to act on an odd-numbered oligosaccharide fragment—probably a heptasaccharide—in which N-acetyl-D-galactosamine 4-sulfate is the terminal nonreducing unit. It therefore depends on the prior action of both hyaluronidase and β-glucuronidase. Further work needs to be done on this enzyme but the possible implications of the data available at present would suggest that degradation of chondroitin 4-sulfate chains in the liver lysosome involves prior attack by hyaluronidase yielding oligosaccharides containing about eight residues, subsequent removal of the terminal uronic acid residue by β-glucuronidase, followed by removal of sulfate from the newly exposed terminal hexosamine residue by the sulfohydrolase. Presumably, lysosomal β-N-acetylhexosaminidase would then have an opportunity to remove the desulfated hexosamine residue. Certainly injected labeled chondroitin 4-sulfate and chondroitin 6-sulfate can be taken up by liver lysosomes and are apparently slowly degraded there (Aronson and Davidson, 1968). In contrast, and in accord with established enzymological knowledge, dermatan sulfate is taken up by lysosomes but persists there for a much longer time.

Strictly speaking, the third known mammalian glycosaminoglycan sulfohydrolase is a sulfamidase, since it removes the N-sulfate groups from heparin. This enzyme was discovered in these laboratories as a result of injecting a semichemically synthesized re-N-^{35}S-sulfated heparin into rats (Lloyd et al., 1966a), a procedure which led to the appearance of considerable amounts of labeled inorganic sulfate in urine. The responsible enzyme has been recognized in extracts of rat, pig, and bovine spleen (Lloyd et al., 1966b, 1968a). The partially purified enzyme removes N-sulfate groups from heparin but does not appear to hydrolyze the O-sulfate groups that are also present. It has no activity toward 2-deoxy-2-sulfoamino-D-glucose, a possible monomeric substrate. It is of some interest that loss of N-sulfate groups from heparin results in the loss of virtually the whole of its anticoagulant activity, so that the sulfamidase may be the enzyme responsible for the rapid loss of this activity which occurs during heparin therapy.

f. Relation between Proteoglycan Biosynthesis and Hydrolysis. Very little is known about the relationship of proteoglycan biosynthesis and hydrolysis but the present view would be that a delicate metabolic balance of proteoglycan synthesis and degradation exists within the tissues. Such a balance is probably subject to control mechanisms the influence of which must alter appreciably with the development and the aging of the tissues in which they are found. At our present state of knowledge, the picture emerges of proteoglycan biosynthesis in the endoplasmic reticulum of the cell being counterbalanced by degradative process taking place within the lysosomes.

Various factors can influence this delicate balance (see Dodgson and Lloyd, 1968) and, in some instances, can give rise to serious clinical disorders. Thus, severe structural changes, accompanied by loss of proteoglycan, occur in cartilage in degenerative arthritis. Other diseases, such as Hurler's syndrome (gargoylism), lead to accumulation of certain glycosaminoglycans in the cell. Much interest is currently being shown in diseases of the latter type, of which several are known. They are of genetic origin and it is becoming clear that they involve excess or absence of certain lysosomal enzymes concerned with proteoglycan degradation. Detailed discussion of these conditions is not relevant to the present work but useful information may be found in recent papers by Fratantoni et al. (1968) and Van Hoof and Hers (1968). In any event, it is certain that a better understanding of them will depend on a better understanding of the relationships between proteoglycan biosynthesis and degradation, and control mechanisms which regulate the activities of the various enzymes involved.

2. Glycolipid Sulfate Esters

a. Occurrence and Structure. Glycolipid sulfoconjugates are important constituents of mammalian organisms. It is probable that they may also occur in lower animals but no investigations of this possibility appear to have been made. In the world of plants a special type of sulfur-containing glycolipid appears to be widely distributed but sulfur is present in the form of sulfonate groups attached to carbohydrate residues (e.g., Benson, 1963), and true glycolipid sulfate esters have not been found. Indeed, the only nonmammalian source of such an ester appears to be the halophilic bacterium, *Halobacterium cutirubrum*, which produces a complex glycolipid containing galactose 3-sulfate residues (Kates et al., 1967).

Mammalian glycolipid sulfoconjugates appear to be of particular importance in nervous tisue and they are abundant in both white and gray matter of brain and in the myelin sheaths of nerves. However, they also feature as membrane components in extraneural tissues and are present in appreciable amounts in kidney, liver, spleen and other organs (e.g., Green and Robinson, 1960; Soper, 1963; Martensson, 1966). The tissue content of these glycolipid sulfates appears to decrease a little with age, but generally speaking the chemical nature of the compounds changes very little apart from minor differences (e.g., Svennerholm and Stalberg-Stenhagen, 1968; O'Brien et al., 1964).

The existence of these sulfoconjugates has been known for many years and was first recognized by Thudichum (1884) who coined the name "sulfatide" to describe them. This trivial name, although quite unsatisfactory from a descriptive point of view, has persisted down the years and hence will be generally used in this section. Further confusing nomenclatures evolved as detail concerning the intimate chemical structures of the compounds became available, but there have been more recent useful attempts at rationalization which now seem to have become generally accepted.

Briefly, two main types of sulfatide exist, with the possibility of a third type occurring in brain.

The basic structure is that of a ceramide mono-or dihexoside monosulfate ester, where ceramide is the term given to sphingosine, the nitrogen of which carries an acyl-linked fatty acid residue. In the mono-hexoside compounds the single hexose residue is galactose which carries an ester sulfate group at position 3 (XIII), while the dihexoside compounds carry an additional glucose residue (XIV). Here, the linkage between galactose and glucose is $\beta 1 \longrightarrow 4$ and the disaccharide component is thus revealed as a lactosyl unit (Stoffyn *et al.*, 1968).

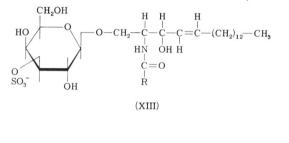

(XIII)

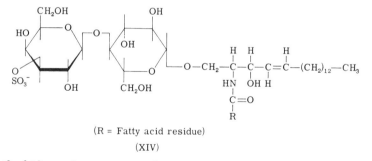

(R = Fatty acid residue)

(XIV)

In the kidney, the two types of compound appear to be distributed in the proportion of about 3 to 1 (e.g., Martensson, 1966). Analogous compounds which carry no ester sulfate group are present in the tissue (i.e., galactosyl ceramides and lactosyl ceramides) and probably constitute the immediate precursors of the sulfate esters. Variations between individual members of the two types of sulfate ester exist and these appear to be principally confined to the fatty acid moiety, although differences have also been recorded for the sphingosine chain (see Martensson, 1966).

The third type of sulfatide has been reported by Hakamori *et al.* (1962) but precise details of the structure have not been reported and the work does not appear to have been confirmed. Briefly, these workers separated two sulfatides from brain and, as a result of chemical studies, concluded that one of them was a hybrid sulfatide composed of galactosyl ceramide 3-sulfate and galactosyl ceramide 6-sulfate portions, possibly linked together by a nontitratable, sulfate diester bridge. This possible type of sulfatide will not be discussed further.

Despite their relatively widespread tissue distribution, their presence in endoplasmic reticular and mitochondrial membranes of brain (e.g., Davison and Gregson, 1966), their obvious relationship to brain and nerve tissue in particular, and the gross disturbances in their metabolism that occur in patients with metachromatic leukodystrophy, no positive physiological role can yet be assigned to these sulfate esters. At present they are therefore generally regarded as "structural glycolipids" whose precise role in membranous structures is quite obscure.

b. Biosynthesis. Surprisingly little is known about the biosynthesis of sulfatides and it is not yet possible to state precisely how and in what sequence the various units are assembled. Indeed, there has been some controversy as to whether the sulfated galactosyl ceramides are precursors of the galactosyl ceramides (e.g., Jatzkewitz, 1960a) or vice versa (e.g., Radin *et al.*, 1957). The implications of the two different views are interesting in that the former implies that sulfation of a small molecule (e.g., UDP-galactose) might be involved, while the latter demands sulfation as the final step in the biosynthesis. The weight of evidence now seems to be heavily in favor of the latter hypothesis. This may be deduced from consideration of the results of incorporation studies using $[^{14}C_6]$ glucose (e.g., Hauser, 1964) and of direct studies of sulfotransferase activity *in vitro* (e.g., Cumar *et al.*, 1968). Incorporation studies reveal that sulfatide synthesis is particularly high during early development (maximum for rat brain at about 20 days, e.g., Davison and Gregson, 1966; Hauser, 1964; McKhann *et al.*, 1965; Balasubramanian and Bachhawat, 1965a) the period of active synthesis coinciding with the onset of the histological appearance of myelin. The specific activity of isolated rat-brain galactosyl ceramide at 20 days after $[^{14}C_6]$ glucose injection is about eight times as high as that of sulfated galactosyl ceramide, indicating the impossibility of the latter being the normal precursor of the former.

Interesting enzymological work on the sulfation of hexosyl ceramides has come from McKhann *et al.* (1965), Balasubramanian and Bachhawat (1965a,b) and Cumar *et al.* (1968). These workers have established the particle-bound nature of the responsible rat and sheep brain sulfotrans-ferases. Rat liver and spleen also contain a similar particle-bound enzyme and rat kidney is an exceptionally good source. In the experiments of Balasubramanian and Bachhawat (1965a) there was no requirement for added acceptor, and sulfate was transferred directly from PAPS to an endogenous acceptor which was firmly attached to protein in the particu-late preparation and which, by indirect evidence, was considered to be a protein-bound galactosyl ceramide.

The experiments of Cumar *et al.* (1968) are particularly interesting because they indicate the presence in rat brain of two different sulfo-transferases. Briefly, all subcellular fractions exhibited varying degrees of sulfotransferase activity toward two distinct types of acceptor. The first type consisted of relatively simple compounds such as galactose, lactose, neuramin lactose and p-nitrophenyl β-D-galactoside, while the second type included galactosyl ceramide, galactosyl sphingosine, and lactosyl ceramide. Separation of the two different sulfotransferase enzymes was not achieved, but a variety of kinetic, cellular-distribution, inhibition, and other experiments clearly indicated the distinct nature of the two activ-ities. The enzyme transferring sulfate to small molecule acceptors will be referred to again in a later section (III, C, 3,b); meanwhile, it is interest-ing that this work not only establishes galactosyl ceramide and lactosyl ceramide as good sulfate acceptors but also indicates that a possible intermediate in the overall biosynthetic pathway to monohexosyl sulfa-tides, galactosyl sphingosine, can readily be sulfated.

With these points in mind the two principal postulated routes to the synthesis of a monohexosyl sulfatide may now be considered. One of these routes [see Eq. (6)] suggests the combination of sphingosine and fatty acid (via an acyl-CoA) to yield ceramide (N-acylsphingosine), followed by the introduction of galactose (presumably via UDP-galactose) and subsequent sulfation. The other suggests that sphingosine and UDP-

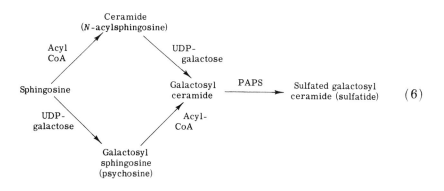

galactose first give galactosyl sphingosine (psychosine) and that this intermediate is then N-acylated and subsequently sulfated.

The work of Cleland and Kennedy (1960) and Brady (1964) suggests that the galactosyl sphingosine route is the favored one. If this is so, then in the light of the fact that galactosyl sphingosine can act as a good sulfate acceptor, the uncertainty about the precise timing of the sulfation step still awaits final clarification. However, the observation (Hauser, 1967) that glucosyl ceramide can act as an acceptor for galactose (from UDP-galactose), with the consequent production of lactosyl ceramide, seems to provide additional support for sulfation as the final step. Meanwhile, the indications are that the final assembly of these relatively large molecules occurs on the endoplasmic reticulum membranes, presumably by an assembly line of enzymes rather analogous to that proposed for proteoglycan biosynthesis.

c. *Enzymic Hydrolysis.* Martensson (1966) has provided some indications of the possible interconversions and metabolic transformations of various glycolipids while enzymes have been described that can release galactose from galactosyl and lactosyl ceramides and glucosyl ceramide (see, e.g., Bowen and Radin, 1968a,b). However, these and other similar degradative transformations will not be discussed further. Of more immediate importance is the probability that sulfated hexosyl ceramides and their nonsulfated isomers are interconvertible. Kidney appears to possess the highest sulfating (Balasubramanian and Bachhawat, 1965a) and desulfating activity (Mehl and Jatzkewitz, 1963, 1964). The sulfohydrolase enzyme has been purified from pig kidney where it is probably present in lysosomes, and it presents some unusual features. In the first instance it appears to consist of two high molecular weight fractions which can be separated by high voltage electrophoresis. One of these fractions is labile but the other is stable and recombination of the two is essential for enzyme activity toward galactosyl ceramide sulfate. Second, as mentioned in Section III, A, 4, *a*, the labile fraction has extremely high arylsulfohydrolase A activity. Further work in which the relationship between arylsulfohydrolase A and the sulfatide sulfohydrolase has constantly been considered (Mehl and Jatzkewitz, 1968) has led the authors to conclude that the heat-labile fraction of the sulfatide sulfohydrolase complex is identical with arylsulfohydrolase A. One or two features of the work perhaps suggest that some oversimplification has been made, particularly the effects of dialysis on the enzyme complex and the exceptionally low activity toward sulfatide as opposed to nitrocatechol sulfate. However, the conclusions must be viewed against the background that arylsulfohydrolase A activity disappears in patients with metachromatic leukodystrophy and further comment on this is made below.

Meanwhile, no one yet seems to have attempted to repeat this work, which is regrettable. Not only is the clinical significance of the work very great but also there are clearly intriguing enzymological problems to be considered, particularly when one recalls the extraordinary anomalous behavior of mammalian arylsulfohydrolase A.

d. Relation between Sulfoconjugation and Sulfohydrolysis. Metachromatic leukodystrophy is a rare inherited metabolic disease of the nervous system and is characterized by the accumulation of sulfatides in the brain, kidney, and other visceral organs (see, e.g., Jatzkewitz, 1958, 1960b; Austin, 1959). A decrease in the tissue content of nonsulfated hexosyl ceramides can also be noticed. The implication of these two factors is that the delicate balance between sulfatide synthesis and degradation, occurring in the endoplasmic reticulum and lysosomes respectively, has been disturbed. The further implication, when considered in relation to certain analogous "storage" diseases such as Gaucher's disease and gargoylism, is that a lysosomal enzyme involved in the degradation of sulfatides is missing. The studies of Austin *et al.* (1963, 1964, 1965) established that lysosomal arylsulfohydrolase A was virtually absent from the urine of patients with metachromatic leukodystrophy and was also low in the tissues. A number of other enzymes, including some of lysosomal origin, were present in normal amounts in the tissue. These observations may now be viewed in relation to those of Mehl and Jatzkewitz concerning arylsulfohydrolase A and sulfatide sulfohydrolase activities. More recent studies by these workers (Jatzkewitz and Mehl, 1969) have extended the numbers of patients examined and have confirmed that both arylsulfohydrolase A and sulfatide sulfohydrolase activities are reduced to the limit of detection in autopsy specimens of renal cortex, liver, and cerebral white matter.

Further disturbances in sulfatide balance are a feature of globoid leukodystrophy, a condition which leads to widespread degeneration of white matter. Sulfatide levels, both in white matter and in cerebral cortex, are markedly decreased (Svennerholm, 1963) but nonsulfated hexosyl ceramides are proportionately elevated. Bachhawat *et al.* (1967) have provided evidence which indicates that the primary enzyme deficiency is that of the sulfotransferase responsible for incorporating sulfate into sulfatide molecules. This enzyme is present in normal amounts in patients with metachromatic leukodystrophy. Deficiency of the enzyme but retention of sulfatide sulfohydrolase activity would ultimately lead to excessive destruction of sulfatide. It is not without interest that these findings also point to sulfation as the last step in sulfatide biosynthesis.

3. Simple Carbohydrate Sulfate Esters

a. General Comments. Mention was made in Section I, B, 3 of some simple carbohydrate sulfate esters which occur naturally and which appear, at the moment, to be biochemical curiosities. One of these, holothurin A, is present in the tropical sea cucumber, *Actinopyga agassizi* (Nigrelli *et al.*, 1955), and a closely related compound, holothurin B, has been recognized in two other sea cucumber species (Yasumoto *et al.*, 1967). Essentially these compounds are steroidal glycosides in which a single sulfate group is present on a sugar residue, although the precise location is not yet known (Fries *et al.*, 1967). The holothurins are potent toxic agents, blocking cholinergic neuromuscular junctions, but they will not be discussed further.

Other simple sugar sulfates that have been detected in mammalian tissues include uridine diphospho-N-acetyl-D-galactosamine 4-sulfate (XV), isolated from hen oviduct by Suzuki and Strominger (1960a); 6-O-sulfato-β-D-galactopyranosyl-(1⟶4)-D-glucopyranose (XVI) (lactose 6-sulfate), isolated from rat mammary gland by Barra and Caputto (1965), and O-α-D-N-acetylneuraminyl-(2⟶3)-O-β-D galactopyranosyl 6-O-sulfate-(1⟶4)-D-glucopyranose (XVII) (neuramin lactose 6-sul-

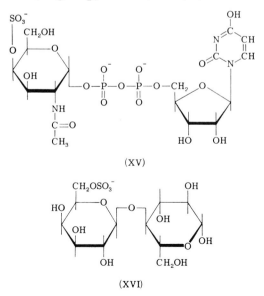

(XV)

(XVI)

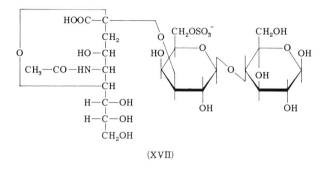

(XVII)

fate) isolated from rat mammary gland by Carubelli *et al.* (1961) and Ryan *et al.* (1965).

b. Sulfoconjugation. In the case of UDP-N-acetyl-D-galactosamine 4-sulfate, Suzuki and Strominger (1960b,c,d) have established that it can be formed by hen oviduct preparations from UDP-N-acetyl-D-galactosamine. It can also be detected in ossifiable cartilage and aorta of rats following injection of inorganic sulfate-^{35}S (Picard and Gardais, 1967). A second and apparently distinct sulfotransferase is present in hen oviduct which can add a second sulfate to position 6 of UDP-N-acetyl-D-galactosamine-4-sulfate but which has no activity toward the desulfated UDP derivative (Harada *et al.*, 1967). It is not yet clear whether these two enzyme systems from hen oviduct are truly soluble or particle bound, as the conditions of centrifuging used would certainly not remove all particles derived from the endoplasmic reticulum. Nor is it yet clear why these sulfate esters should be synthesized at all (cf. Section III, C, 1, *d*, *ii*).

The precise mechanism whereby neuramin lactose 6-sulfate and lactose 6-sulfate receive their sulfate groups is not yet known. It will be recalled (Section III, C, 2, *b*) that Cumar *et al.* (1968) were able to achieve the sulfoconjugation of neuramin lactose, lactose, and galactose by rat brain preparations and the sulfotransferase enzyme was probably different from the one responsible for the sulfation of hexosyl ceramides. However, the rates of acid hydrolysis of the simple sugar sulfates formed in these experiments were much greater than would have been obtained with C_6-sulfated sugars and, for the enzymically synthesized lactose sulfate, hydrolysis proceeded at twice the rate of that for authentic lactose 6-sulfate. Presumably, in view of the initial isolation of neuramin lactose 6-sulfate and lactose 6-sulfate from rat mammary gland, it would seem more appropriate to use this tissue as the likely source of the 6-sulfating system. It is of interest that both Harada *et al.* (1967) and Cumar *et al.*

(1968) have observed the presence of low molecular weight, water-soluble, endogenous sulfate acceptors in their oviduct and brain preparations respectively. In brain preparations, transfer of sulfate to these endogenous acceptors appears to be very efficient. It would certainly be worthwhile to examine these endogenous acceptors more closely in order to establish whether they are related in any way to the biosynthesis of proteoglycans, sulfated mucins or sulfatides, or whether they perhaps represent carbohydrate sulfate esters of types that have not previously been suspected.

c. Sulfohydrolysis. A number of nonmammalian sources of sulfohydrolases that can use relatively simple sugar sulfates have been reported and the reader is referred to the reviews of Lloyd (1966), Dodgson (1966), and Dodgson and Lloyd (1968) for many of the details. Whether such enzymes are present in mammals is less certain. Injection of ^{35}S-labeled hexose and hexosamine sulfates into experimental animals (Lloyd *et al.*, 1964; Lloyd, 1961, 1962) maintained on an antibiotic regimen in order to suppress gut flora, has always been followed by an extremely rapid excretion of the major part of the injected material in an unchanged form. Nevertheless there has always been a small amount of radioactivity (usually less than 10% of the administered dose) appearing in the urine as inorganic sulfate. This does indicate the possible existence of mammalian glycosulfohydrolases. A number of searches for such an enzyme have been made, generally employing glucose 6-sulfate, N-acetylglucosamine 6-sulfate or N-acetylgalactosamine 6-sulfate as potential substrates, assuming that like other glycosulfohydrolases (see Dodgson, 1961; Lloyd *et al.*, 1968b), activity toward glucose 6-sulfate would also imply activity toward galactose 6-sulfate. These searches have been singularly unsuccessful.

Recently, Mehl and Jatzkewitz (1968) found that their pig kidney sulfatide sulfohydrolase preparations were able to hydrolyze galactose 3-sulfate but not galactose 6-sulfate, implying a high substrate specificity similar to that noted for the glycosulfohydrolase of *Trichoderma viride* (Lloyd *et al.*, 1968b). Such an enzyme may perhaps be involved in some (as yet unknown) way in the breakdown of sulfatides but can obviously not be concerned with the degradation of such compounds as lactose 6-sulfate.

Clearly the relationship between the biosynthesis and hydrolysis of these relatively simple carbohydrate sulfate esters in mammals is still a mystery and the earlier reference to these esters as biochemical curiosities seems fully justified at the present time.

IV. Conjugates with N—SO₃⁻ Linkages

A. OCCURRENCE

The principal examples of naturally occurring sulfoconjugates of this type are, of course, heparin and heparan sulfates in mammals and the mactins in the surf clam. These compounds have already been discussed in Section III, but it is worthwhile recalling that in the case of heparin the presence of the N-sulfate (sulfamate) groupings is essential for its anticoagulant activity and the chemical or enzymic removal leads to an immediate loss of that activity.

In addition to these compounds, one or two examples of analogous but much simpler sulfamates have come to light during the course of drug metabolism work. Thus Boyland *et al.* (1957) and Parke (1960) have shown that simple arylamines such as aniline and 2-naphthylamines were excreted in rat or rabbit urine to a small extent as the corresponding phenylsulfamate (XVIII) and naphthyl 2-sulfamate (XIX), respectively.

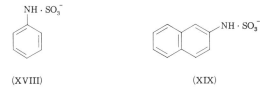

(XVIII)   (XIX)

Smith (1962) has also detected *p*-carboxyphenylsulfamate in the excreta of spiders following administration of *p*-aminobenzoic acid. This conjugation mechanism does not appear to be a significant one and has been relatively little studied. However, it has one or two features of interest which might be mentioned. These sulfamates, like the sulfamate groups of heparin and heparan sulfate, are relatively unstable compared to the normal C-O-SO₃⁻ structures, and indeed N-desulfated heparin can readily be prepared from heparin by mild acid hydrolysis under conditions that apparently do not affect the O-sulfate ester groups that are present.

B. BIOSYNTHESIS

Most of the biosynthetic work on these compounds can be attributed to Roy (e.g., 1960b, 1961b, 1962) using rat or guinea pig liver preparations. Briefly, the rat liver enzyme is able to transfer sulfate from PAPS to acceptors such as 1- and 2-naphthylamines and aniline but not to glucosamine or to benzylamine, where the nitrogen is separated from the aromatic ring by a —CH$_2$ grouping. In common with several other sulfo-transferases the enzyme is activated by Mg^{++} ions and is apparently dependent on intact thiol groups for its activity. Very little work has been done on the natural distribution of this enzyme but toad liver is a potent source of the enzyme as well as of many other sulfotransferases (unpublished results, these laboratories). In contrast, the opossum shows little or no transfer activity toward 2-naphthylamine or, indeed, toward estrone, androstenolone, or testosterone (Roy, 1963) while it is only feebly active toward p-nitrophenol. The contrast between these two species is certainly interesting from an evolutionary point of view.

During the course of his studies Roy observed some interesting effects of 17-oxo steroids on crude preparations of the sulfotransferase enzymes from rat and guinea pig. For example, the rat enzyme was greatly activated by 3β-methoxyandrost-5-en-17-one whereas the guinea pig enzyme was strongly inhibited. Conclusions were drawn from these studies which were certainly not warranted in view of the crude enzyme preparations that were employed, both to produce PAPS *in situ* and to transfer its sulfate to the acceptor. However, further work (Roy, 1964; Banerjee and Roy, 1966) led to a modification of the earlier conclusions and to the suggestion that the guinea pig sulfotransferase was an allosteric enzyme which was inhibited by very low concentrations of 17-oxosteroids, and which may therefore have some important part to play in the overall control of the activity of the enzyme.

A further surprising thing emerged from these studies (Banerjee and Roy, 1966) namely, the conclusion on the basis of enzyme purification, fractionation, and other studies that a true arylamine sulfotransferase does not exist, but that this type of transfer activity may be an additional feature of estrone and androstenolone sulfotransferases. This emerges as such an extraordinary conclusion in terms of specificity that it seems essential to put the conclusion to further tests. In relation to this, it is of some interest that the highly purified estrone sulfotransferase of bovine adrenal glands is completely devoid of transfer activity toward 2-naphthylamine (Adams and Poulos, 1967).

It is now pertinent to enquire about the possible relationship between arylamine sulfotransferase activity and sulfotransferase activity toward heparin amino groups. Briefly, there would appear to be none since the former activity appears to be present in the soluble fraction of the cell while the latter is largely associated with the endoplasmic reticulum. Indirect support for this view might be derived from the observations of Carrol and Spencer (1965) that fetal rat liver preparations are devoid of arylamine sulfotransferase activity but the fetus is still able to produce sulfated glycosaminoglycans, presumably including heparin.

C. Sulfohydrolysis

There have been no reports of sulfohydrolases active toward arylamine sulfamates, but the spleen enzyme which liberates the N-sulfate groups from heparin is without activity toward phenyl sulfamate, although the latter compound inhibits the activity of the enzyme toward heparin (Lloyd et al., 1968a).

D. Conclusions

The present state of knowledge on the importance of arylamine sulfo-conjugates and their biosynthesis is clearly in an unsatisfactory state. Other effective methods for the disposal of arylamines are available to mammalian organisms and the mechanism, at the very best, appears to be only marginally important to the problem of "detoxication." This particular field of endeavor still does not attract the attention of investigators, yet there always remains the possibility that the mechanism is concerned with sulfoconjugation of well-known nitrogen-containing compounds of biological importance, for example, the ring nitrogens of bile pigments or the side-chain nitrogens of compounds such as noradrenaline. Alternatively, the mammalian mechanism may be an evolutionary relic relating to a period of evolution in which there was a need for such activity. However, to the student of drug metabolism such a proposition is unlikely to carry much favor.

V. General Comment

In this chapter an attempt has been made to indicate the changes that have occurred in the parts played by "active" forms of sulfate during the course of evolution. Both APS and PAPS have been important factors in the evolution of pathways of sulfur metabolism and have been involved in such pathways in individual and collective ways. The present indications are that their functions on the primitive Earth were mainly concerned with dissimilatory and assimilatory reductive processes as well as with oxidative pathways of sulfur metabolism and their emergence as sulfate sources for sulfoconjugation processes may have been a relatively late evolutionary phenomenon. With the developing complexity of living organisms, their role in sulfoconjugation has become more sophisticated, until in the higher mammals we find a bewildering multiplicity of relatively specific sulfotransferase enzymes concerned with the production of many different types of sulfoconjugates. Some of these enzymes are present in the cell sap in soluble form, while others are attached to the membranous endoplasmic reticulum of the cell. All of them are difficult to resolve from each other and to purify and, in the few cases that have been closely studied, there are indications of structural and mechanistic similarities. Many of these enzymes are involved in the biosynthesis of compounds with obvious biological functions and/or importance while others produce compounds of obscure significance. In addition, arylsulfotransferases and possibly arylamine sulfotransferases appear to fulfill useful roles in enabling the animal to remove unwanted (and sometimes toxic) materials from the body. From the point of view of evolution, and with the evidence available at present, it is tempting to assume that this particular role may have been a comparatively recent development— perhaps not achieving real significance until the emergence of terrestial forms of animal life.

Also present in living organisms are hydrolytic enzymes of differing specificity which can liberate sulfate from sulfoconjugates. In many instances they appear to bear little relationship to the sulfotransferase enzymes and there are indications that organisms which possess the latter need not necessarily possess the former and vice versa. In some cases (i.e., bacteria and fungi) this situation is clearly established but in others it is possible that the apparent lack of coexisting enzymes might merely reflect the lack or inadequacy of searches that have been made for them. In the mammalian world the sulfohydrolases appear in several cellular loci ranging from the soluble APS-sulfohydrolase of the cell sap, through the lysosomal enzymes, to the highly insoluble steroid, estrogen, and arylsulfohydrolases of endoplasmic reticulum. Only in one case (APS and PAPS breakdown and synthesis) is it clearly established that both biosynthetic and degradative enzymes occur, at least in part, in the same cell locus. This cellular compartmentalization of sulfohydrolases and sulfotransferases would seem to imply that sulfoconjugates can and do penetrate to all regions of the cell and that a complex functional relationship exists between these regions in terms of biosynthesis and degradation of the conjugates. In most instances it is impossible to say with certainty precisely what this relationship is. However, the encouraging developments that have occurred in relation to cerebroside sulfate metabolism in metachromatic leukodystrophy might lead one to believe that the elucidation of the relationship might be of considerable importance and value.

The different sulfohydrolases, together with their biosynthetic counterparts, constitute an important field for evolutionary and comparative study. However, some time will probably elapse before such studies can become reality and some new approaches to the problems of separation and purification of the various enzymes will first need to be made.

Meanwhile, general interest in the field continues to increase rapidly and what was formerly held to be a quiet backwater of biochemistry is now assuming considerable importance. It is probably safe to prophesy that, parallel with the growth of interest will be the emergence of new sulfoconjugates and of enzymes that are concerned with their biosynthesis and degradation and that such discoveries will provide further fascinating fields of study for the future.

ACKNOWLEDGMENTS

The authors of this chapter wish to express their appreciation to the U.S. Public Health Service for support extending over several years which has enabled our own laboratories to contribute to progress in the field described.

REFERENCES

Abraham, A., and Bachhawat, B. K. (1964). *Indian J. Biochem.* 1, 192.

Acheson, R. M., and Hands, A. R. (1961). *J. Chem. Soc.* p. 746.

Adams, J. B. (1962). *Biochim. Biophys. Acta* 62, 17.

Adams, J. B. (1964). *Biochim. Biophys. Acta* 83, 127.

Adams, J. B. (1967). *Biochim. Biophys. Acta* 146, 522.

Adams, J. B., and Chulavatnatol, M. (1967). *Biochim. Biophys. Acta* 146, 509.

Adams, J. B., and Edwards, A. M. (1968). *Biochim. Biophys. Acta* 167, 122.

Adams, J. B., and Meaney, M. F. (1961). *Biochim. Biophys. Acta* 54, 592.

Adams, J. B., and Poulos, A. (1967). *Biochim. Biophys. Acta* 146, 493.

Ali, S. Y., Evans, L., Stainthorpe, E., and Lack, C. H. (1967). *Biochem. J.* 105, 549.

Allen, E., and Roy, A. B. (1968). *Biochim. Biophys. Acta* 168, 243.

Anast, C., Kennedy, R., Volk, G., and Adamson, L. (1965). *J. Lab. Clin. Med.* 65, 903.

Anastasi, A., Bertaccini, G., and Erspamer, V. (1966). *Pharmacol. Chemotherap.* 27, 479.

Anastasi, A., Erspamer, V., and Endean, R. (1968). *Arch. Biochem. Biophys.* 125, 57.

Anderson, B., Hoffman, P., and Meyer, K. (1963). *Biochim. Biophys. Acta* 74, 309.

Anderson, B., Hoffman, P., and Meyer, K. (1965). *J. Biol. Chem.* 240, 156.

Anderson, I. G., and Haslewood, G. A. D. (1967). *Biochem. J.* 104, 1061.

Anderson, S. O. (1959a). *Acta Chem. Scand.* 13, 120.

Anderson, S. O. (1959b). *Acta Chem. Scand.* 13, 884.

Anderson, S. O. (1959c). *Acta Chem. Scand.* 13, 1671.

Antonopolous, C. A., Engfeldt, B., Gardell, S., Hjertquist, S. O., and Solheim, K. (1965). *Biochim. Biophys. Acta* 101, 150.

Arcos, M., and Lieberman, S. (1967). *Biochemistry* 6, 2032.

Aronson, N. N., and Davidson, E. A. (1965). *J. Biol. Chem.* 240, 3222 PC.

Aronson, N. N., and Davidson, E. A. (1967). *J. Biol. Chem.* **242**, 437.

Aronson, N. N., and Davidson, E. A. (1968). *J. Biol. Chem.* **243**, 4494.

Ashai, T. (1964). *Biochim. Biophys. Acta* **82**, 58.

Astudillo, M. D., Espliguero, M. S., Zumel, C. L., and Sanz, F. (1962). *Actas Soc. Espan. Cienc. Fisiol.* **7**, 45.

Astudillo, M. D., Espliguero, M. S., Zumel, C. L., and Sanz, F. (1964). *Revta Espan. Fisiol.* **20**, 113.

Austin, J. H. (1959). *Proc. Soc. Exptl. Biol. Med.* **100**, 361.

Austin, J. H., Balasubramanian, A. S., Pattabiraman, T. N., Saraswathi, S., Basu, D. K., and Bachhawat, B. K. (1963). *J. Neurochem.* **10**, 805.

Austin, J. H., McAfee, D., Armstrong, D., O'Rourke, M., Shearer, L., and Bachhawat, B. K. (1964). *Biochem. J.* **93**, 15C.

Austin, J. H., McAfee, D., and Shearer, L. (1965). *Arch. Neurol.* **13**, 593.

Bachhawat, B. K., Austin, J. H., and Armstrong, D. (1967). *Biochem. J.* **104**, 15C.

Baddiley, J., Buchanan, J. G., and Letters, R. (1957). *J. Chem Soc.* p. 1067.

Baddiley, J., Buchanan, J. G., Letters, R., and Sanderson, A. R. (1959). *J. Chem. Soc.* p. 1731.

Bailey-Wood, R., Dodgson, K. S., and Rose, F. A. (1969). *Biochem. J.* **112**, 2C.

Balakhovski, S. D., and Kuznetsova, I. V. (1958). *Dokl. Akad. Nauk SSSR* **118**, 331.

Balasubramanian, A. S., and Bachhawat, B. K. (1962). *Biochim. Biophys. Acta* **59**, 389.

Balasubramanian, A. S., and Bachhawat, B. K. (1965a). *Biochim. Biophys. Acta* **106**, 218.

Balasubramanian, A.S., and Bachhawat, B.K. (1965b). *Indian J. Biochem.* **2**, 212.

Balasubramanian, A. S., Spolter, L., Rice, L. I. Sharon, J. B., and Marx, W. (1967). *Anal. Biochem.* **21**, 22.

Banerjee, R. K., and Roy, A. B. (1966). *Mol. Pharmacol.* **2**, 56.

Banerjee, R. K., and Roy, A. B. (1967). *Biochim. Biophys. Acta* **137**, 211.

Banerjee, R. K., and Roy, A. B. (1968). *Biochim. Biophys. Acta* **151**, 573.

Barra, H. S., and Caputto, R. (1965). *Biochim. Biophys. Acta* **101**, 367.

Basford, J. M., Jones, J. G., Rose, F. A., and Dodgson, K. S. (1966). *Biochem. J.* **99**, 534.

Baulieu, E. E., and Corpéchot, C. (1965). *Bull. Soc. Chim. Biol.* **47**, 443.

Baum, H., and Dodgson, K. S. (1957). *Nature* **179**, 312.

Baum, H., and Dodgson, K. S. (1958a). *Biochem. J.* **69**, 573.

Baum, H., and Dodgson, K. S. (1958b). *Nature* **181**, 115.

Baum, H., Dodgson, K. S., and Spencer, B. (1958). *Biochem. J.* **69**, 567.

Baum, H., Dodgson, K. S., and Spencer, B. (1959). *Clin. Chim. Acta* **4**, 453.

Baumann, E. (1876). *Ber. Deut. Chem. Ges.* **9**, 54.

Benson, A. A. (1963). *Advan. Lipid Res.* **1**, 3.

Bernstein, S., and McGilvery, R. W. (1952a). *J. Biol. Chem.* **198**, 195.

Bernstein, S., and McGilvery, R. W. (1952b). *J. Biol. Chem.* **199**, 745.

Bertaccini, G., deCaro, G., Endean, R., Erspamer, V., and Impicciatore, M. (1968a). *Brit. J. Pharmacol.* **34**, 291.

Bertaccini, G., Endean, R., Erspamer, V., and Impicciatore, M. (1968b). *Brit. J. Pharmacol.* **34**, 311.

Bettelheim, F. R. (1954). *J. Am. Chem. Soc.* **76**, 2838.

Bhavanandan, V. P., and Meyer, K. (1968). *J. Biol. Chem.* **243**, 1052.

Bidwell, R. G. S., and Gosh, N. R. (1963). *Can. J. Botany* **41**, 209.

Bleszynski, W. (1967). *Enzymologia* **32**, 169.
Bollet, A. J., Bonner, W. M., and Nance, J. L. (1963). *J. Biol. Chem.* **238**, 3522.
Bolte, E., Mancuso, S., Eriksson, G., Wiqvist, N., and Diczfalusy, E. (1964). *Acta Endocrinol.* **45**, 535.
Boström, H., and Åqvist, S. (1952). *Acta Chem. Scand.* **6**, 1557.
Boström, H., and Vestermark, A. (1959). *Nature* **183**, 1593.
Boström, H., and Vestermark, A. (1960). *Scand. J. Clin. Lab. Invest.* **12**, 323.
Boström, H., and Wengle, B. (1964). *Acta Soc. Med. Upsalien.* **69**, 41.
Boström, H., and Wengle, B. (1967). *Acta Endocrinol.* **56**, 691.
Boström, H., Gustafsson, B. E., and Wengle, B. (1963). *Proc. Soc. Exptl. Biol. Med.* **114**, 742.
Bowen, D. M., and Radin, N. S. (1968a). *Biochim. Biophys. Acta* **152**, 587.
Bowen, D. M., and Radin, N. S. (1968b). *Biochim. Biophys. Acta* **152**, 599.
Boyland, E., Manson, D., and Orr, S. F. D. (1957). *Biochem. J.* **65**, 417.
Brady, R. O. (1964). *In* "Metabolism and Physiological Significance of Lipids" (R. M. C. Dawson and D. N. Rhodes, eds.), p. 95. Wiley, New York.
Bray, B. A., Lieberman, R., and Meyer, K. (1967). *J. Biol. Chem.* **242**, 3373.
Bray, H. G., Humphris, B. G., Thorpe, W. V., White, K., and Wood, P. B. (1952a). *Biochem. J.* **52**, 419.
Bray, H. G., Thorpe, W. V., and White, K. (1952b). *Biochem. J.* **52**, 419.
Brimacombe, J. S., and Webber, J. M. (1964). "Mucopolysaccharides," Elsevier, Amsterdam.
Brunngraber, E. G. (1958). *J. Biol. Chem.* **233**, 472.
Bryant, J. H., Leder, I. G., and Stetten, D. W. (1958). *Arch. Biochem. Biophys.* **76**, 122.
Buddecke, E., and Platt, D. (1965). *Z. Physiol. Chem.* **343**, 61.
Buddecke, E., Kroz, W., and Lanka, E. (1963). *Z. Physiol. Chem.* **331**, 196.
Burkhardt, G. N., Ford, W. G. K., and Singleton, E. (1936). *J. Chem. Soc.* p. 17.
Burson, S. L., Fahrenbach, M. J., Fromhagen, L. H., Riccarda, B. A., Brown, R. A., Brockman, J. A., Lewry, H. V., and Stockstad, E. L. R. (1956). *J. Am. Chem. Soc.* **78**, 5874.
Carroll, J., and Spencer, B. (1965). *Biochem. J.* **94**, 20P.
Carubelli, R., Ryan, L. C., Trucco, R. E., and Caputto, R. (1961). *J. Biol. Chem.* **236**, 2381.
Chapeville, F., and Fromageot, P. (1957). *Biochim. Biophys. Acta* **26**, 538.
Cherniak, R., and Davidson, E. A. (1964). *J. Biol. Chem.* **239**, 2986.
Chu, T. M., and Slaunwhite, W. R. (1968). *Steroids* **12**, 309.
Cifonelli, J. A. (1968). *In* "The Chemical Physiology of Mucopolysaccharides" (G. Quintarelli, ed.), p. 91. Churchill, London.
Cifonelli, J. A., and Dorfman, A. (1961). *Biochem. Biophys. Res. Commun.* **4**, 328.
Cifonelli, J. A., and Dorfman, A. (1962). *Biochem. Biophys. Res. Commun.* **7**, 41.
Cifonelli, J. A., and Mathews, M. B. (1968). *Federation Proc.* **27**, No. 2, Abstr. 3339.
Cleland, W. W., and Kennedy, E. P. (1960). *J. Biol. Chem.* **235**, 45.
Conchie, J., and Hay, A. J. (1963). *Biochem. J.* **87**, 354.
Creange, J. E., and Szego, C. M. (1967). *Biochem. J.* **102**, 898.
Cumar, F. A., Barra, H. S., Maccioni, H. J., and Caputto, R. (1968). *J. Biol. Chem.* **243**, 3807.
Curtis, C. G (1966). The Transport and Metabolism of Sulfuric Acid Esters in Mammalian Organisms. Ph.D. Thesis, Univ. of Wales.

Dalgliesh, C. E. (1955). *Biochem. J.* **61**, 334.

Davidson, E. A., and Meyer, K. (1954). *J. Biol. Chem.* **211**, 605.

Davidson, E. A., and Riley, J. G. (1960). *J. Biol. Chem.* **235**, 3367.

Davis, V. E., Huff, J. A., and Brown, H. (1966). *Clin. Chim. Acta* **13**, 380.

Davison, A. N., and Gregson, N. A. (1966). *Biochem. J.* **98**, 915.

de Duve, C. (1959). *Exptl. Cell Res. Suppl.* **7**, 169.

de la Haba, G., and Holtzer, H. (1965). *Science* **149**, 1263.

DeLuca, S., and Silbert, J. E. (1968). *J. Biol. Chem.* **243**, 2725.

DeMeio, R. H. (1945). *Arch. Biochem.* **7**, 323.

DeMeio, R. H., and Arnolt, R. I. (1944). *J. Biol. Chem.* **156**, 577.

DeMeio, R. H., Wizerkaniuk, M., and Fabiani, E. (1953). *J. Biol. Chem.* **203**, 257.

DeMeio, R. H., Lewycka, C., Wizerkaniuk, M., and Salciunas, O. (1958). *Biochem. J.* **68**, 1.

Denner, W. H. B., Olavesen, A. H., Powell, G. M., and Dodgson, K. S. (1969). *Biochem. J.* **111**, 43.

Derrien, M. (1911). *Bull. Soc. Chim. France* **9**, 110.

Deyrup, I. J. (1963). *Federation Proc.* **22**, 332.

Diczfalusy, E., Cassmer, O., Alonso, C., and de Micquel, M. (1961). *Acta Endocrinol.* **38**, 31.

Dingle, J. T. (1961). *Biochem. J.* **79**, 509.

Dodgson, K. S. (1956). *In* "Colloque sur la Biochimie du Soufre" (C. Fromageot, ed.), p. 123. R. S., Paris.

Dodgson, K. S. (1959). *Proc. 4th Intern. Congr. Biochem., Vienna, 1958* **13**, 3.

Dodgson, K. S. (1961). *Biochem. J.* **78**, 324.

Dodgson, K. S. (1966). *In* "The Amino Sugars" (R. W. Jeanloz, and E. A. Balazs, eds.,), Vol. 2B, p. 201. Academic Press, New York.

Dodgson, K. S., and Lloyd, A. G. (1957). *Biochem. J.* **66**, 53.

Dodgson, K. S., and Lloyd, A. G. (1968). *In* "Carbohydrate Metabolism and its Disorders" (F. Dickens, P. J. Randle, and W. J. Whelan, eds.), Vol. 1. p. 169. Academic Press, New York.

Dodgson, K. S., and Powell, G. M. (1959a). *Biochem. J.* **73**, 666.

Dodgson, K. S., and Powell, G. M. (1959b). *Biochem. J.* **73**, 672.

Dodgson, K. S., and Rose, F. A. (1966). *Nutr. Abstr. Rev.* **36**, 327.

Dodgson, K. S., and Spencer, B. (1953a). *Biochem. J.* **53**, 444.

Dodgson, K. S., and Spencer, B. (1953b). *Biochem. J.* **55**, 315.

Dodgson, K. S., and Spencer, B. (1956a). *Biochem. J.* **62**, 30P.

Dodgson, K. S., and Spencer, B. (1956b). *Biochim. Biophys. Acta* **21**, 175.

Dodgson, K. S., and Spencer, B. (1957a). *Rept. Progr. Chem.* **53**, 318.

Dodgson, K. S., and Spencer, B. (1957b). *Methods Biochem. Analy.* **4**, 211.

Dodgson, K. S., and Spencer, B. (1957c). *Biochem. J.* **65**, 668.

Dodgson, K. S., and Spencer, B. (1957d). *Clin. Chim. Acta* **1**, 478.

Dodgson, K. S., and Tudball, N. (1960). *Biochem. J.* **74**, 154.

Dodgson, K. S., and Wynn, C. H. (1958). *Biochem. J.* **68**, 387.

Dodgson, K. S., Spencer, B., and Thomas, J. (1953a). *Biochem. J.* **53**, xxxvi.

Dodgson, K. S., Lewis, J. I. M., and Spencer, B. (1953b). *Biochem. J.* **55**, 253.

Dodgson, K. S., Spencer, B., and Thomas, J. (1954a). *Biochem. J.* **57**, xxi.

Dodgson, K. S., Spencer, B., and Thomas, J. (1954b). *Biochem. J.* **56**, 177.

Dodgson, K. S., Melville, T. H., Spencer, B., and Williams, K. (1954c). *Biochem. J.* **58**, 182.

Dodgson, K. S., Rose, F. A., and Spencer, B. (1955a). *Biochem. J.* **60**, 346.
Dodgson, K. S., Spencer, B., and Thomas, J. (1955b). *Biochem. J.* **59**, 29.
Dodgson, K. S., Spencer, B., and Williams, K. (1955c). *Biochem. J.* **61**, 374.
Dodgson, K. S., Spencer, B., and Wynn, C. H. (1956a). *Biochem. J.* **62**, 500.
Dodgson K. S., Spencer, B., and Williams, K. (1956b). *Biochem. J.* **64**, 216.
Dodgson, K. S., Rose, F. A., and Spencer, B. (1957a). *Biochem. J.* **66**, 357.
Dodgson, K. S., Rose, F. A., Spencer, B., and Thomas, J. (1957b). *Biochem. J.* **66**, 363.
Dodgson, K. S., Rose, F. A., and Tudball, N. (1959). *Biochem. J.* **71**, 10.
Dodgson, K. S., Lloyd, A. G., and Tudball, N. (1961a). *Biochem. J.* **79**, 111.
Dodgson, K. S., Powell, G. M., Rose, F. A., and Tudball, N. (1961b). *Biochem. J.* **79**, 209.
Dodgson, K. S., Gatehouse, P. W., Lloyd, A. G., and Powell, G. M. (1965). *Biochem. J.* **95**, 18P.
Dodgson, K. S., Basford, J. M., Jones, J. G., and Mattock, P. (1967). *Indian J. Biochem.* **4**, Suppl. 2, 7.
Doolittle, R. F., and Blombäck, B. (1964). *Nature* **202**, 147.
Doyle, J. (1967). *Biochem. J.* **103**, 325.
Drayer, N. M., and Giroud, C. J. P. (1964). *Steroids* **5**, 289.
Drayer, N. M., and Lieberman, S. (1965). *Biochem. Biophys. Res. Commun.* **18**, 126.
Drayer, N. M., Roberts, K. D., Bandi, L., and Lieberman, S. (1964). *J. Biol. Chem.* **239**, PC3113.
Dzialoszysnski, L. M., and Gniot-Szulzycka, J. (1967). *Clin. Chim. Acta* **15**, 381.
Dziewiatkowski, D. D. (1954). *J. Exptl. Med.* **100**, 11.
Dziewiatkowski, D. D., and Di Ferrante, N. (1957). *J. Biol. Chem.* **227**, 347.
Dziewiatkowski, D. D., Tourtellotte, C. D., and Campo, R. D. (1968). In "The Chemical Physiology of Mucopolysaccharides" (G. Quintarelli, ed.), p. 63. Churchill, London.
Egami, F., and Takahashi, N. (1962). In "Biochemistry and Medicine of Mucopolysaccharides" (F. Egami and Y. Oshima, eds.), p. 53. Maruzen, Tokyo.
Eisenman, R. A., Balasubramanian, A. S., and Marx, W. (1967). *Arch. Biochem. Biophys.* **119**, 387.
Esterly, J. S., Standen, A. C., and Pearson, B. (1968). *J. Histochem. Cytochem.* **16**, 489.
Fell, H. B. (1965). *Proc. Nutr. Soc. (Engl. Scot.)* **24**, 166.
Fell, H. B., and Dingle, J. T. (1963). *Biochem. J.* **87**, 403.
Fell, H. B., Mellanby, E., and Pelc, S. R. (1956). *J. Physiol. (London)* **134**, 179.
Fisher, D., Higham, M., Kent, P. W., and Pritchard P. (1966). *Biochem. J.* **98**, 46P.
Fisher, D., Whitehouse, M. W., and Kent, P. W. (1967). *Nature* **213**, 204.
Flynn, T. G., Dodgson, K. S., Powell, G. M., and Rose, F. A. (1967). *Biochem. J.* **105**, 1003.
Ford, E. A., and Ruoff, P. M. (1965). *Chem. Commun.* p. 630.
Fowler, L. R., and Rammler, D. H. (1964). *Biochemistry* **3**, 230.
Franek, M. D., and Dunstone, J. R. (1967). *J. Biol. Chem.* **242**, 3460.
Fransson, L. A. (1968a). *J. Biol. Chem.* **243**, 1504.
Fransson, L. A. (1968b). *Biochim. Biophys. Acta* **156**, 311.
Fransson, L. A., and Rodén, L. (1967a). *J. Biol. Chem.* **242**, 4161.
Fransson, L. A., and Rodén, L. (1967b). *J. Biol. Chem.* **242**, 4170.

Frantantoni, J. C., Hall, C. W., and Neufield, E. F. (1968). *Proc. Natl. Acad. Sci. U.S* **60**, 699.

French, A. P., and Warren, J. C. (1965). *Steroids* **6**, 865.

French, A. P., and Warren, J. C. (1966). *Steroids* **8**, 79.

French, A. P., and Warren, J. C. (1967). *Biochem. J.* **105**, 233.

Fries, S. L., Durant, R. C., Chanley, J. D., and Fash, F. J. (1967). *Biochem. Pharmacol.* **16**, 1617.

Fujiwara, T., and Spencer, B. (1962). *Biochem J.* **85**, 19P.

Garton, G. A., and Williams, R. T. (1948). *Biochem. J.* **43**, 206.

Garton, G. A., and Williams, R. T. (1949). *Biochem. J.* **44**, 234.

Geison, R. L., Rogers, W. E., Jr., and Johnson, B. C. (1968). *Biochim. Biophys. Acta* **165**, 448.

Gibian, H., and Bratfisch, G. (1956). *Z. Physiol. Chem.* **305**, 265.

Goldberg, J. H., and Delbruck, A. (1959). *Federation Proc.* **18**, 235.

Grebner, E. E., Hall, C. W., and Neufeld, E. F. (1966). *Arch. Biochem. Biophys.* **116**, 391.

Green, J. P., and Robinson, J. D. (1960). *J. Biol. Chem.* **235**, 1621.

Gregory, H., Hardy, P. M., Jones, D. S., Kenner, G. W., and Sheppard, R. C. (1964). *Nature* **204**, 931.

Gregory, J. D., and Lipmann, F. (1957). *J. Biol. Chem.* **229**, 1081.

Gregory, J. D., and Robbins, P. W. (1960). *Ann. Rev. Biochem.* **29**, 347.

Gregory, J. D., Laurent, T. C., and Rodén, L. (1964). *J. Biol. Chem.* **239**, 3312.

Gregory, R. A. (1968). *Proc. Roy. Soc. (London)* **B170**, 81.

Grimes, A. J. (1959). *Biochem. J.* **73**, 723.

Gurpide, E., Roberts, K. D., Welch, M. T., Bandi, L., and Lieberman, S. (1966). *Biochemistry* **5**, 3352.

Guthrie, R. D., Ferrier, R. J., and How, M. J. (1968). *In* "Carbohydrate Chemistry" (R. D. Guthrie, ed.), Periodical Rept., Vol. 1, p. 254. Chem. Soc., London.

Hakamori, S., Ishimoda, T., and Nakamura, K. (1962). *J. Biochem. (Tokyo)* **52**, 468.

Hanahan, D. J., Everett, N. B., and Davis, C. D. (1949). *Arch. Biochem.* **23**, 501.

Harada, T. (1954). *Nippon Nogeikagaku Kaishi* **28**, 840.

Harada, T. (1959). *Bull. Agr. Chem. Soc. Japan* **23**, 222.

Harada, T., and Spencer, B. (1960). *J. Gen. Microbiol.* **22**, 520.

Harada, T., and Spencer, B. (1962). *Biochem. J.* **82**, 148.

Harada, T., Shimizu, S., Nakanishi, Y., and Suzuki, S. (1967). *J. Biol. Chem.* **242**, 2288.

Hasegawa, E., Debruck, A., and Lipmann, F. (1961). *Federation Proc.* **20**, 86.

Haslewood, G. A. D. (1967). *J. Lipid Res.* **8**, 535.

Hauser, G. (1964). *Biochim. Biophys. Acta* **84**, 212.

Hauser, G. (1967). *Biochem. Biophys. Res. Commun.* **28**, 502.

Havinga, E., Dejongh, R. O., and Dorst, W. (1956). *Rec. Trav. Chim.* **75**, 290.

Hawkins, J. B., and Young, L. (1954). *Biochem. J.* **56**, 166.

Heacock, R. A., and Mahon, M. E. (1964). *Can. J. Biochem.* **42**, 813.

Hearse, D. J., Powell, G. M., Olavesen, A. H., and Dodgson, K. S. (1969a). *Biochem. Pharmacol.* **18**, 181.

Hearse, D. J., Powell, G. M., Olavesen, A. H., and Dodgson, K. S. (1969b). *Biochem. Pharmacol.* **18**, 205.

Heinegard, D., and Gardell, S. (1967). *Biochim. Biophys. Acta* **148**, 164.

Held, E., and Buddecke, E. (1967). Z. Physiol. Chem. 348, 1047.
Helting, T., and Rodén, L. (1968). Biochem. Biophys. Res. Commun. 31, 786.
Higaki, M., Takahashi, M., Suzuki, T., and Sahashi, Y. (1965). J. Vitaminol. (Kyoto) 11, 266.
Hilz, H., and Lipmann, F. (1955). Proc. Natl. Acad. Sci. U.S. 41, 880.
Hoffman, P., Mashburn, T. A., and Greenberg, J. (1967a). Federation Proc. 26, 282.
Hoffman, P., Mashburn, T. A., and Meyer, K. (1967b). J. Biol. Chem. 242, 3805.
Holcenberg, J. S., and Rosen, S. W. (1965). Arch. Biochem. Biophys. 110, 551.
Hori, K., and Cormier, M. J. (1966). Biochim. Biophys. Acta 130, 420.
Horwitz, A. L., and Dorfman, A. (1968). J. Cell Biol. 38, 358.
Hunt, S., and Jevons, F. R. (1965). Biochim. Biophys. Acta 101, 214.
Hunt, S., and Jevons, F. R. (1966). Biochem. J. 98, 522.
Huovinen, J. A., and Gustaffsson, B. E. (1967). Biochim. Biophys. Acta 136, 441.
Hussey, C. A., Orsi, B. A., Scott, J., and Spencer, B. (1965). Nature 207, 632.
Hutterer, F. (1966). Biochim. Biophys. Acta 115, 312.
Immers, J. (1962). "Investigations on Macromolecular Sulfated Polysaccharides in Sea Urchin Development." Almqvist & Wiksell, Uppsala.
Inoue, S. (1965). Biochim. Biophys. Acta 101, 16.
Ishihara, K. (1968). Exptl. Cell Res. 51, 473.
Itahashi, M. (1961). J. Biochem. (Tokyo) 50, 52.
Jarrige, P. (1962). D. S. N. Thesis, Univ. of Paris.
Jatzkewitz, H. (1958). Z. Physiol. Chem. 311, 279.
Jatzkewitz, H. (1960a). Z. Physiol. Chem. 318, 265.
Jatzkewitz, H. (1960b). Z. Physiol. Chem. 320, 134.
Jatzkewitz, H., and Mehl, E. (1969). J. Neurochem. 16, 19.
Jeanloz, R. W. (1963a). In "Comprehensive Biochemistry" (M. Florkin and E. H. Stotz, eds.), Vol. 5, p. 262. Elsevier, Amsterdam.
Jeanloz, R. W. (1963b). Advan. Enzymol. 25, 433.
Jerfy, A., and Roy, A. B. (1969). Biochim. Biophys. Acta 175, 355.
Jevons, F. R. (1963). Biochem. J. 89, 621.
Jirku, H., Hogsander, U., and Levitz, M. (1967). Biochim. Biophys. Acta 137, 558.
John, R. A., Rose, F. A., Wuseman, F. S., and Dodgson, K. S. (1966). Biochem. J. 100, 278.
Jones, J. G., and Dodgson, K. S. (1965). Biochem. J. 94, 331.
Jones, J. G., Dodgson, K. S., Powell, G. M., and Rose, F. A. (1963). Biochem. J. 87, 548.
Jones, J. G., Scotland, S. M., and Dodgson, K. S. (1966). Biochem. J. 98, 138.
Kates, M., Palameta, B., Perry, M. P., and Adams, G. A. (1967). Biochim. Biophys. Acta 137, 213.
Katsura, N., and Davidson, E. A. (1966a). Biochim. Biophys. Acta 121, 120.
Katsura, N., and Davidson, E. A. (1966b). Biochim. Biophys. Acta 121, 128.
Kenner, G. W., and Sheppard, R. C. (1968). Proc. Roy. Soc. (London) B170, 89.
Kent, P. W., and Marsden, J. C. (1963). Biochem. J. 87, 38P.
Kent, P. W., Ackers, J., and Marsden, J. C. (1967). Biochem. J. 105, 24P.
Kittredge, J. S., Simonsen, D. G., Roberts, E., and Jelinek, B. (1962). In "Amino Acid Pools" (J. T. Holden, ed.), p. 176. Elsevier, Amsterdam.
Kleine, T. O., Kirsig, H. J., and Hilz, H. (1968). Z. Physiol. Chem. 349, 1037.
Knecht, J., Cifonelli, J. A., and Dorfman, A. (1967). J. Biol. Chem. 242, 4652.
Koizumi, T., Suematsu, T., Kawasaki, A., Hiramatsu, K., and Abe, H. (1969).

Biochim. Biophys. Acta **184**, 106.

Kornfeld, S., Kornfeld, R., Neufeld, E. F., and O'Brien, P. J. (1964). *Proc. Natl. Acad. Sci. U.S.* **52**, 371.

Krajewski, T., and Blombäck, B. (1968). *Acta Chem. Scand.* **22**, 1339.

Laatikainen, T., Peltokallio, P., and Vihko, R. (1968). *Steroids* **12**, 407.

Layton, L. L., and Frankel, D. R. (1951). *Arch. Biochem. Biophys.* **31**, 161.

Leon, Y. A., Bulbrook, R. D., and Corner, E. D. S. (1960). *Biochem. J.* **75**, 612.

Levi, A. S., Geller, S., Root, D. M., and Wolf, G. (1968). *Biochem. J.* **109**, 69.

Levitz, M., and Katz, J. (1968). *J. Clin. Endocrinol. Metab.* **28**, 862.

Levitz, M., Condon, G. P., Money, W. L., and Dancis, J. (1960). *J. Biol. Chem.* **235**, 973.

Levitz, M., Condon, G. P., and Dancis, J. (1961). *Endocrinology* **68**, 825.

Lewis, M. H. R., and Spencer, B. (1962). *Biochem. J.* **85**, 18P.

Lindahl, U. (1966). *Biochim. Biophys. Acta* **130**, 368.

Lindahl, U. (1967). *Arkiv Kemi* **26**, 101.

Lindahl, U. (1968). *Biochim. Biophys. Acta* **156**, 203.

Lindahl, U., and Rodén, L. (1965). *J. Biol. Chem.* **240**, 2821.

Lindahl, U., and Rodén, L. (1966). *J. Biol. Chem.* **241**, 2113.

Lindahl, U., Cifonelli, J. A., Lindahl, B., and Rodén, L. (1965). *J. Biol. Chem.* **240**, 2817.

Linker, A., and Hovingh, P. (1968). *Biochim. Biophys. Acta* **165**, 89.

Lloyd, A. G. (1961). *Biochem. J.* **80**, 572.

Lloyd, A. G. (1962). *Biochim. Biophys. Acta* **58**, 1.

Lloyd, A. G. (1966). *In* "Methods in Enzymology. Vol. 8: Complex Carbohydrates" (V. Ginsburg and E. F. Neufeld, eds.), p. 663. Academic Press, New York.

Lloyd, A. G., Large, P. J., James, A. M., and Dodgson, K. S. (1964). *J. Biochem. (Tokyo)* **55**, 669.

Lloyd, A. G., Embery, G., Wusteman, F. S., and Dodgson, K. S. (1966a). *Biochem. J.* **98**, 33P.

Lloyd, A. G., Embery, G., Powell, G. M., Curtis, C. G., and Dodgson, K. S. (1966b). *Biochem. J.* **98**, 33P.

Lloyd, A. G., Fowler, L. J., Embery, G., and Law, B. A. (1968a). *Biochem. J.* **110**, 54P.

Lloyd, A. G., Large, P. J., Davies, M., Olavesen, A. H., and Dodgson, K. S. (1968b). *Biochem. J.* **108**, 393.

McCluskey, R. T., and Thomas, L. (1959). *Am. J. Pathol.* **35**, 819.

Machlin, J. L., Pearson, P. B., and Denton, C. A. (1955). *J. Biol. Chem.* **212**, 469.

McKenna, J., Menini, E., and Norymberski, J. K. (1961). *Biochem. J.* **79**, 11P.

McKhann, G. M., Levy, R., and Ho, W. (1965). *Biochem. Biophys. Res. Commun.* **20**, 109.

Malawista, I., and Schubert, M. (1958). *J. Biol. Chem.* **230**, 535.

Manley, G., Severn, M., and Hawksworth, J. (1968). *J. Clin. Pathol.* **21**, 339.

Martensson, E. (1966). "Glycolipids of Human Kidney." Elanders, Gothenberg, Sweden.

Mason, M., and Gullekson, E. H. (1960). *J. Biol. Chem.* **235**, 1312.

Mathews, M. B., and Decker, L. (1968). *Biochim. Biophys. Acta* **156**, 419.

Mattock, P. (1967). Some Aspects of Sulfate Activation and Transfer in Biological Systems. Ph.D. Thesis, Univ. of Wales.

Matukas, V. J., and Krikos, G. A. (1968). *J. Cell Biol.* **39**, 43.

Mayers, G. L., and Haines, T. H. (1967). *Biochemistry* **6**, 1665.

Mayers, D. F., and Kaiser, E. T. (1968). *J. Am. Chem. Soc.* **90**, 6192.

Meezan, E., and Davidson, E. A. (1967). *J. Biol. Chem.* **242**, 4965.

Mehl, E., and Jatzkewitz, H. (1963). *Z. Physiol. Chem.* **331**, 292.

Mehl, E., and Jatzkewitz, H. (1964). *Z. Physiol. Chem.* **339**, 260.

Mehl, E., and Jatzkewitz, H. (1965). *Biochem. Biophys. Res. Commun.* **19**, 407.

Mehl, E., and Jatzkewitz, H. (1968). *Biochim. Biophys. Acta* **151**, 619.

Menini, E., and Diczfalusy, E. (1961). *Endocrinology* **68**, 492.

Meyer, K., and Anderson, B. (1965). *Exptl. Eye Res.* **4**, 346.

Migeon, C., and Plager, J. E. (1955). *J. Clin. Endocrinol. Metab.* **15**, 702.

Milazzo, F. H., and Fitzgerald, J. W. (1967). *Can. J. Microbiol.* **13**, 659.

Milsom, D. W., Rose, F. A., and Dodgson, K. S. (1968). *Biochem. J.* **109**, 40P.

Morley, J. S. (1968). *Proc. Roy. Soc. (London)* **B170**, 97.

Moser, H. W., Moser, A. B., and Orr, J. C. (1966). *Biochim. Biophys. Acta* **116**, 146.

Muir, H. (1958). *Biochem. J.* **69**, 195.

Muir, H. (1964). *Intern. Rev. Connective Tissue Res.* **2**, 101.

Mukherji, B., and Bachhawat, B. K. (1967). *Biochem. J.* **104**, 318.

Mumma, R. O. (1968). *Biochim. Biophys. Acta* **165**, 571.

Neuberg, C., and Kurono, K. (1923). *Biochem. Z.* **140**, 295.

Neuberg, C., and Simon, E. (1932). *Ergeb. Physiol.* **34**, 896.

Ney, K. H. (1959). *Z. Vitamin-, Hormon-, Fermentforsch.* **10**, 139.

Ney, K. H., and Ammon, R. (1959). *Z. Physiol. Chem.* **315**, 415.

Nichol, L. W., and Roy, A. B. (1964). *J. Biochem. (Tokyo)* **55**, 643.

Nichol, L. W., and Roy, A. B. (1965). *Biochemistry* **4**, 386.

Nichol, L. W., and Roy, A. B. (1966). *Biochemistry* **5**, 1379.

Nigrelli, R. F., Chanley, J. D., Kohn, S. K., and Sobotka, H. (1955). *Zoologica* **40**, 47.

Noir, B. A., Groszman, R. J., and de Walz, A. T. (1966). *Biochim. Biophys. Acta* **117**, 297.

Nose, Y., and Lipmann, F. (1958). *J. Biol. Chem.* **233**, 1348.

O'Brien, J. S., Fillerup, D. L., and Mead, J. F. (1964). *J. Lipid Res.* **5**, 109.

Öckerman, P. A. (1968). *Clin. Chim. Acta* **20**, 1.

Oertel, G. W. (1961). *Biochem. Z.* **334**, 431.

Oertel, G. W. (1963). *Biochem. Z.* **339**, 125.

Oertel, G. W., and Kaiser, E. (1962). *Clin. Chim. Acta* **7**, 463.

Oertel, G. W., and Treiber, L. (1966). *Z. Physiol. Chem.* **344**, 163.

Oertel, G. W., Kaiser, E., and Bruhl, P. (1962). *Biochem. Z.* **336**, 154.

Oertel, G. W., Menzel, P., and Wenzel, F. (1968). *Z. Physiol. Chem.* **349**, 1551.

Olsson, I., Gardell, S., and Thunell, S. (1968). *Biochim. Biophys. Acta* **165**, 309.

Omachi, A. (1964). *Science* **145**, 1449.

Orii, T. (1968). *Z. Physiol. Chem.* **349**, 816.

Orsi, B. A., and Spencer, B. (1964). *J. Biochem. (Tokyo)* **56**, 81.

Pardee, A. B. (1966). *J. Biol. Chem.* **241**, 5886.

Pardee, A. B., Prestidge, L. S., Whipple, M. B., and Dreyfuss, J. (1966). *J. Biol. Chem.* **241**, 3962.

Parke, D. V. (1960). *Biochem. J.* **77**, 493.

Pasqualini, J. R. (1964). *Arch. Biochem. Biophys.* **106**, 15.

Pasqualini, J. R., and Jayle, M. F. (1961). *Biochem. J.* **81**, 147.

Pasqualini, J. R., and Jayle, M. F. (1962). *J. Clin. Invest.* **41**, 981.

Pasqualini, J. R., Cedard, L., Nguyen, B. L., and Alsatt, E. (1967). *Biochim. Biophys. Acta* **139**, 177.

Pasternak, C. A. (1960). *J. Biol. Chem.* **235**, 438.

Pasternak, C. A., and Pirie, A. (1964). *Exptl. Eye Res.* **3**, 365.

Pasternak, C. A., Humphries, S. K., and Pirie, A. (1963). *Biochem. J.* **86**, 382.

Payne, A. H., and Mason, M. (1963). *Biochim. Biophys. Acta* **71**, 719.

Peck, H. D. (1962). *Bacteriol. Rev.* **26**, 67.

Picard, J., and Gardais, A. (1967). *Bull. Soc. Chim. Biol.* **49**, 1689.

Pinter, G. C. (1967). *Experientia* **23**, 100.

Poux, N. (1966). *J. Histochem. Cytochem.* **14**, 932.

Powell, G. M., Rose, F. A., and Dodgson, K. S. (1963). *Biochem. J.* **87**, 545.

Powell, G. M., Rose, F. A., and Dodgson, K. S. (1964). *Biochem. J.* **91**, 6P.

Powell, G. M., Curtis, C. G., and Dodgson, K. S. (1967). *Biochem. Pharmacol.* **16**, 1997.

Pulkinnen, M. O. (1957). *Acta Physiol. Scand.* **41**, Suppl. 145, p. 115.

Pulkinnen, M. O. (1961). *Acta Physiol. Scand.* **52**, Suppl. 180, p. 9.

Pulkinnen, M. O., and Hakarainen, H. (1965). *Acta Endocrinol.* **48**, 313.

Pulkinnen, M. O., and Paunio, I. (1963). *Ann. Med. Exptl. Biol. Fenniae (Helsinki)* **41**, 283.

Purdy, R. H., Engel, L. L., and Oncley, J. L. (1961). *J. Biol. Chem.* **236**, 1043.

Radhakrishnamurthy, B., Dalferes, E. R., and Berenson, G. S. (1968). *Anal. Biochem.* **24**, 397.

Radin, N. S., Martin, F. B., and Brown, J. R. (1957). *J. Biol. Chem.* **224**, 499.

Rambaut, P. C., and Miller, S. A. (1965). *Federation Proc.* **24**, 373.

Rammler, D. H., Grado, C., and Fowler, L. R. (1964). *Biochemistry* **3**, 224.

Rappaport, C. (1966). *Proc. Soc. Exptl. Biol. Med.* **121**, 1025.

Raud, H. R., and Hobkirk, R. (1966). *Can. J. Biochem.* **44**, 657.

Raud, H. R., and Hobkirk, R. (1968). *Can. J. Biochem.* **46**, 749.

Reichard, P., and Ringerts, N. R. (1957). *J. Am. Chem. Soc.* **79**, 2025.

Rice, L. I., Spolter, L., Tokes, Z., Eisenman, R., and Marx, W. (1967). *Arch. Biochem. Biophys.* **118**, 374.

Rice, L. I., Rice, E. H., Spolter, L., Marx, W., and O'Brien, J. S. (1968). *Arch. Biochem. Biophys.* **127**, 37.

Riggs, T. R., and Walker, L. M. (1964). *Endocrinology* **74**, 483.

Robbins, P. W. (1962). *In* "The Enzymes" (P. D. Boyer, H. Lardy, and K. M. Myrbäck, eds.), 2nd Ed., Vol. 6, p. 469. Academic Press, New York.

Robbins, P. W. (1963). *In* "Methods in Enzymology" (S. P. Colowick and N. O. Kaplan, eds.), Vol. 6, p. 766. Academic Press, New York.

Robbins, P. W., and Lipmann, F. (1958a). *J. Biol. Chem.* **233**, 681.

Robbins, P. W., and Lipmann, F. (1958b). *J. Biol. Chem.* **233**, 686.

Robert, L., and Dische, Z. (1963). *Biochem. Biophys. Res. Commun.* **10**, 209.

Roberts, K. D., Vande Wiele, R. L., and Lieberman, S. (1961). *J. Biol. Chem.* **236**, 2213.

Roberts, K. D., Bandi, L., Calvin, H. I., Drucker, W. D., and Lieberman, S. (1964). *Biochemistry* **3**, 1983.

Robinson, D., and Abrahams, H. E. (1967). *Biochim. Biophys. Acta,* **132**, 212.

Robinson, H. C. (1965). *Biochem. J.* **94**, 687.

Robinson, H. C., and Pasternak, C. A. (1964). *Biochem. J.* **93**, 487.

Robinson, H. C., Telser, A., and Dorfman, A. (1966). *Proc. Natl. Acad. Sci. U.S.* **56**, 1859.

Roche, J., Michel, R., Closon, J., and Michel, O. (1959). *Biochim. Biophys. Acta* **33**, 461.

Roche, J., Michel, R., Closon, J., and Michel, O. (1960). *Biochim. Biophys. Acta* **38**, 325.

Rodén, L., and Armand, G. (1966). *J. Biol. Chem.* **241**, 65.

Rodén, L., and Smith, R. (1966). *J. Biol. Chem.* **241**, 5949.

Rose, F. A., Flanagan, T. H., and John, R. A. (1966). *Biochem. J.* **98**, 168.

Rosenberg, L., Schubert, M., and Sandson J. (1967). *J. Biol. Chem.* **242**, 4691.

Roy, A. B. (1953). *Biochem. J.* **53**, 12.

Roy, A. B. (1954). *Biochim. Biophys. Acta* **14**, 149.

Roy, A. B. (1956a). *Biochem. J.* **62**, 35P.

Roy, A. B. (1956b). *Biochem. J.* **63**, 294.

Roy, A. B. (1956c). *Biochem. J.* **62**, 41.

Roy, A. B. (1957). *Biochem. J.* **66**, 700.

Roy, A. B. (1958). *Biochem. J.* **68**, 519.

Roy, A. B. (1960a). *Advan. Enzymol.* **22**, 205.

Roy, A. B. (1960b). *Biochem. J.* **74**, 49.

Roy, A. B. (1960c). *Biochem. J.* **77**, 380.

Roy, A. B. (1961a). *J. Histochem. Cytochem.* **10**, 106.

Roy, A. B. (1961b). *Biochem. J.* **79**, 253.

Roy, A. B. (1962). *Biochem. J.* **82**, 66.

Roy, A. B. (1963). *Australian J. Exptl. Biol. Med. Sci.* **41**, 331.

Roy, A. B. (1964). *J. Mol. Biol.* **10**, 176.

Roy, A. B., and Banerjee, R. K. (1966). *Proc. 2nd Intern. Congr. Hormonal Steroids, Milan. Excerpta Med. Intern. Congr. Ser.* **132**, 397.

Ryan, L. C., Carubelli, R., Caputto, R., and Trucco, R. E. (1965). *Biochim. Biophys. Acta* **101**, 252.

Salomon, L. L. (1957). *Texas Rept. Biol. Med.* **15**, 925.

Sato, T., Suzuki, T., Fukuyama, T., and Yoshikawa, H. (1956). *J. Biochem. (Tokyo)* **43**, 421.

Sato, T., Yamada, M., Suzuki, T., Fukuyama, T., and Yoshikawa, H. (1959). *J. Biochem. (Tokyo)* **46**, 79.

Saunders, A. M., and Silverman, L. (1967). *Nature* **214**, 194.

Sayers, G., Sayers, M. A., Fry, E. G., White, A., and Long, C. N. H. (1944). *Yale J. Biol. Med.* **16**, 361.

Scardi, V., Iaccarino, M., and Scarano, E. (1962). *Biochem. J.* **83**, 413.

Schacter, B., and Marrian, G. F. (1938). *J. Biol. Chem.* **126**, 663.

Schanberg, S. M., Breese, G. R., Schildkraut, J. J., Gordon, E. K., and Kopin, I. J. (1968). *Biochem. Pharmacol.* **17**, 2006.

Schneider, J. J., and Lewbart, M. L. (1956). *J. Biol. Chem.* **222**, 787.

Schrager, J., and Oates, M. D. (1968). *Biochem. J.* **106**, 523.

Scott, J., and Spencer, B. (1965). *Biochem. J.* **95**, 50P.

Segal, H. L. (1955). *J. Biol. Chem.* **213**, 161.

Segal, H. L., and Mologne, L. A. (1959). *J. Biol. Chem.* **234**, 909.

Seno, N., Meyer, K., Anderson, B., and Hoffman, P. (1965). *J. Biol. Chem.* **240**, 1005.

Serafini-Fracassini, A. (1968). *Biochim. Biophys. Acta* **170**, 289.

Serafini-Fracassini, A., and Duward, J. J. (1968). *Biochem. J.* **109**, 693.

Shatton, J., and Schubert, M. (1954). *J. Biol. Chem.* **211**, 565.

Silbert, J. E. (1967a). *J. Biol. Chem.* **242**, 2301.

Silbert, J. E. (1967b). *J. Biol. Chem.* **242**, 5146.

Silbert, J. E. (1967c). *J. Biol. Chem.* **242**, 5153.

Smith, J. N. (1955). *Biol. Rev. Cambridge Phil. Soc.* **30**, 455.

Smith, J. N. (1962). *Nature* **195**, 399.

Solheim, K. (1966). *J. Oslo City Hosp.* **16**, 17.

Soper, R. (1963). *Comp. Biochem. Physiol.* **10**, 325.

Spencer, B. (1958). *Biochem. J.* **69**, 155.

Spencer, B. (1959a). *Biochem. J.* **71**, 500.

Spencer, B. (1959b). *Biochem. J.* **69**, 442.

Spencer, B. (1960). *Biochem. J.* **77**, 294.

Spencer, B., and Raftery, J. (1966). *Biochem. J.* **99**, 35P.

Sperber, I. (1948). *Kgl. Lantbruksakad. Tidskr.* **15**, 317.

Stoffyn, A., Stoffyn, P., and Martensson, E. (1968). *Biochim. Biophys. Acta* **152**, 353.

Subba Rao, K., and Ganguly, J. (1964). *Biochem. J.* **90**, 104.

Subba Rao, K., and Ganguly, J. (1966). *Biochem. J.* **98**, 693.

Subba Rao, K., Seshadri Sastry, P., and Ganguly, J. (1963). *Biochem. J.* **87**, 312.

Sundaresan, P. R. (1966). *Biochim. Biophys. Acta* **113**, 95.

Sundaresan, P. R., Elford, R. M., and Wolf, G. (1964). *Federation Proc.* **23**, 479.

Suzuki, S., and Strominger, J. L. (1960a). *J. Biol. Chem.* **235**, 257.

Suzuki, S., and Strominger, J. L. (1960b). *J. Biol. Chem.* **235**, 267.

Suzuki, S., and Strominger, J. L. (1960c). *J. Biol. Chem.* **235**, 274.

Suzuki, S., and Strominger, J. L. (1960d). *J. Biol. Chem.* **235**, 2768.

Suzuki, S., Takahashi, N., and Egami, F. (1957). *Biochim. Biophys. Acta* **24**, 444.

Suzuki, S., Threnn, R. H., and Strominger, J. L. (1961). *Biochim. Biophys. Acta* **50**, 169.

Svennerholm, L. (1963). *In* "Brain Lipids and Lipoproteins and the Leucodystrophies" (J. Folch-Pi and H. Bauer, eds.), p. 104. Elsevier, Amsterdam.

Svennerholm, L., and Stalberg-Stenhagen, S. (1968). *J. Lipid Res.* **9**, 215.

Szent-Györgyi, A. J. (1928). *Biochem. J.* **22**, 1386.

Takahashi, N. (1960). *J. Biochem.* (Tokyo) **47**, 230.

Tallan, H. H., Bella, S. T., Stein W. H., and Moore, S. (1955). *J. Biol. Chem.* **217**, 703.

Tan, Y. H., and Bowness, J. M. (1968). *Biochem. J.* **110**, 9.

Telser, A., Robinson, H. C., and Dorfman, A. (1965). *Proc. Natl. Acad. Sci. U.S.* **54**, 912.

Telser, A., Robinson, H. C., and Dorfman, A. (1966). *Arch. Biochem. Biophys.* **116**, 458.

Thomas, L. (1956). *J. Exptl. Med.* **104**, 245.

Thudichum, J. L. W. (1884). "A Treatise on the Chemical Constitution of the Brain." Ballière, London.

Tominga, F., Oka, K., and Yoshida, H. (1965). *J. Biochem.* (Tokyo) **57**, 717.

Touchstone, J. C., Greene, J. W., McElroy, R. C., and Murawee, T. (1963). *Biochemistry* **2**, 653.

Troen, P., Nilsson, B., Wiqvist, N., and Diczfalusy, E. (1961). *Acta Endocrinol.* **38**, 361.

Tsaltas, T. T. (1966). *Federation Proc.* **25**, 663.

Tsuyuki, H., and Idler, D. R. (1957). *J. Am. Chem. Soc.* **79**, 1771.

Tudball, N., and Davidson, E. A. (1969). *Biochim. Biophys. Acta* **171**, 113.

Tudball, N., and Thomas, J. H. (1965). *Abstr. 2nd Meeting Federation European Biochem. Soc., Vienna* p. 201.

Twombly, G. H., and Levitz, M. (1960). *Am. J. Obstet. Gynecol.* **80**, 889.

Vaes, G. (1967). *Biochem. J.* **103**, 802.

Van Hoof, F., and Hers, H. G. (1968). *European J. Biochem.* **7**, 34.

Varandani, P. P., Wolf, G., and Johnson, B. C. (1960). *Biochem. Biophys. Res. Commun.* **3**, 97.

Vestermark, A., and Boström, H. (1959). *Exptl. Cell Res.* **18**, 174.

Vestermark, A., and Boström, H. (1960). *Experientia* **9**, 408.

Viala, R., and Gianetto, R. (1955). *Can. J. Biochem. Physiol.* **33**, 839.

Waley, S. G. (1959). *Biochem. J.* **71**, 132.

Warren, J. C., and French, A. P. (1965). *J. Clin. Endocrinol. Metab.* **25**, 278.

Warren, J. C., and Timberlake, C. E. (1962). *J. Clin. Endocrinol. Metab.* **22**, 1148.

Webb, E. C., and Morrow, P. F W. (1959). *Biochem. J.* **73**, 7.

Webb, E. C., and Morrow, P. F. W. (1960). *Biochim. Biophys. Acta* **39**, 542.

Wedding, R. T., and Black, M. K. (1960). *Plant Physiol.* **35**, 72.

Wengle, B. (1964). *Acta Soc. Med. Upsalien.* **69**, 105.

Wengle, B., and Boström, H. (1963). *Acta Chem. Scand.* **17**, 1203.

Williams, R. T. (1938). *Biochem. J.* **32**, 878.

Williams, R. T. (1959). "Detoxication Mechanisms." Chapman & Hall, London.

Wilson, L. G. (1962). *Ann. Rev. Plant Physiol.* **13**, 201.

Wilson, L. G., and Bandurski, R. S. (1956). *Arch. Biochem. Biophys.* **62**, 503.

Wilson, L. G., and Bandurski, R. S. (1958). *J. Biol. Chem.* **233**, 975.

Woolley, D. W., and Peterson, W. H. (1937). *J. Biol. Chem.* **122**, 213.

Wortman, B. (1962). *Arch. Biochem. Biophys.* **97**, 70.

Wortman, B. (1963). *Biochim. Biophys. Acta* **77**, 65.

Yagi, T. (1966). *J. Biochem. (Tokyo)* **59**, 495.

Yasumoto, T., Nakamura, K., and Hashimoto, Y. (1967). *Agr. Biol. Chem. (Tokyo)* **31**, 7.

GLYCOPROTEIN AND MUCOPOLYSACCHARIDE HYDROLYSIS

(Glycoprotein and Mucopolysaccharide Hydrolysis in the Cell)

EUGENE A. DAVIDSON

I. Introduction

The number and types of potentially available glycoprotein or mucopolysaccharide substrates found in mammalian systems are unusually broad. In addition, both endo- or exoglycosidases as well as peptidases may act on such substrates. The discussion in this chapter will be limited to those carbohydrate-containing polymers in which a covalent linkage is present between the polypeptide or protein moiety and the saccharide component. The term "polymer," at least in this sense, merely indicates

TABLE I

MONOSACCHARIDE OCCURRENCE IN GLYCOPROTEINS AND MUCOPOLYSACCHARIDES[a]

Sugar	Source
D-Xylose	Mucopolysaccharides
D-Galactose (Gal)	Mucopolysaccharides + glycoproteins
D-Mannose (Man)	Glycoproteins
N-Acetyl D-glucosamine (GNAc)	Mucopolysaccharides + glycoproteins
N-Acetyl D-galactosamine (GalNAc)	Mucopolysaccharides + glycoproteins
L-Fucose	Glycoproteins
D-Glucuronic acid (GA)	Mucopolysaccharides
L-Iduronic acid	Mucopolysaccharides
N-Acetyl neuraminic acid (NAN)	Glycoproteins and mucins

[a] D-Glucose may also be present in some tissue-specific glycoproteins but is quite restricted in occurrence and not usually found.

the size of the intact substrate since the molecular weight range of the carbohydrate portion of such compounds may vary from disaccharide units to high molecular weight chains. The number and type of sugars found in such compounds is considerable although usually restricted within a given polymer. A representative list of sugars present in these macromolecules is given in Table I.

Systems of nomenclature or classification are usually arbitrary, and terms adopted in this area are frequently confusing. In general, a glycoprotein is defined as such in terms of the percentage contribution of the carbohydrate moiety to the total structure. Representative examples might include ovalbumin (Nuenke and Cunningham, 1961), ribonuclease B (Plummer and Hirs, 1964), α-1-acid glycoprotein (Kamiyama and Schmid, 1962; Satake *et al.*, 1965), fetuin (Spiro, 1962), transferrin (Jamieson, 1965), and γ-globulin (Clamp and Putnam, 1964) (Table II).

The term "mucin" is more or less operationally descriptive and refers

TABLE II

CARBOHYDRATE CONTENT OF REPRESENTATIVE GLYCOPROTEINS

Glycoprotein	CHO component	
	Sugars present	Amount (% of total)
Ribonuclease B	Man, GNAc	3
γ-Globulin	Fucose, Gal, Man, GNAc, NAN	3.1
Ovalbumin	Man, GNAc	3.2
Transferrin	Gal, Man, GNAc, NAN	5.6
Fetuin	Gal, NAN, GNAc	22.0
α-1 Acid glycoprotein	Fucose, Gal, Man, GNAc, NAN	41.0

primarily to secretions of epithelial cells such as those elaborated by the submaxillary or sublingual glands. These have rather characteristic physical properties which serve to distinguish them from most of the glycoproteins, but there are some features in common.

Finally, the term "mucopolysaccharide" or "mucopolysaccharide-protein complex" is generally employed when discussing components where the majority of the molecular mass is attributable to the carbohydrate portion of the molecule.[1]

It can readily be recognized that the areas of distinction between glycoproteins, mucins, and mucopolysaccharides are rather blurred and that operational classification of one or another macromolecule might result in ambiguity. It seems likely that families of compounds exist which tend to bridge these rather artificially constructed barriers. However, distinctions are frequently applicable either on chemical or physical bases. For example, the properties of a molecule such as ribonuclease B or γ-globulin are most readily reflected in terms of the protein conformation and its structure rather than in the nature or type of carbohydrate covalently attached to the polypeptide. This does not imply that the carbohydrate moiety has no physiological function even though it may only represent a relatively small proportion of the total macromolecule. Areas such as antibody specificity and immunological determinants frequently fall within the physiological role of such carbohydrate groups.

Many properties of the mucins are readily ascribable to the carbohydrate moieties and, in particular, to the presence of N-acetyl neuraminic acid as a characteristic terminal nonreducing end of the carbohydrate oligomers. Finally, the properties of the mucopolysaccharides, at least in terms of their physical behavior, are ascribable primarily to the carbohydrate units and, to a lesser degree, to the protein structure.

With these distinctions in mind, the following discussion will indicate the nature of the substrates to be encountered in dealing with hydrolytic problems of glycoproteins and mucopolysaccharides. It should be recognized that many of the potential substrates are infrequently, if at all, exposed to cellular degradative processes. However, there is a growing body of evidence which indicates that macromolecules circulating in the serum can be cleared by liver or kidney cells at a reasonable rate and subjected to degradative processes within these cells. Unfortunately,

---

[1] A wide variety of terminology has been utilized with regard to these compounds. Terms such as "protein polysaccharide," "glycosaminoglycan," and "proteoglycan" have been employed by various authors, and there is certainly no unanimity of agreement as to the most appropriate nomenclature. As much as possible, the terms employed in this review will be within fairly discrete limits.

there is relatively little data on discrete representatives of these groups. The few examples that we have indicate that the process may be of a somewhat general nature. Certainly studies in the area of protein turnover would indicate the presence in the whole animal of degradative systems capable of attacking not only the protein but also the carbohydrate portion of any of these macromolecules.

II. Substrates

A. GLYCOPROTEINS

The number and type of glycoproteins thus far known to be present in serum are considerable. The carbohydrate content of such glycoproteins ranges from 1 to 3% in the case of ovalbumin to nearly 40% for substrates such as α-1-acid glycoprotein. It is convenient to subdivide the glycoproteins into two broad classes—those which contain N-acetyl neuraminic acid and those which do not. This distinction is frequently operational in terms of the fractionation properties of these molecules and can serve as a convenient point of departure.

The data summarized in Table II indicate some of the glycoproteins commonly encountered and their carbohydrate components. It should be noted that for almost all of the glycoproteins that are known, there are certain characteristic carbohydrate units which are found. These include either N-acetyl glucosamine or N-acetyl galactosamine, D-mannose, D-galactose, and L-fucose. D-Glucose is rarely found as a constituent of glycoproteins and has only been identified in the case of collagen and a glycoprotein isolated from aorta. In the study of such glycoprotein substrates there are two primary chemical problems which need to be answered. The first is the exact structure, linkage, and configuration of the monosaccharide groups; and the second is the mode of covalent attachment between the oligosaccharide portion and the polypeptide chain.

The structure of the carbohydrate oligomers can be established by conventional techniques usually following exhaustive proteolytic digestion to remove the bulk of the amino acids, gel filtration to isolate glycopeptides thus produced, and sequential enzymic or chemical degradations to establish the order of the carbohydrates, their linkage positions, and configurations. In most cases the configuration of the carbohydrate linkage is established by the use of stereospecific glycosidases. Techniques such as periodate susceptibility, release of monosaccharides on treatment with purified glycosidases, Smith degradation, and others have proven effective in determining the order of residues and points of linkage attachment. Typical oligosaccharide structures are indicated in Fig. 1.

The modes of linkage between the carbohydrate oligomers and the polypeptide chain can, at least on theoretical grounds, be of several types. These would include glycosidic ester, glycosidic amide, and glycosidic

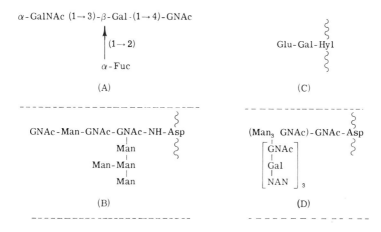

FIG. 1. Some typical oligosaccharide structures. (A) Blood group A tetrasaccharide. (B) Ovalbumin oligosaccharide. (C) Disaccharide linked to hydroxylysyl residue in collagen. (D) Carbohydrate moiety of fetuin; three such moles are present per mole of glycoprotein. Human chorionic gonadotropin has a very similar oligosaccharide component.

ether linkages to serine or threonine and, in the case of collagen, to hydroxyproline or hydroxylysine. In the case of the first two, either glutamic or aspartic acid may participate in the attachment. Representative structures are indicated in Fig. 2. Thus far, for the glycoproteins in which the linkage group has been established, the glycosidic amide linkage to asparagine is the only one that has been documented. In almost all cases examined, the sugar immediately attached to the asparagine residue is N-acetyl glucosamine or N-acetyl galactosamine. The mode of biosynthesis of this linkage can be visualized as involving the transfer of an N-acetyl hexosaminyl residue from a uridine diphospho N-acetyl amino sugar to the polypeptide acceptor as illustrated in Fig. 3. In general, sequential addition of the other monosaccharide units from the appropriately activated carbohydrate derivative takes place in a stepwise fashion. There is relatively little known about the specificity of the enzymes involved in these transfer reactions, but it is presumed that they

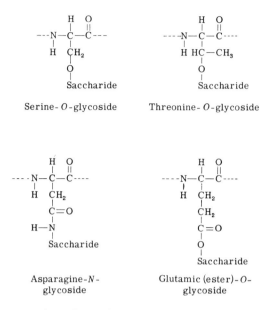

FIG. 2.   Types of covalent linkages between carbohydrate and protein.

recognize the specific protein substrate and attached carbohydrate units. It is entirely reasonable to assume that different glycosyl transferases are involved for linkages of different configuration and position.

The electrophoretic and physical behavior of a number of the glyco-proteins can readily be accounted for by the presence in them of an unusual 9-carbon sugar acid, N-acetyl neuraminic acid. This compound arises biologically by means of an aldol condensation between N-acetyl mannosamine 6-phosphate and phosphoenol pyruvate to form the N-acetyl neuraminic acid 9-phosphate from which the phosphate group is removed by a specific phosphatase (Roseman, 1962). N-Acetyl neuraminic acid is activated prior to transfer to an acceptor molecule by the forma-

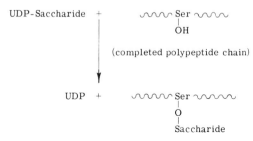

FIG. 3.   Synthesis of glycoprotein linkage.

tion of a cytidine monophosphate derivative of the ketal hydroxyl (Kean and Roseman, 1966). The relatively low pK of the carboxyl group of the acetyl neuraminic acid (approximately 2.7) confers charged properties on the glycoproteins in which it is found and this, coupled with the extreme lability of the ketosidic linkage involved, gives rise to extensive polymorphism.[2]

Structure	Protein or source
NAN-(2→6)-GalNAc-Protein	Sheep submaxillary
NAN-Gal-GNAc-Protein	Goat colostrum
Man — Man — Man-Man-GNAc-Protein	Gonadotropin
(NAN)$_2$-Gal$_2$-Man$_4$-GNAc$_4$-Protein	Transferrin
(NAN)$_2$-Gal$_4$-Man$_3$-GNAc$_5$-Protein	Thyroglobulin

For the gonadotropin structure, a bracketed side chain appears beneath the first Man:
$$\left[\begin{array}{c} \text{GNAc} \\ | \\ \text{Gal} \\ | \\ \text{NAN} \end{array}\right]_2 \quad \text{GNAc} \quad \text{GNAc}$$

FIG. 4. Representative structures of N-acetyl neuraminic acid containing glycoproteins.

It is noteworthy that the N-acetyl neuraminic acid in such glycoproteins exists invariably as a terminal nonreducing end. Schematic structures for several N-acetyl neuraminic acid-containing carbohydrate units are indicated in Fig. 4. Since essentially all of the degradative glycosidases found in mammalian systems are exoenzymes, that is, they attack only the terminal nonreducing carbohydrate moiety of an oligosaccharide structure, it can be seen that for this class of macromolecules the action of a neuraminidase is an absolute prerequisite to degradation of the carbohydrate moiety.

B. MUCINS

In contrast to the family of glycoproteins, the chemical architecture of epithelial secretions is far less well documented. The most thoroughly

---

[2] A variation in the N-acetyl neuraminic acid content is not the only explanation for polymorphism. Most studies on glycoproteins are conducted with pooled material regardless of the type of animal, and small variations in the polypeptide backbone as well as variations in the amount of N-acetyl neuraminic acid present can give rise to this phenomenon. Preparations frequently appear homogeneous by several physical criteria but can be fractionated by techniques such as starch gel electrophoresis (Schmid et al., 1962).

studied mucin, that elaborated by the submaxillary gland, contains approximately 40% carbohydrate. The vast majority is present as disaccharide units of N-acetyl neuraminic acid linked to N-acetyl galactosamine which in turn is joined in ether linkage to the hydroxyl groups of either serine or threonine (Tanaka *et al.*, 1964). The molecular weight of the polypeptide is apparently quite large since the entire macromolecule has a particle size in the range of 800,000. Based on these figures, this would leave a figure of approximately 480,000 for the protein moiety, which suggests that a subunit structure is involved. However, detailed

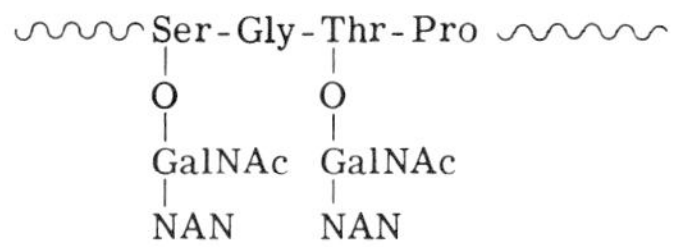

$$
\begin{array}{ccc}
\text{\textcolor{black}{\char126\char126\char126}Ser-Gly-Thr-Pro \char126\char126\char126} \\
| \qquad\quad\; | \\
O \qquad\quad\; O \\
| \qquad\quad\; | \\
\text{GalNAc} \quad \text{GalNAc} \\
| \qquad\quad\; | \\
\text{NAN} \qquad \text{NAN}
\end{array}
$$

Fig. 5. Schematic structure of submaxillary mucin. Approximately 2000 oligosaccharide units are present per mole although the possibility of a subunit structure for the protein moiety exists.

information is not available at this time. The less well-studied mucins elaborated by the gastric mucosa appear to have more complicated structures in terms of the number and type of carbohydrate units, and several of these appear to contain ester sulfate groupings as well. A typical structure for the submaxillary mucins is indicated in Fig. 5.

C. MUCOPOLYSACCHARIDES

The number and type of mucopolysaccharides found in the connective tissue of animals is rather limited, and these can be conveniently subdivided into sulfated and nonsulfated structures. There are certain general similarities for all of these compounds which can be indicated as follows:

(1) The compounds contain negative charges either on every monosaccharide residue or on every other monosaccharide residue. These charges are either carboxylate, arising from the carboxyl function of either D-glucuronic or L-iduronic acid, or ester sulfate, the ester sulfate groups most often being attached to N-acetyl amino sugar residues.

(2) All representatives of this class contain either N-acetyl glucosamine or N-acetyl galactosamine as one of the component sugars.

(3) The polysaccharide portions of these macromolecules are straight-chain polymers; although the molecular weight ranges may vary to a con-

siderable degree, branching of the carbohydrate chains has thus far not been detected.

(4) The polysaccharides can be visualized as having a repeating disaccharide unit containing an N-acetyl amino sugar and, in most cases, a uronic acid. The one exception is keratan sulfate which contains galactose instead of uronic acid.

(5) The mode of linkage of the polysaccharide portions to the polypeptide core proteins appears to be fairly uniform within this group and consists of an unusual trisaccharide: galactosyl-galactosyl-xylose linked via the reducing position of the xylose unit to the hydroxyl group of a serine in the polypeptide chain (Roden and Smith, 1966).

The major nonsulfated polysaccharide found in connective tissue, hyaluronic acid, contains alternating D-glucuronic acid and N-acetyl-D-glucosamine residues. The polysaccharide is found in skin, heart valves, vitreous humor, synovial fluid, and umbilical cord, and in addition, may be present in smaller amounts in other areas of the connective tissue. The molecular weight of hyaluronic acid appears to be very large, on the order of 500,000 to 50,000,000 depending upon source, method of isolation, and most probably, method of measurement. In any case, there is general agreement that the molecule is very large, probably polydisperse in molecular weight distribution, and that these features appear to be directly related to its physiological function. The mode of linkage to protein, if any, has not been conclusively established. Biosynthesis appears to take place by transfer of activated monosaccharide units from nucleoside diphospho derivatives to an acceptor which appears to be membrane bound. The exact mechanism of this transfer reaction, which requires inversion of configuration of the monosaccharide linkages, has not been established nor has the nature of the acceptor. Preparations isolated from synovial fluid by electrodeposition appear to be substantially protein free; but, for a molecule of molecular weight of several millions, the presence of a small amount of covalently linked peptide could easily escape detection. In any case, there appears to be little size restriction during the biosynthetic process, and any covalently attached protein probably represents a biosynthetic remnant rather than an essential physiological feature of the molecule.

A number of years ago, a polysaccharide in cornea was described which differed from hyaluronic acid only in the configuration of the amino sugar (Davidson and Meyer, 1954). Although the molecule isolated was not totally free of sulfate, the sulfate content was approximately one per eight disaccharide units. This polymer, chondroitin, has not been found in other areas of connective tissue; and its biological relationship to the more fully sulfated representatives is obscure. It may well be restricted to the cornea since even in embryonic material, fractions with this low sulfate content are generally not found in other tissues.

There are a number of sulfated polysaccharides found in the connective tissue, and the basic structures are indicated in Table III.[3] It

TABLE III

BASIC STRUCTURES OF ACID POLYSACCHARIDES FOUND IN CONNECTIVE TISSUE

Polysaccharide	Sugars present and linkage
Hyaluronic acid	D-Glucuronic acid (β 1 $\to$ 3)
	N-Acetyl-D-glucosamine (β 1 $\to$ 4)
Chondroitin	D-Glucuronic acid (β 1 $\to$ 3)
	N-Acetyl-D-galactosamine (β 1 $\to$ 4)
Chondroitin 4-sulfate	D-Glucuronic acid (β 1 $\to$ 3)
	N-Acetyl-D-galactosamine-4-O-sulfate (β 1 $\to$ 4)
Chondroitin 6-sulfate	D-Glucuronic acid (β 1 $\to$ 3)
	N-Acetyl-D-galactosamine-6-O-sulfate (β 1 $\to$ 4)
Dermatan sulfate	L-Iduronic acid (α 1 $\to$ 3)[a]
	N-Acetyl-D-galactosamine-4-O-sulfate (β 1 $\to$ 4)
Keratan sulfate	D-Galactose (β 1 $\to$ 3)
	N-Acetyl-D-glucosamine-6-O-sulfate (β 1 $\to$ 4)
Heparin sulfate	D-Glucuronic acid (linkage details still incomplete)
	N-Sulfo-D-glucosamine
	N-Acetyl-D-glucosamine

[a] Same conformation as D-glucuronic acid. Linkage is equatorial, carboxyl group axial.

should be noted that although there are relatively minor differences in the chemical architecture of these macromolecules, their physical properties and association with protein are clearly different; and one can presume that the physiological role played by each of these is distinctly different. It should be noted that the fine structure of heparitin

[3] A polysaccharide generally included in this group, although improperly so, is heparin. The inclusion of heparin among the acid mucopolysaccharides is based on its highly anionic structure, presence of ester sulfate groups, the presence of amino sugar, and the presence of uronic acid. However, although these areas of similarity are considerable, the areas that serve to differentiate heparin from the other mucopolysaccharides are more compelling. These include the configuration of the glycosidic linkages as well as their position, the presence of N-sulfate groups, and the fact that heparin is restricted primarily to the mast cells and, as such, cannot be considered a structural component of the connective tissue. Finally, the known biological roles for heparin, those of anticoagulant activity and lipoprotein lipase activator, are quite different from any presumed to exist for the other polysaccharides.

sulfate and keratan sulfate, especially with regard to the nature of the polypeptides present or associated with these molecules, has not yet been delineated.

The chondroitin sulfates, especially chondroitin 4-sulfate, exist in tissues as highly specific protein polysaccharide macromolecules with a well-

TABLE IV

TYPICAL AMINO ACID COMPOSITION

Amino acid	Content[a]	
	2	1
Lysine	0.12	0.21
Histidine	0.082	0.24
Arginine	0.35	0.32
Cystine, methionine	Nil	Nil
Aspartic	0.50	0.69
Threonine	0.49	0.50
Serine	1.01	0.97
Glutamic	1.02	1.21
Proline	0.79	1.00
Glycine	1.00	1.00
Alanine	0.57	0.76
Valine	0.34	0.47
Isoleucine	0.19	0.24
Leucine	0.32	0.59
Tyrosine	0.27	0.19
Phenylaline	0.39	0.38
Glucosamine	0.69	1.35
Galactosamine	10.42	18.60

[a] Molar ratio to glycine = 1.00. Total peptide content is 8–12%. Preparation 1 is from pig embryo costal cartilage and preparation 2 is from mature pig costal cartilage. Hydroxyproline was not detected in either fraction.

defined structure (Marler and Davidson, 1965). It is worthy of note that the molecular weight of the polysaccharide chains, at least where these have been determined, is smaller than that of hyaluronic acid by at least one order of magnitude and probably two or three, that the polydispersity is different among the groups, and that relatively little biosynthetic information is available. The nature of the polypeptides involved does deserve some comment. In addition to a high percentage of serine, there are relatively large amounts of glycine, aspartic acid, alanine, and glutamic acid (Katsura and Davidson, 1969; Hranisavljevic and Davidson, 1970). A typical amino acid composition is indicated in Table IV. In the absence of detailed sequence information, the fine structure of these molecules cannot be reconstructed, and relatively little information is available about the susceptibility of intact structures to proteolytic degradation, especially by tissue cathepsins.

III. Enzymes

A. LOCALIZATION OF DEGRADATIVE ENZYMES

The variety of carbohydrate-containing macromolecules, ranging from proteins containing small oligosaccharide units to highly complex systems

TABLE V
In Vivo TURNOVER OF CONNECTIVE TISSUE POLYSACCHARIDES

Polysaccharide	Source	Half-life (days)
Hyaluronic acid	Skin	4–6
Dermatan sulfate	Skin	10–14
Chondroitin 4-sulfate	Cartilage	30
Keratan sulfate	Nucleus pulposus	>120

containing 90% or more carbohydrate, suggests a multiplicity of degradative pathways. Studies conducted in numerous laboratories have established that a fairly active turnover exists in living systems (Schiller *et al.*, 1954; Katsura and Davidson, 1963). This is not restricted to the carbohydrate moieties but also includes the ester sulfate group. Relatively less information is available about the turnover of the associated protein cores, and fewer studies have been done on the half-lives of glycoproteins and mucins. Typical data are indicated in Table V. The fact that these macromolecules apparently undergo regular degradation and resynthesis points to the existence of a series of degradative enzymes capable of extensive fragmentation of the carbohydrate chains. There are several problems in the study of such degradative enzymes; primarily they are concerned with the exact nature of the substrate attacked by such enzymes, the localization of these for a given cell type, and the presence of a given enzyme system in differentiated cell systems. In general, most studies have indicated that the primary sites of degradation are the liver, kidney, and spleen although the failure to detect specific degradative enzymes in skin, cartilage, or bone may be more a reflection of the methodology employed than of the presence or absence of a particular activity.

Studies conducted in a number of laboratories have clearly delineated the lysosome concept as originally defined by deDuve (Weissmann, 1965; deDuve and Wattiaux, 1966). The fact that upward of 30 hydrolytic activities, all with acid pH optima and comparable sedimentation properties, can be demonstrated in cells strongly suggests the presence of a highly organized system specifically geared to the degradation of macromolecular substrates. In general, the isolation of subcellular particles and the localization of degradative enzymes follows fairly standard techniques. A typical procedure for subcellular fractionation is as follows (Trouet, 1966).

The isolation of purified lysosomes takes advantage of the fact that the detergent Triton WR1339 is specifically taken up by these subcellular organelles but is not degraded. The uptake results in an associated lowering of the buoyant density of the lysosomes so that they can readily be separated from mitochondria by sucrose density-gradient centrifugation. Animals are pretreated with the detergent 3–4 days prior to sacrifice. The livers are homogenized in 0.25 M sucrose using an all-glass homogenizer and the nuclear fraction removed by centrifugation at 1000 g for 10 minutes. A combined mitochondrial and lysosomal pellet is isolated by centrifugation at 25,000 rpm in a Spinco Model L2 centrifuge using the 30 rotor for a period of 8 minutes and 20 seconds. The supernatant remaining from this fraction can be centrifuged at 144,000 g for 2 hours to yield a microsomal pellet and a final supernatant fraction. The combined mitochondrial and lysosomal pellet is resuspended in sucrose and subjected to density-gradient centrifugation using a discontinuous sucrose gradient consisting of sucrose, density 1.21, at the bottom of the tube and layers of sucrose, density 1.155 and 1.06, going toward the top. After centrifugation in the swinging bucket rotor at 24,000 rpm for 2 hours, the lysosomes are found at the interface between the 1.06 and 1.155 density sucrose and can be removed by displacement.

Routine examination of the lysosomal fraction for mitochondrial activity indicates less than 3% contamination with mitochondrial marker enzymes, a finding which is confirmed by electron microscopy. Examination of a wide variety of hydrolytic activities after such a fractionation procedure shows 80% or more of the total hydrolytic capacity of the cell in the lysosomal fraction. This estimate is probably conservative since not all of the organelles may take up the detergent, and there are obviously some losses during the fractionation procedure. However, it has not yet been conclusively demonstrated that comparable hydrolytic activities are totally absent from the nuclear fraction or the endoplasmic reticulum. Nevertheless, these appear to be of small, if any, quantitative significance.

In general, there are two approaches which can be employed to study the hydrolytic capacity of such organelles. Most, if not all, of the enzymes can be solubilized from the lysosomes by disruption of the lysosomal membrane using either detergents, freeze-thawing, or disintegration by ultrasonic vibration. The various enzymes present can then be subjected to a variety of purification procedures conventionally employed in enzymology and the properties of the individual isolated enzymes assessed. These can then be reconstructed to provide a reasonable estimate of the total degradative capacity of the organelle and should indicate the range of substrates which are capable of being degraded. However, under many circumstances this can prove to be a deceptive picture. Certainly under physiological conditions it is reasonable to presume that a substrate taken up by such an organelle will be simultaneously subjected to hydrolysis by the entire enzyme complement present. Accordingly, a more appropriate action pattern can be obtained, perhaps, if the mixed lysosomal content is permitted to attack the substrate under study. The rate of release of degradation products and the nature of the end products accumulated under such conditions may then provide a more realistic appraisal of degradative capacities. It should be noted that without exception, all of the lysosomal enzymes have a distinctly acid pH optimum with little or no activity at the conventional physiological pH of 7.4. However, this very fact would seem to argue for the combining of these hydrolytic activities in a subcellular particle which may easily maintain a pH gradient between its interior and the cytoplasm of the cell in which it is found.

Thus far, one or another type of lysosome activity has been demonstrated in every cell type that has been examined, including cells grown in tissue culture. It is not unreasonable to presume that such subcellular organelles are present in virtually every cell of higher animals, but this does not necessarily imply that the enzyme complement found within a given differentiated tissue is identical to that found in another. In fact, quite the contrary may be the case.

In addition to the complement of lysosomal enzymes which are probably responsible for the vast majority of hydrolase activities occurring, at least one enzyme is apparently present in a nonlysosomal state; this is testicular hyaluronidase. The possible involvement of hyaluronidase in the fertilization process has been the subject of speculation for quite a number of years, but the actual mode of action is still uncertain. In any case, the activity present appears to be restricted to the testis, although it is not fundamentally different from the comparable lysosomal enzyme found in the liver.

B. UPTAKE OF SUBSTRATES BY CELLS

In view of the enormous breadth of degradative enzymes present in a single subcellular organelle, it is reasonable to ask whether there is any uptake specificity for macromolecules associated with the lysosomal fraction; that is, do macromolecules which may encounter a given cellular system find their way to the lysosome directly; is the process random or organized in some specific way? Relatively few studies have been made in this area, but the following data may well be representative (Aronson and Davidson, 1968).

TABLE VI

ASSIMILATED RADIOACTIVITY IN VARIOUS SUBCELLULAR FRACTIONS[a]

Fraction	Chondroitin-^{35}S 4-sulfate (% of total in liver)
Mitochondria	0.9
Nuclei	10.1
Lysosomes	68.3
Microsomes	3.6
Supernatant	17.1

[a] 10^5 cpm of chondroitin-^{35}S 4-sulfate injected intravenously 15 minutes prior to sacrifice. Liver fractions were isolated as described by Aronson and Davidson (1968). Approximately 1% of the injected dose was recovered.

Intravenous injection of rats with radioactively labeled chondroitin 4-sulfate results in the rapid uptake of the label by liver lysosomes—approximately 1% of the injected dose can be isolated from the lysosomes within 15 minutes. When the various subcellular fractions are isolated from liver and the uptake specificity examined, it is found that virtually all of the assimilated radioactivity is present in the lysosomal fraction with negligible amounts found in the nuclear, mitochondrial, and microsomal pellets (see Table VI). These data imply a very specific relationship between the uptake process of the liver cell and the nature of the lysosomal membrane and possibly, the origin of the lysosomal particles themselves. It can be visualized that the initial uptake mechanism may be endocytotic, resulting in the formation of an organelle with its outer membrane derived from the plasma membrane of the cell. The fusion of this particle with a lysosome may occur through a type of membrane recognition phenomenon built into the architecture of the membrane structures. A similar recognition may not be present, for example, for nuclear or mitochondrial membranes. Once such a fusion has taken place,

the degradative processes can then proceed. This is illustrated schematically in Fig. 6. The same study also demonstrated that after 96 hours,
polysaccharides which had degradative enzymes present in the lysosomes
could no longer be detected; on the other hand, the lysosome does not
appear to contain the requisite degradative enzymes for dermatan sulfate

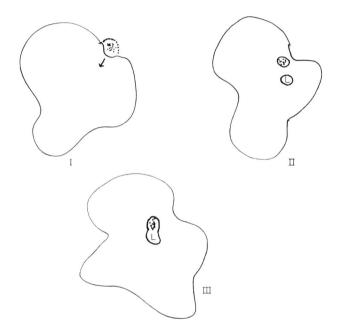

FIG. 6.   Schematic representation of endocytosis leading to the formation of a
vacuole containing entrapped macromolecules derived from the extracellular space.
Fusion with a primary lysosome (L) yields a particle in which degradation may
proceed.

and it was still present after this extended time period. Thus, it would
appear that there is some specificity in terms of the uptake by subcellular
organelles of macromolecular substrates present in the extracellular space.
However, the somewhat larger problem of specificity for cell type has not
as yet received any serious study. The accessibility of a given cell to an
extracellular substrate will certainly be reflected in any study of this
nature, and it is reasonable to expect that this would vary widely from
tissue to tissue. Liver, spleen, and kidney which are highly cellular,
highly vascular tissues, would be expected to participate in continual
clearing of degradable macromolecules from the extracellular compart-

ments of these tissues. However, connective tissues, cartilage, dermis, and bone which are relatively acellular and avascular may participate in such a process but would be expected to do so at a much lower level. There are as yet no data on this particular point.

C. Enzyme Types and Action

1. Glycosidases

As indicated above, an enormous variety of hydrolytic activities are present in the lysosome. In general, these are exclusively of the exo na-

TABLE VII
GLYCOSIDASES IDENTIFIED IN LYSOSOMES

Glycosidase	Reference
β-Xylosidase	Robinson and Abrahams (1967)
α-Glucosidase	Lejeune et al. (1963)
β-Glucosidase	Beck and Tappel (1968)
β-Galactosidase	Sellinger et al. (1960)
α-Mannosidase	Sellinger et al. (1960)
α-N-Acetyl glucosaminidase	Weissmann and Friederici (1966)
β-N-Acetyl glucosaminidase	Sellinger et al. (1960)
α-N-Acetyl galactosaminidase	Weissmann et al. (1967)
β-N-Acetyl galactosaminidase	Woolen et al. (1961)
β-Glucuronidase	de Duve et al. (1955)
α-Fucosidase	Fisher et al. (1966)
N-Acetyl neuraminidase	Mahadevan et al. (1967)
Hyaluronidase	Aronson and Davidson (1967)
Lysozyme	Shibko and Tappel (1965)

ture; that is, the enzymes act only on terminal nonreducing carbohydrate moieties, and to this date only one endoglycosidase, hyaluronidase, has been demonstrated. A partial listing of these glycosidases is given in Table VII. Several of these deserve more detailed commentary.

a. N-Acetyl Neuraminidase. N-Acetyl neuraminidase is, as has been mentioned, a prerequisite to degradative activities on any N-acetyl neuraminic acid containing glycoprotein substrate. The N-acetyl neuraminidase that has been described in lysosomal preparations appears to be unusually labile after solubilization, and this may account for the apparent failure of lysosomal preparations to extensively digest typical glycoprotein substrates such as the α-1-acid glycoprotein (Mahadevan et al., 1967). However, when freshly prepared lysosomal preparations are employed in such studies and repeated additions of the enzymes are made, it is found that nearly complete degradation can be achieved (Aronson and deDuve, 1968). It is also possible, although not yet documented by complete studies, that some linkages, particularly those which may be involved in branched substrates, may be resistant to glycosidase action.

  b. α-L-*Fucosidase*. α-L-Fucosidase is of similar interest to N-acetyl neuraminidase since L-fucose generally also occupies terminal nonreducing ends in those glycoproteins in which it is found. There is relatively little information about the properties of the lysosomal fucosidase, but it also appears to be quite labile, losing nearly 70% of its activity in 24 hours.

$$2 \text{ GA-GNAc-(GA-GNAc)}_n\text{-GA-GNAc}$$

$$\downarrow$$

$$\text{GA-GNAc-GA-GNAc)}_{n+1}\text{-GA-GNAc}$$

$$\text{(GA-GNAc)}_n\text{-GA-GNAc, etc.}$$

FIG. 7. Transglycosidation reaction catalyzed by testicular or lysosomal hyaluronidase. Tetrasaccharide ($n=1$) is inactive and accumulates as the major end product. A small amount of disaccharide is also formed as a result of hexasaccharide cleavage. The affinity of the enzyme increases with increasing n but transfer reactions with high molecular weight substrates ($n > 6$) have not been studied in detail.

  c. *Hyaluronidase*. The presence of an enzyme active against hyaluronic acid and chondroitin 4- and 6-sulfates in extracts of testis has been known for some considerable time. Presumably the testis enzyme is involved to some degree in the fertilization process, but its exact role is unknown. In addition to the testis enzyme, there is present in liver and other lysosomal preparations a hyaluronidase with remarkably similar specificity and action pattern. The existence of this enzyme was first suggested by Bollet *et al.* (1963) who observed that extensive incubation of hyaluronic acid with liver preparations resulted in release of N-acetyl hexosamine end groups which he suspected were due to the action of a liver hyaluronidase. More extensive recent studies have demonstrated that the hyaluronidase present in liver is associated with the lysosomal fraction and has characteristics very similar to those of the testis enzyme, including the ability to catalyze transglycosylation reactions (Aronson and Davidson, 1968) (see Fig. 7). Thus, the digestion of hyaluronic acid by liver lysosomes could be expected to proceed in the following manner (Fig. 8).

The endoglycosidase acting on the macromolecular substrate would yield cleavage products of even-numbered oligosaccharides ranging from tetra- to perhaps decasaccharide or higher. As these products are released and the lower numbered oligosaccharides accumulate, the exo-enzymes, β-glucuronidase, and N-acetyl-β-glucosaminidase, can act

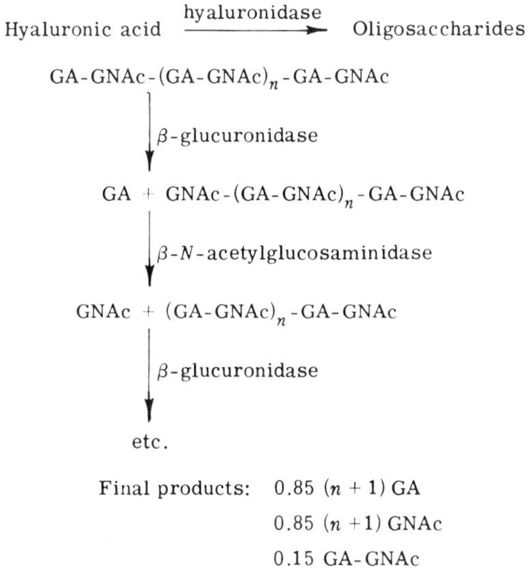

$$
\text{Hyaluronic acid} \xrightarrow{\text{hyaluronidase}} \text{Oligosaccharides}
$$

$$
\text{GA-GNAc-(GA-GNAc)}_n\text{-GA-GNAc}
$$

$$
\downarrow \beta\text{-glucuronidase}
$$

$$
\text{GA} + \text{GNAc-(GA-GNAc)}_n\text{-GA-GNAc}
$$

$$
\downarrow \beta\text{-}N\text{-acetylglucosaminidase}
$$

$$
\text{GNAc} + \text{(GA-GNAc)}_n\text{-GA-GNAc}
$$

$$
\downarrow \beta\text{-glucuronidase}
$$

$$
\text{etc.}
$$

Final products: 0.85 $(n + 1)$ GA

0.85 $(n + 1)$ GNAc

0.15 GA-GNAc

FIG. 8.   Hyaluronic acid degradation by lysosomal enzymes.

sequentially from the terminal nonreducing end of the saccharides to yield glucuronic acid, N-acetyl glucosamine, and shorter saccharide units. The digestion apparently reaches completion when approximately 85% of the sugar residues initially present have been liberated as monosaccharides and the rest accumulated as disaccharide units (Aronson and deDuve, 1968). It is of interest that the repeating disaccharide N-acetyl hyalobiuronic acid appears to be resistant to the action of the lysosomal β-glucuronidase.

d. General Features of Cellular Glycosidases.   It should be emphasized at this point that the majority of the lysosomal enzymes have not been purified to a significant degree primarily because they are very similar in their fractionation characteristics, but also because they occur in relatively small amounts and the processing of unusually large amounts of tissue would be required to obtain any one of these enzymes in reasonable quantities. Nevertheless, the stereospecificity of all of these seems to be sufficiently high so that studies carried out with mixed enzyme preparations are indicative of the properties of any one of the components.

TABLE VIII

DIGESTION OF GLYCOPROTEINS BY LYSOSOMAL ENZYMES

Substrate	% Carbohydrate released (24 hours)
Fetuin	8
Fetuin glycopeptides	18
Ovalbumin glycopeptides	60
Submaxillary mucin	80
Orosomucoid[a]	40

[a] Nearly 70% of the N-acetyl neuraminic acid and 25% of the hexosamine are released with smaller amounts of galactose and mannose.

The action of lysosomal glycosidases *en masse* on both glycoprotein and mucopolysaccharide substrates has been studied by several laboratories (Mahadevan *et al.*, 1969; Langley, 1968). In general, two types of measurements have been made, reflecting either the extent of release of the carbohydrate portion of the substrates or the extent of proteolytic breakdown achieved by the cathepsins present in the lysosomes. The latter subject will be discussed subsequently. Both natural substrates and substrates in which the terminal N-acetyl neuraminic acid residues have been removed by mild acid hydrolysis have been examined. Table VIII summarizes the results achieved by incubation of freshly prepared lysosomes with a variety of glycoproteins and also with glycopeptides derived from ovalbumin by extensive prior proteolytic digestion and fractionation. It can be noted that those sugar residues which appear to be peripheral are nearly quantitatively released from the glycoprotein substrates while residues which may occur closer to the peptide core or in a branched sequence may be appreciably more resistant. The lack of extensive release of inner core saccharides may also reflect the concomitant proteolytic digestion occurring under these circumstances since the action of glycosidases may be restricted when, for example, the amino acid involved in linkage to the carbohydrate has either a free amino or a free carboxyl terminus. In this connection, it should be pointed out that xylosyl serine itself is resistant to the action of the β-xylosidase present in lysosomes whereas the same residue is cleaved when the serine is part of a larger peptide unit. In some earlier experiments, the extreme lability of the lysosomal N-acetyl neuraminidase was not recognized; and the failure of such preparations to carry out extensive degradation of substrates such as the α-1-acid glycoprotein was incompletely understood. However, the use of freshly prepared material and, if necessary, repeated additions of enzyme, have overcome this experimental problem.

The time course of carbohydrate release has also been examined, at least in part, to obtain information about the arrangement of saccharide residues peripheral to the polypeptide chain. Information obtained from such experiments is at best suggestive since the rates of cleavage of individual monosaccharide units may vary according to the specificity of the enzyme involved and its absolute content in a given preparation. Concurrent studies of the effect of proteolysis on the release of carbohydrates from glycoprotein substrates suggest that some proteolytic digestion or denaturation promotes carbohydrate release; whereas, as mentioned above, extensive proteolysis may result in peptides still containing linked carbohydrates which are resistant to further degradation.

As is common with most of the lysosomal enzymes, the pH optimum for saccharide release is in the neighborhood of 4.0. This will vary somewhat, depending on the particular enzymes involved since, for example, the lysosomal β-glucosidase has a pH optimum closer to 5. Because almost none of the enzymes has been purified to homogeneity, certain aspects of their substrate specificity remain undetermined. In particular, the β-xylosidase activity observed in such preparations may be due to the broader specificity of the β-glucosidase and not to the presence of a separate enzyme; this point remains to be clarified. Furthermore, the specificity of these preparations with regard to hydrolysis of sugars involved in branch-point linkages has also not been thoroughly documented. Preliminary results have indicated that certain residues, particularly galactose and mannose units, appear to be resistant when part of a branched chain, but the configuration and linkage specificity of this apparent resistance is not known.

Studies comparable to those with hyaluronidase (e.g., Section III, C, 1,c) on the digestion of chondroitin 4- and 6-sulfates, which are also substrates for the lysosomal hyaluronidase, have been carried out but not quite as extensively. Initial cleavage of the sulfated polysaccharides will follow a pattern very similar to that of hyaluronic acid, yielding intermediate-sized oligosaccharides varying from tetra to decasaccharides, each with a terminal N-acetyl galactosamine sulfate on the reducing end and a D-glucuronic acid residue on the nonreducing end. The oligosaccharides thus liberated can be degraded by β-glucuronidase to yield odd-numbered oligosaccharides, but owing to the presence of the sulfate ester group, the hexosaminidic linkage is not cleaved by the galacto-saminidase that is present. A recent report has documented the presence in lysosomes of a sulfatase which can act on a terminal nonreducing N-acetyl galactosamine 4-O-sulfate residue, liberating inorganic sulfate and thus making the terminal saccharide unit subject to further degradation

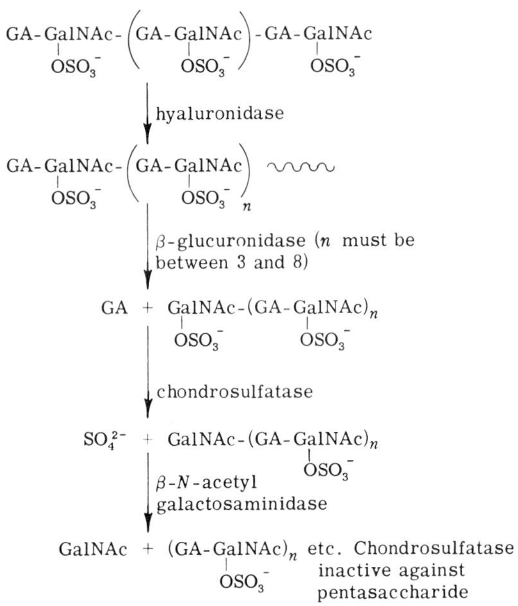

$$GA\text{-}GalNAc\text{-}\left(\underset{OSO_3^-}{GA\text{-}GalNAc}\right)\text{-}\underset{OSO_3^-}{GA\text{-}GalNAc}$$
$$\underset{OSO_3^-}{\big|}$$

$\downarrow$ hyaluronidase

$$GA\text{-}GalNAc\text{-}\left(\underset{OSO_3^-}{GA\text{-}GalNAc}\right)_n \sim\!\sim\!\sim$$
$$\underset{OSO_3^-}{\big|}$$

$\downarrow$ β-glucuronidase (n must be between 3 and 8)

$$GA \;+\; \underset{OSO_3^-}{GalNAc}\text{-}\underset{OSO_3^-}{(GA\text{-}GalNAc)_n}$$

$\downarrow$ chondrosulfatase

$$SO_4^{2-} \;+\; GalNAc\text{-}\underset{OSO_3^-}{(GA\text{-}GalNAc)_n}$$

$\downarrow$ β-N-acetyl galactosaminidase

$$GalNAc \;+\; \underset{OSO_3^-}{(GA\text{-}GalNAc)_n} \;\text{etc. Chondrosulfatase}$$
inactive against pentasaccharide

FIG. 9.   Chondroitin 4-sulfate degradation by lysosomal enzymes.

by glycosidase action. The substrate specificity of the sulfatase is such that prior removal of the glucuronic acid residue is necessary for the sulfatase to act, and the action is effective on products at the hexa- or octasaccharide level but not on tetrasaccharide and not on polymeric material (Tudball and Davidson, 1969). This entire degradative scheme is summarized schematically in Fig. 9.

Mucopolysaccharide substrates such as keratan sulfate, dermatan sulfate, and heparitin sulfate apparently are resistant to the action of the lysosomal glycosidases—at least to those known to be present in liver or kidney. All of these substrates are hyaluronidase resistant as well, and it is possible that the failure to demonstrate their degradation may reflect either a true lack of the appropriate enzyme or inappropriate experimental conditions. It should be noted that keratan sulfate has a rather low biological turnover, and higher animals appear to be in a positive balance throughout their life span—that is, keratan sulfate accumulates in progressively larger quantities with increasing age. Dermatan sulfate, however, especially in skin, where the most extensive studies have been carried out, appears to be rather active metabolically but little is known about enzymes involved in its degradation. Neither endo- nor exoglycosidases which act on dermatan sulfate or its oligosaccharides have been described. It should be noted, however, that extensive studies have not been carried out with skin lysosomes which would appear to be the most likely source for such enzymes in view of the distribution of the polymer and the documented active turnover in this tissue. Relatively little is known about the biological half-lives of heparitin sulfate or heparin.

e. Other Polysaccharide Hydrolases. In a number of studies that have been conducted, the presence of specific glycosidases has been inferred from the ability of various preparations to cleave synthetic glycoside substrates, most frequently *p*-nitrophenyl or *o*-nitrophenyl glycosides. This may give a particularly misleading conclusion regarding the ability of these enzymes to act on macromolecular substrates and, more specifically, on glycopeptides. It does not seem likely that a large family of β-galactosidases might exist, each of which has a different specificity for linkage configuration and position. Nevertheless, this possibility has not been excluded, and the presence or absence of activities against synthetic substrates should not be interpreted as reflecting the ability of the enzymes to act on natural polymers. In view of the known capacity of lysosomes to accumulate macromolecular substrates, whether by endocytosis or autophagy, and in view of the failure in most normal systems for such substrates to progressively accumulate within these particles, it seems likely that such organelles play a significant physiological role in the degradation of entrapped substrates. A natural corollary of this conclusion is that the internal pH of these organelles must permit effective hydrolytic action. This may require either active proton transport by the lysosomal membrane or may be the result of a Donnan equilibrium effect due to the storage of acid mucopolysaccharides within the lysosome. In any event, there is an obvious advantage to the cell in having such hydrolytic activities regulated by a single organelle with a pH–activity relationship that makes these enzymes normally inactive in the cytoplasm.

2. Cathepsins

It has been known for a number of years that lysosomal preparations are capable of carrying out extensive proteolysis, and several specific proteolytic activities have been demonstrated in such preparations. In addition to the activities associated with these organelles, several neutral proteinases and peptidases have been reported; but much less documentation is available regarding their specificity and action pattern. The naturally acid pH which appears to prevail within the lysosomal particle may facilitate digestion of specific substrates by providing conditions under which one or another protein may be denatured sufficiently to render it susceptible to a lysosomal attack. Clearly, the enzymes present within the lysosome do not digest one another, and one would therefore expect that proteins which are stable under acid pH conditions would not be attacked by this system.

Cathepsins are a class of proteolytic enzymes obtained from animal tissues which, because of their tissue occurrence, can be distinguished from those enzymes, such as pepsin and trypsin, which are secreted into the digestive tract. To this date, none of these enzymes have been obtained in a satisfactory state of homogeneity so that their exact properties can be satisfactorily defined, and the classification of the various cathepsins is frequently based on their activity toward artificial substrates. Nevertheless, in view of the acid pH optimum of the majority of the described catheptic activities and their association with lysosomal particles, it is not an unreasonable conclusion that the physiological function of these enzymes is akin to the other acid hydrolases found in the lysosome and thus is associated with normal cellular degradative processes. The cathepsins appear to be ubiquitous in their distribution among animal tissues and have been found not only in higher animals but also in tissues of lower vertebrates and invertebrates and may be considered a component of most animal cells.

Four different cathepsins have been described and the distinction between these is as follows:

The assay for cathepsin A depends on the ability of this enzyme to hydrolyze carbobenzoxy-L-glutamyl-L-tyrosine at pH 5 without the requirement for prior or concomitant activation by sulfhydryl compounds. The assay for cathepsin B is based on the ability to deamidate benzoyl-L-argininamide also at an acid pH, but, in this case, activation by sulfhydryl compounds is necessary. Cathepsin C deamidates glycyl-L-tyrosinamide and is also SH activated. The assays for cathepsin D are frequently based on proteolysis of hemoglobin which has previously been denatured, and this enzyme is distinguished from the other cathepsins primarily in reactivity toward SH compounds and the failure of cathepsin D to cleave either of the substrates routinely employed for demonstration of cathepsin B or C activity.

In general, the ability of lysosomal preparations to degrade protein substrates will be influenced by the susceptibility of such substrates to denaturation under acid pH conditions. Those proteins which are readily denatured at pH values near 4 are undergoing rapid catheptic proteolysis by lysosomal enzymes. Examples include globin and serum albumin. However, ferritin or invertase, for example, appear to be resistant to proteolysis in the native state and are stable under the acid pH conditions. More complex substrates, such as those containing covalently bound carbohydrate, will be acted upon to the degree that the carbohydrate substituents do not interfere with the action of the proteolytic enzymes. Since the action pattern of most cathepsins appears to involve either basic residues such as arginine or aromatic residues and since neither of these groups is involved directly in covalent linkage with carbohydrate units, fairly extensive digestion of glycoproteins may be expected to take place. However, in at least one case, it has been documented that mucins which contain a high degree of substitution by N-acetyl-neuraminic acid residues may be resistant to extensive proteolytic cleavage until the bulk of the anionic groups are removed by prior neuraminidase action. A detailed study on the end products of the lysosomal digestion of typical protein substrates has indicated that hydrolysis to small peptides and amino acids takes place (Coffey and deDuve, 1968). Thus, nearly 50% of the protein substrate appeared as free amino acid after exposure to the lysosomal complement of enzymes, indicating the presence of both endo- and exopeptidases, possibly, in addition to the cathepsins previously described. To this date, very little detailed work has been reported that is concerned with the exact nature and specificity of these enzymes.

In addition to the more apparent catheptic activities, persistent reports have appeared in the literature of the capability of other cell organelles to catalyze protein hydrolysis (Marks and Lajtha, 1963). In particular, the presence of neutral proteinases associated with the mitochondrial fraction has received significant attention. However, to this date, no activity associated with these fractions has been purified sufficiently so that detailed studies can be made. It seems reasonable to conclude, in view of the uptake studies discussed previously, that should such activity exist in the mitochondrial fraction, for example, it probably is of small significance in relation to the ability of the cells to hydrolyze glycoprotein or mucopolysaccharide substrates since such substrates will never penetrate the mitochondrial membrane and thus be exposed to these degradative activities.

IV. Unsolved Problems

There seems little doubt, in view of the accumulated evidence, that the lysosomal fraction of all animal cells possesses the requisite enzyme complement to cause extensive degradation of most, if not all, glycoprotein substrates. In addition, those mucins whose detailed structure is chemically known also appear to be susceptible to enzymes known to be associated with lysosomes. However, in spite of the fact that a number of the acid polysaccharides have been shown by turnover studies to be metabolically active, there still have not been described activities, whether of lysosomal or of other origin, which are capable of degrading substrates such as dermatan sulfate or keratan sulfate. It is noteworthy that in a number of pathological conditions excessive accumulation of acid polysaccharides appears to occur within specific cell types. In particular, storage of such materials may take place within the lysosomes, as is evidenced in the case of Hurler's disease in which at least one aspect of the defect appears to be associated with the failure of the normal cellular processes to degrade the polysaccharides rather than a loss in the regulation of their biosynthesis (Fratantoni *et al.*, 1968). It is possible that a number of other pathological conditions reflect the lack of a specific degradative enzyme normally associated with lysosomes.

REFERENCES

Aronson, N. N., and Davidson, E. A. (1967). *J. Biol. Chem.* **242**, 437.
Aronson, N. N., and Davidson, E. A. (1968). *J. Biol. Chem.* **243**, 4494.
Aronson, N. N., and de Duve, C. (1968). *J. Biol. Chem.* **243**, 4564.
Beck, C., and Tappel, A. (1968). *Biochim. Biophys. Acta* **151**, 159.
Bollet, A. J., Bonner, W. M., and Nance, J. L. (1963). *J. Biol. Chem.* **238**, 3522.
Clamp, J. R., and Putnam, F. H. (1964). *J. Biol. Chem.* **239**, 3233.
Coffey, J. W., and de Duve, C. (1968). *J. Biol. Chem.* **243**, 3255.
Davidson, E. A., and Meyer, K. (1954). *J. Biol. Chem.* **211**, 605.
de Duve, C., and Wattiaux, R. (1966). *Ann. Rev. Physiol.* **28**, 435.
de Duve, C., Pressman, B. C., Gianetto, R., Wattiaux, R., and Applemans, F. (1955). *Biochem. J.* **60**, 604.
Fisher, D., Higham, M., Kent, P. W., and Pritchard, P. (1966). *Biochem. J.* **98**, 46P.
Fratantoni, J. C., Hall, C. W., and Neufeld, E. F. (1968). *Science* **162**, 570.
Hranisavljevic, J., and Davidson, E. A. (1970). Unpublished observations.
Jamieson, G. A. (1965). *J. Biol. Chem.* **240**, 2914.
Kamiyama, S., and Schmid, K. (1962). *Biochim. Biophys. Acta* **58**, 80.
Katsura, N., and Davidson, E. A. (1963). *Biochim. Biophys. Acta* **69**, 453.
Katsura, N., and Davidson, E. A. (1969). *Biochim. Biophys. Acta* **184**, 503.
Kean, E. L., and Roseman, S. (1966). *J. Biol. Chem.* **241**, 5643.
Langley, T. J. (1968). *Arch. Biochem. Biophys.* **128**, 304.
Lejeune, N., Thines-Sempoux, D., and Hers, H. G. (1963). *Biochem. J.* **86**, 16.
Mahadevan, S., Nduaguba, J. C., and Tappel, A. L. (1967). *J. Biol. Chem.* **242**, 4409.
Mahadevan, S., Dillard, C. J., and Tappel, A. L. (1969). *Arch. Biochem. Biophys.* **129**, 525.
Marks, N., and Lajtha, A. (1963). *Biochem. J.* **89**, 438.

Marler, E., and Davidson, E. A. (1965). *Proc. Natl. Acad. Sci. U.S.* **54**, 648.

Nuenke, R. H., and Cunningham, L. W. (1961). *J. Biol. Chem.* **236**, 2452.

Plummer, T. H., and Hirs, C. H. W. (1964). *J. Biol. Chem.* **239**, 2530.

Robinson, D., and Abrahams, H. E. (1967). *Biochim. Biophys. Acta* **132**, 212.

Roden, L., and Smith, R. (1966). *J. Biol. Chem.* **241**, 5949.

Roseman, S. (1962). *Federation Proc.* **21**, 1075.

Satake, M., Okuyama, T., Ishihara, K., and Schmid, K. (1965). *Biochem. J.* **95**, 749.

Schiller, S., Mathews, M. B., Jefferson, H., Ludowieg, J., and Dorfman, A. (1954). *J. Biol. Chem.* **211**, 717.

Schmid, K., Binette, J. P., Kamiyama, S., Pfister, V., and Takahashi, S. (1962). *Biochemistry* **1**, 959.

Sellinger, O. Z., Beaufay, H., Jacques, P., Doyen, A., and de Duve, C. (1960). *Biochem. J.* **74**, 450.

Shibko, S., and Tappel, A. (1965). *Biochem. J.* **95**, 731.

Spiro, R. G. (1962). *J. Biol. Chem.* **237**, 382.

Tanaka, K., Bertalini, M., and Pigman, W. (1964). *Biochem. Biophys. Res. Commun.* **16**, 404.

Trouet, A. (1966). *Arch. Intern. Physiol. Biochim.* **72**, 698.

Tudball, N., and Davidson, E. A. (1969). *Biochim. Biophys. Acta* **171**, 113.

Weissmann, G. (1965). *New Engl. J. Med.* **273**, 1084.

Weissmann, B., and Friederici, D. (1966). *Biochim. Biophys. Acta* **117**, 496.

Weissmann, B., Rowin, G., Marshall, J. M., and Friederici, D. (1967). *Biochemistry* **6**, 207.

Woolen, J. W., Walker, P. G., and Heyworth, R. (1961). *Biochem. J.* **79**, 294.

HYDROLASES AND CELLULAR DEATH

J. L. VAN LANCKER[1]

O, that this too too solid flesh would melt,
Thaw, and resolve itself into a dew!

Hamlet (I, ii, 129–130)

[1] The research done in this laboratory was supported by grants of the USPHS AM-06556 and USPHS GM-14395 and a grant from the American Cancer Society E-411.

I. Introduction

A. ANABOLISM AND CATABOLISM

Nature has built within itself cycles of life and death. Outside of the tropics, every autumn almost every leaf dies slowly and it is no wonder that in approaching death men refer to their old age as the winter of their life. The growth and maintenance of the animal organism results from the delicate balance of cellular proliferation, maturation, and elimination. During embryogenesis (Saunders and Fallon, 1966) millions of cells are killed or commit suicide in preselected areas. After elimination, the dead cells are replaced by other cells, often of a different type.

Morphogenesis of bone tissue results from repeated destruction of preexisting histological structures with replacement by new ones. The red cells, the polymorphonuclears, and the lymphocytes of the blood all have limited life spans, yet the total circulating number of these cells remains constant. Superficial layers of skin and mucosae are shed periodically but never are surfaces denuded in the healthy organism. These few examples suggest that cellular death is as important to organismal health and survival as cellular proliferation. Moreover, it is likely that the very life of cells depends upon the balanced elimination and replacement of intracellular organelles. Consequently, in normal cells one finds side by side catabolic and anabolic enzyme systems. However, the mysteries of the segregation and the delicate regulation of catabolic and anabolic activities continue to elude us in spite of much research in the field.

Many of the catabolic enzymes are hydrolases. Our catalog of the hydrolases found in mammals is not likely to be complete, yet an impressive number of mammalian hydrolases have been identified with respect to their catalytic properties, their specificity, their function, etc. Some hydrolases function inside the cell (e.g., cathepsins), some are excreted (pepsin, chemotrypsin). Some act on small molecules, others attack the ends (exopeptidases, exonucleases), others the inside of large molecules (endopeptidases, endonucleases).

The specificity of the enzyme may be broad or it may be very restricted. Conditions for activity *in vitro* vary considerably even for enzymes which act on identical substrates. Some deoxyribonucleases require magnesium for activity, others do not. Some ribonucleases, deoxyribonucleases or proteases, etc., function optimally at basic or neutral pH and others at acidic pH. Thus the cell has stored within it a vast and varied arsenal of weapons that can be triggered into action in many situations compatible with life and possibly after death. To protect itself against reckless destruction of its building blocks, its chemical sources of energy, its multimolecular bioenergetic pathways, its replicating, transcribing, and translating macromolecular systems, its delicate array of membranes with selective permeabilities, the cell often muzzles the activity of the hydrolases.

In some cases the mechanism by which the cell is protected against hydrolytic activity is known at least in part. At present, three different types of protection against the breakdown of specific substrates by their respective hydrolases can be distinguished. We shall call them intramolecular, intermolecular, and subcellular protection. The proteases of pancreas and blood provide remarkable examples of "intramolecular protection." In pancreas, chymotrypsin and trypsin exist in the form of "zymogen" (Desnuelle, 1960). During their elaboration on ribosomes, the enzymes are made longer by a few amino acids. The lengthened chain is inactive (for reasons that cannot be discussed here) and it is only after splitting off the protective polypeptide segments that enzymes and substrate interact (Desnuelle, 1960; Ottesen, 1967).

Intermolecular types of protection against hydrolase activity can be of various kinds, including *strapping* of the active center on membranes, *masking* of the active center by inhibitors, and *emasculation* of the active center because of the absence of activators. One ATPase is believed to be maintained inactive by virtue of its attachment to other protein through S—S bonds and it is only after reduction of the S—S bonds that the enzyme splits ATP. Protein synthesis would be impossible if messenger and ribosomal RNA were not constantly protected against the hydrolytic action of the almost ubiquitous ribonuclease. Nothing is known of the regulation of ribonuclease activity before, during, or after protein synthesis. Whether or not the inhibitors (Roth and Bachmurski, 1957; Roth, 1956a,b, 1958a) of ribonuclease play a role in this regulation is not certain, but their presence is indicative.

Similarly, DNA synthesis is self-defeating if it must take place in presence of the numerous DNase's found in different cell fractions and functioning optimally at different pH's (Dabrowska *et al.*, 1949; Cooper *et al.*, 1950; Cruz-Coke *et al.*, 1951; Henstell *et al.*, 1952; Kurnick *et al.*, 1953). Again there is not conclusive evidence that inhibitors protect DNA against the hydrolytic action of DNase's but it is relevant that a DNase inhibitor (a protein other than histones) was found and partially purified from the nucleus (Van Lancker and Holtzer, 1964).

The mode of protection against hydrolytic activity that has been known for the longest time is the subcellular. Pancreatic enzymes that are made in the acinar cells for export are enclosed within the boundaries of the membrane of the zymogen granule which is excreted into the pancreatic ducts, from there into the intestines where it soon ruptures. As a result, enzymes and zymogens are released and activated. Pancreatic hydrolases which are elaborated for excretion are believed to be contained in solution within the zymogen granule; but it is likely that the relationship between pancreatic hydrolases and zymogen granules is much more complex than just described, because portions of the enzyme activities cannot readily be released from their granular support (Holtzer *et al.*, 1963).

That structural segregation of the enzymes protects the cell, however, is illustrated in the lipid necrosis that takes place when lipases are freed (H. Busch, 1959). Pathologists have long been familiar with the unusual types of necrosis that follow the release of pancreatic lipase.

B. INJURY, RESPONSE TO INJURY, AND REVERSIBILITY OF INJURIES

The duration of the three stages of cell life, proliferation, maturation, and elimination, varies with the cell type. In some cells, such as lymphocytes, the full life span is less than a day, for others (red cells) it is several months long. Some tissues, such as liver, have few dividing cells in the absence of special stimuli, yet the hepatic cell maintains its proliferating ability and after partial hepatectomy the cell divides rapidly. Except for glia, most of the cells found in the central nervous system lose their ability after maturation to proliferate and therefore they are irreplaceable when eliminated.

Knowledge of the molecular events in mammalian cell proliferation and maturation has been reviewed (Gross, 1968; Van Lancker, 1969). Our understanding of the molecular sequence of changes is meager except for the fact that maturation probably results in part from selective transcription of the genome and that proliferation results from selective transcription and faithful and integral replication of the genome. Little is known of the structure of interreaction of the molecules involved in triggering, regulating, and stopping proliferation or maturation.

Although it seems likely that in some cases programmed progressive and irreversible repression of the genome is the prelude to cellular death (e.g., red cells, see below), it is at present impossible to describe the sequence of molecular events that lead to cellular death in as simple a situation as ischemia (Farber *et al.*, 1968), in spite of numerous attempts to study the mechanisms triggering the sequence and the point of irreversibility in the sequence of molecular events that lead to cellular death. It is important that the cause or causes of cellular death be identified because such knowledge is at the roots of our understanding of cellular pathology and may be the source of new therapeutic procedures. Whenever the internal milieu or the environment of the cell is so suddenly or so drastically modified that smooth adjustment of the intracellular events to the insult is impossible, the cell is said to be injured. Few insults, with the possible exception of fixation, stop all cellular reactions at once. Most injuries are compatible with at least temporary persistence of many if not most intracellular functions. However, as a result of injury, membrane permeabilities are modified, bioenergetic pathways are ineffective, mitosis is blocked, lipid or other intra- or extracellular material accumulates, and the delicate balance between catabolic and anabolic pathways is tilted one way or the other. Because the cell presents such changes after injury, it is often said to "react to injury." The expression "reaction to injury" implies that the changes that took place in the injured cell are not passive consequences of the insult but rather dynamic response to it either to secure survival or to accelerate death. Among responses to injury, one must distinguish primary from secondary response. Special injurious agents have affinities for specific molecules. For example, cyanide blocks the activity of cytochrome oxidase, and many antimetabolites interfere with the active site of specific enzymes, thereby preventing metabolism of normal substrates. These are primary injuries.

The primary biochemical insult is usually followed by a cascade of secondary metabolic alterations. A case in point is the mutation in the DNA molecules, which is responsible for the absence of uridyl transferase and thereby leads to the disease called galactosemia, which is believed to result in the accumulation of galactose 1-phosphate, an inhibitor of a number of enzymes. Inhibition of these enzymes could explain the development of hypoglycemia, jaundice, and cataracts and possibly the mental retardation that develops in galactosemic patients (Sidbury, 1961).

The primary injury is not always restricted to a single molecule, and it may produce damage in several molecular systems at the same time. For example, although there seems to be no doubt that much of the damage produced by ultraviolet light results from the formation of thymine dimers in the DNA molecule (Wacker, 1963), it cannot be excluded that the ultraviolet light also damages other macromolecular systems (Errera, 1968). Similarly X-irradiation of bacteria or mammalian tissues interferes with DNA synthesis. This interference is likely to result from the insult to the DNA molecule (Van Lancker, 1969). Yet because of the very nature of X-radiation, it seems unlikely that the primary injury affects only DNA (Errera, 1968) although it is quite possible that only damage to the DNA molecule is critical for cell survival (Van Lancker, 1969). In spite of the need for distinguishing between primary and secondary, and between critical and noncritical injuries, by "reaction to injury" pathologists mean all intracellular changes that take place following the insult. Yet in some cases it is obvious that the changes are an inevitable consequence of the primary molecular injury. The accumulation of glycogen in glycogenosis or of lipids after carbon tetrachloride intoxication is simply the inevitable consequence of the insult to other molecules. Although it is likely that purposeful reaction to injury exists, it is seldom if ever (with present methods of investigation) possible to distinguish dynamic from passive responses to injury. Therefore, whenever the expression "reaction to injury" is used in this paper, it should not be construed to mean a dynamic purposeful response, but simply a response that may be either a passive or a dynamic reaction to injury.

If programmed cellular death may be considered to be a part of the organism's life cycle, premature death of even a few cells can have drastic consequences for the entire organism. For example, the destruction of a few cells of the anterior horns in the spinal cord by the poliomyelitis virus paralyzes and sometimes kills the victim. The occlusion of relatively small vessels, the coronary or the cerebral arteries, is responsible for the destruction of a varied population of cells in heart and brain. The infarcts either kill the victims or make invalids of them.

Depending upon the nature of the injurious agent, a number of molecular events have been described. Such events include interference with transcription of the genome (La Farge et al., 1966; Van Lancker, 1966; Kirk, 1960), interference of translation of RNA templates into polypeptide chains (Gottlieb et al., 1964), changes in cellular permeability (King *et al.*, 1959; Judah *et al.*, 1964, 1965, 1966), uncoupling of oxidative phosphorylation (Fonnesu, 1960; Stoner and Threlfall, 1960), trapping of ATP (Farber *et al.*, 1964), lipid peroxidation (Recknagel, 1967), interference of the biosynthesis of specific biochemical pathways, such as cholesterol (Love *et al.*, 1965), block of the electron transport chain (Recknagel, 1967; Trump and Bulger, 1967, 1968; Malamed, 1966), and modification in the intracellular distribution or activity of hydrolases.

Even if the catalog of molecular alterations that follow injury were complete, it could provide only a partial explanation of the mechanism of cellular death. A comprehensive understanding of such mechanism or mechanisms should include at least three other different types of information: a complete description of the sequence of events in time; a clear distinction between events that are critical and those that are uncritical for cell survival; a clear distinction between those events that are the cause and those that are the consequence of cellular death.

Among the events that take place during the time interval between the application of the injurious agent and the disappearance of the cell as a result of cellular death, it is likely that only a few are directly responsible for cell death. Furthermore, it is not impossible that within the chain of causal events, a point can be identified which separates reversible from irreversible changes. Thus if the nefarious agent can be withdrawn before that crucial moment, the cell will survive. Consequently, identification of the point of no return is obviously of considerable importance in human pathology, if prevention of cellular death, for example, as a result of ischemia, is to be achieved. Accurate identification of the point of no return requires determining what takes place and when.

C. Necrosis, Necrobiosis, Necrocytosis, and Autodigestion

In Greek, the word νεχρος means dead body or dead. From it are derived the terms necrobiosis, necrocytosis, and necrosis. In this discussion all three terms will be used with specific meaning. Necrocytosis will be used to refer to the death of individual cells, necrobiosis to that of a large number of cells under physiological conditions, and necrosis to the death of a large number of cells under pathological conditions. By definition, the terms necrosis and necrobiosis exclude organismal death or that of excised organs. The term "autolysis" will be used in these two cases. Several authors have used the term "autophagia" to refer to small intracellular foci of self-digestion of the cytoplasm. Inasmuch as "autophagia" implies a dynamic process resulting from the movement of intracellular organelles and thereby prejudges the interpretation of available fact, we prefer the term autodigestion.

Because the cells which die in necrobiosis and necrosis are by definition surrounded by living cells, reaction of surrounding necrotic tissue should be anticipated. Among other events, these reactions involve vascular changes and cellular exudation at the site of necrosis or necrobiosis. We shall not be concerned with these reactions except in some indirect fashion, when phagocytosis of death cells by either macrophages or polymorphonuclears is discussed.

There are many situations in which the injury to the cell leaves most of its biosynthetic or bioenergetic machinery intact, but still causes severe disturbances which result, for example, in accumulation of excessive normal or abnormal metabolites. Intoxication with carbon tetrachloride or orotic acid (Novikoff et al., 1966) leads to the deposition of lipids in the liver cells (Lombardi, 1966; Smuckler, 1966; Dianzani et al., 1966; Magee, 1966). Cerebrosides (Fredrickson, 1966), gangliosides (Fredrickson and Trams, 1966), sphingomyelin (Fredrickson, 1966), mucopolysaccharides (Dorfman, 1960), and glycogen (Hers, 1964; Field, 1966) accumulate in cells in a number of inborn errors of metabolism. In absence of a better term, we will refer to all forms of cytoplasmic alterations resulting from injury (other than cytoplasmic autodigestion) by the words cytoplasmic degeneration. Admittedly, these definitions have no intrinsic merit or are not even new. Our only purpose in presenting them here is to prevent confusion.

In this discussion of the role of hydrolases in necrobiosis, autolysis, and cellular degeneration, we shall consider that role in the context of modern knowledge of cellular death. Such an attempt should help to determine whether the hydrolases participate in the triggering event or are among last steps of necrocytosis, and whether changes in hydrolase activity are part of the reversible or the irreversible sequence of events.

II. The Development of the Lysosomal Hypothesis

A. Biochemistry of Lysosomes

An interesting sequence of discoveries was at the origin of the notion of lysosomes. A painstaking tissue fractionation study by Novikoff et al. (1953) proved to be prophetic. Liver homogenates were separated into a large number of fractions by differential centrifugation and each fraction was assayed for activities of a number of enzymes. In this study the choice of enzymes was most remarkable. Among the enzymes selected by

Novikoff, acid phosphatase and uricase later proved to be good markers for at least two cytoplasmic structures unknown at the time. The study further revealed that differential centrifugation partially dissociated the activity of acid phosphatase and uricase from that of succinoxidase and cytochrome oxidase. The authors believed that these results simply reflected the heterogeneity of the population of traditional intracytoplasmic organelles, namely mitochondria and microsomes. Studies on homogenates and traditional mitochondrial fractions (which are now known to contain both mitochondria and lysosomes) had revealed an important property of the acid phosphatase bound to these particles (Berthet and de Duve, 1951; Appelmans and de Duve, 1955; Berthet et al., 1951). The activity of acid phosphatase of liver homogenate prepared at 0.25 M was found to increase when the osmolarity of the incubation medium was reduced. Although these findings were reminiscent of previous observations made with glutamic dehydrogenase (a mitochondrial enzyme), a comparison of the osmotic activation of glutamic dehydrogenase to that of acid phosphatase revealed marked differences. Acid phosphatase is gradually released between concentration of sucrose of 0.15 M and 0.005 M. Glutamic dehydrogenase remains associated with the organelle up to concentration of sucrose as low as 0.005 M and then is suddenly released when the sucrose concentration of the medium is further reduced.

An ingenious and simple method (Wattiaux and de Duve, 1956) for measuring enzyme activities bound to organelles versus that released from organelles was devised and the study on acid phosphatase was extended in three ways: (1) by attempting to release enzymes by methods other than osmotic shock; (2) by attempting to prepare cleaner preparations of lysosomes; and (3) by studying the release of a number of acid hydrolases other than acid phosphatase. The methods used to release the enzymes were rather drastic. They included the use of detergent, repeated freezing and thawing, shearing in the Waring blender, and incubation with lipase (Gianetto and de Duve, 1955; de Duve et al., 1953; Viala and Gianetto, 1955; MacFarlane and Datta, 1954; Beaufay, 1957) and it is, therefore, not surprising that under such conditions all acid hydrolases studied were released at similar rates. The test for free and bound activity of lysosomal enzymes was based on the assumption that the substrates of lysosomal enzymes do not enter intact lysosomes. More recent studies from Rosenberg and Janoff (1968) challenge these views in that they showed penetration of phenolphthalein glucuronide within the granules. If the latter experiments can be confirmed, not only will it be necessary to reconsider interpretation of early work on free and bound lysosomal enzyme, but the concept of a granule surrounded by a membrane impermeable to substrate must be reexamined. Instead of considering the lysosome as a bag containing hydrolytic enzyme that ruptures when needed, one would have to look at them rather as a sieve with selective permeabilities for the substrate.

Attempts from our laboratory to demonstrate a penetration and hydrolysis of substrate using phenolphthalein glucuronide in presence of liver lysosomes were unsuccessful (therefore unpublished). In contrast, the substrate and the product of rhodanese, another structurally latent enzyme possibly found in mitochondria, freely entered the granule, indicating that conversion of the substrate to the product occurs inside of the organelle.

Appelmans *et al.* (1955) worked out a new fractionation scheme which concentrated acid phosphatase and other hydrolases in a pellet with sedimentation properties intermediate between those of mitochondria and microsomes. However, the separation was far from complete. The new pellet contained only 40% of the total acid hydrolase activity.

Beaufay *et al.* (1958) applied the method of Kuff *et al.* (1957) of centrifugation in linear sucrose gradient to the study of the distribution of acid hydrolase and mitochondrial enzyme and established that the particles with which these two groups of enzymes were associated differed in their mean sedimentation constant (5000 S for acid phosphatase-containing organelles, 10,000 S for glutamic dehydrogenase and cytochrome oxidase-containing organelles). They calculated sizes of 0.4 µ for lysosomes and 0.8 µ for mitochondria. Thus there was a separation between the smaller lysosomes and the larger mitochondria but middle-sized particles of both sources overlapped broadly.

Results of these studies were confirmed by sedimentation of cytoplasmic organelles in heavy water (D_2O) gradients (de Duve, 1959). There was a sharp distribution of cytochrome oxidase and malic dehydrogenase between density values of 1.21–1.23, a broad distribution of acid hydrolase (between 1.21 and 1.27) with a small peak at 1.24 for acid phosphatase, cathepsin, and β-glucuronidase, and peaks at 1.25 and 1.26 for acid deoxyribonuclease and acid ribonuclease, respectively. A striking observation was the appearance of a new sharp peak associated with uricase activity, a finding that de Duve would ably exploit to demonstrate the existence of a separate class of organelles, the "peroxisomes," or microbodies (Baudhuin *et al.*, 1964; for review, see Hruban and Rechcigl, 1970).

The biochemical findings led to the establishment of the lysosomal concept (de Duve, 1964, 1965; de Reuck and Cameron, 1963) which proposed that the cell contains granules limited by a single lipoprotein membrane, and the granules hold only acid hydrolases in solution which are released under osmotic shock. The concept rests on a number of assumptions: (1) the granules contain no other enzymes than acid hydrolases; (2) acid hydrolases found in other cell fractions result from contamination by enzymes released from lysosomes during the preparation of the homogenate; (3) all hydrolases present in lysosomes are either in solution or attached in a similar loose fashion to the organelle ultrastructure; (4) the enzymes contained in the organelle become functional only after rupture of the membranes and release of enzyme in the medium. Although all four assumptions were only partially correct, they formed the basis of a working hypothesis which was at the source of innumerable studies on the role of hydrolases in physiological conditions.

When attempts are made to separate lysosomes in tissues other than liver, although latent acid phosphatase activity is always detectable, separation of granules containing acid phosphatase from traditional mitochondria and microsomal fractions is always incomplete. Fractions rich in acid hydrolases exhibiting structural latency have been prepared in pancreas (Van Lancker and Holtzer, 1959a), bone (Vaes, 1965a; Vaes and Jacquest, 1965; Vaes and Nichols, 1961, 1962), mammary gland (Slater, 1961a), kidney (Strauss, 1954, 1956, 1957, 1959, 1962a, 1964a,b), prostate (Harding and Samuels, 1961), polymorphonuclear (Cohn and Hirsch, 1960), and lymphoid tissue (Bowers and de Duve, 1967).

Van Lancker and Holtzer (1959b) studied the intracellular distribution of biochemical markers in cell fractions obtained from mouse pancreas. The homogenate was submitted to increasing centrifugal forces and pellets were prepared. Only the distribution of DNA and that of cytochrome oxidase satisfied requirements for adequate markers. An ideal marker for biochemical characterization of cell fractions should be stable under the conditions of the assay, and there should be no interference with its quantitative determination. If the marker is an enzyme, the conditions of the assay should measure all its potential activity and a high proportion of the constituents should be isolated in a single organelle. Amylase has a clear-cut bimodal distribution, one fraction being associated with the zymogen granules and another with the microsomes. The plot of the activity of acid phosphatase and deoxyribonuclease versus the centrifugal forces used to sediment each pellet yielded spectra of distribution even broader than those described by Novikoff in similar studies on liver (Novikoff et al., 1953).

Strauss isolated kidney droplets that appear in the brush border cells of the convoluted tubules by a combination of centrifugation and filtration, demonstrated that they were rich in acid hydrolases, and proposed that these granules also were lysosomes (Strauss, 1963).

Elegant studies of a group of investigators at the Rockefeller Institute (Cohn and Hirsch, 1960; Hirsch, 1965) established that clean preparation of polymorphonuclear and macrophage granules were rich in lysosomal enzymes. However, there are important differences between the preparation of polymorphonuclear granules and that of liver lysosomes isolated by de Duve. The polymorphonuclear preparations contain phagocytin, a bacteriocidic basic protein which kills both gram-positive and gram-negative microorganisms. A number of enzymes not usually found in the lysosomal pellets obtained from other tissues are associated with polymorphonuclear granules. Among them one should include lysozyme, peroxidase, alkaline phosphatase, lipase, and NADH oxidase. The stories behind the alkaline phosphatase, the lipase, and the NADH oxidase are of particular interest. Classic liver lysosomes were assumed to contain only acid hydrolases and, therefore, the finding of alkaline phosphatase in the polymorphonuclear granules was puzzling. Bainton and Farquhar (1968a,b) examined polymorphonuclear granules histochemically and after combination of histochemistry and electron microscopy, they claim that there are at least two types of granules in the polymorphonuclear: the lysosomelike granule, which contains acid phosphatase and other acid hydrolases, and a specific granule which contains alkaline phosphatase and lysozyme. Somewhat disturbing are the facts that the peroxidase is found in lysosomelike granules and that small amounts of acid phosphatase activity are associated with the special granules. Zeya and Spitznagel (1966, 1969) using zonal centrifugation separated from the lysosomelike and the special granule another type of granule, poor in enzyme, containing six cationic proteins that are not found in macrophages or in liver cells.

Evans and Karnovsky (1961) have demonstrated the presence of NADH oxidase in polymorphonuclear granules prepared from guinea pig lymphocyte and have implicated the enzyme in the phagocytic process. Woodin (1961) was unable to confirm Evans and Karnovsky's findings in rabbit leukocytes. Tappel *et al.* (1962) have claimed that they purified liver lysosomal preparations containing an NADH dehydrogenase which is different from the enzyme found in mitochondria, but which shares some of the properties of the microsomal enzyme. Unless it is conclusively established that the presence of NADH dehydrogenase in polymorphonuclear granules and liver lysosomes is the result of contamination of the cell fraction, either by mitochondria or endoplasmic reticulum, the lysosomal concept must be expanded to include the existence of granules containing NADH dehydrogenase. Although the presence of the above enzyme in polymorphonuclear granules is not at all surprising, the lipase has not been found in lysosomes from other sources.

Eisenberg et al. (1968) attempted to study the distribution of phosphatide acyl-hydrolase and lysophosphatide acyl-hydrolase activity in human and rat arteries. It is of interest that these two hydrolases which function at optimal pH's of 8 were not activated by Triton X-100, in contrast to sphingomyelin phosphoryl choline hydrolase, for which the presence of Triton X-100 in the incubation mixture was indispensable for activity.

It would serve no purpose to list all tissues in which the intracellular distribution of acid hydrolases has been studied by biochemical methods. The examples just described will serve to illustrate the problem.

The methods used by de Duve to release acid hydrolases from their granule support were drastic ones and, therefore, they could hardly discriminate the mode of attachment of various enzymes. When gentler methods were used to release the hydrolases, such as incubation at 0°C, selective enzyme release clearly took place (Walkinshaw et al., 1964). Selective enzyme release has also been demonstrated with the aid of detergent (Ugazio and Pani, 1963) and in some in vivo conditions (Reid and Stevens, 1958). Osmotic shock releases acid DNA selectively in rat kidney. Moreover, Nosman et al. (1966) have claimed that after X-radiation release of lysosomal enzyme is quite selective. In fact, the authors have divided the lysosomal enzymes into four categories with respect to their ability to be released after exposure to total body doses of X-radiation.

Structural latencies of lysosomal enzymes might not be the result of their enclosure within a bag surrounded by a lipoprotein membrane. Koenig (1962) and Koenig et al. (1964) proposed that latency is due to cationic binding to the structure of the granules. Not only would such a view go far in explaining selective release of the enzyme, but it would also relate to the mode of formation of lysosomal enzyme which seems to be derived from enzymes elaborated and tightly bound to the membranes of the endoplasmic reticulum.

Any attempt to interpret the role of lysosomes in necrosis or necrobiosis must take into account the possibility of the existence of enzymes in other cell fractions. Even in liver, which has yielded the best lysosomal preparations obtained from a parenchymal tissue, only small fractions of hydrolases are found in association with the lysosomal pellet, as indicated by a plot of hydrolase activity (in ordinate) versus the centrifugal forces (in abscissa). When such plots are made for cytochrome oxidase or glucose-6-phosphatase, a sharp curve for enzyme activities clearly appears in association with either mitochondrial or microsomal pellet. In contrast, the plot for acid phosphatase activity is broad and widespread and substantial amounts of acid hydrolase activities are associated with mitochondrial, nuclear, and microsomal pellet as well as with the cytosol. de Duve's method of plotting the results shows clearly that the activity of the acid hydrolase per milligram of nitrogen is highest in the lysosomal fraction but it does not exclude the possibility that at least part of the acid hydrolases found in other cell fractions are genuine.

The work of Walkinshaw and Van Lancker (1964), Van Lancker (1964), Van Lancker and Lentz (1970), as well as that of Fishman *et al.* (1967), Frieden *et al.* (1964), and Pettengill and Fishman (1962), leaves no doubt that at least a substantial amount of the β-glucuronidase activity found in the liver cell is genuine to the endoplasmic reticulum and the cytosol. The evidence for and significance of the association of the acid hydrolases with the endoplasmic reticulum will be discussed in the section devoted to the origin of the acid hydrolases that appear in areas of cytoplasmic autodigestion.

For the purpose of studying the intracellular distribution of glycolytic enzymes and especially their concentration in the nucleus, Siebert developed various methods of preparation of the nuclei, some of which yielded high concentration of acid hydrolases. Although it is simple to disregard these observations as an artifact of the preparation resulting from contamination by enzyme released from the lysosome, a careful reexamination of the problem is warranted. Substantial amounts of acid phosphatase, cathepsin, and β-glucuronidase were found in some of the nonaqueous preparation of nuclei free of mitochondria and uricase-bearing particles.

These results, however, are not in agreement with Chauveau *et al.* (1957) who found little contamination of the nuclear pellet. Studies of the effect of X-radiation and the release of spleen hydrolases from their granular support might be relevant (Fausto *et al.*, 1964). Six hours after the administration of 800 rads, the total activity of deoxyribonuclease and β-glucuronidase is unchanged, but the amount of enzyme associated with the cytosol is almost doubled. The increase in soluble enzyme is associated with a decrease in lysosomal enzyme but there is no change in the levels of β-glucuronidase and DNase in nuclear and microsomal pellets, suggesting that the mode of binding of the enzyme to these organelles might be different from their mode of binding to lysosomes. Such a conclusion, however, will not be justified until absorption of released enzymes has been excluded.

B. MORPHOLOGY OF LYSOSOMES

To substantiate the lysosome concept, morphological identification of the granule was imperative. Novikoff was the first to examine lysosomal pellets prepared in de Duve's laboratory with the aid of the electron microscope (Novikoff et al., 1956; Novikoff, 1958). These early preparations were far from clean and contained mitochondria, microbodies, and dense peribiliary bodies. Yet the findings suggested that it was dense peribiliary bodies that contained the lysosomal enzymes. These pioneering findings later developed in two different paths of investigation: further examination of lysosomal pellets by light microscopy, histochemistry, and electron microscopy of lysosomes.

There are few fields of biology which have benefitted as much from histochemistry as lysosomal research (Gahan, 1967). As pointed out by Scarpelli and Kanczak (1965), it is ironical that the method of Gomori (1952) for acid phosphatase was the one that provided most of the information on the intracellular distribution of acid phosphatase. Using the Gomori method for acid phosphatase on frozen tissues fixed in cold formol calcium, Novikoff was able to demonstrate high acid phosphatase activity in liver peribiliary bodies and in the droplets of the cytoplasm of proximal tubules (Novikoff, 1958). When histochemical methods for acid phosphatase were extended to electron microscopy by Sheldon et al. (1955), several laboratories soon demonstrated intense acid phosphatase activity in the pericanalicular dense bodies of liver and in a heterogeneous population of dense bodies of other intact cells (Goldfischer et al., 1963; Novikoff, 1963).

The limitations of ultrastructural histochemistry should, however, not be underestimated. The difficulties that one encounters in exactly localizing enzymes after fixation is illustrated by the case of the acetyl esterase. Tissue fractionation studies of rodent liver indicate that esterase is found in the microsomal fraction. The intracellular distribution of the esterase thus parallels that of glucose-6-phosphatase.

On electron microscopic sections, esterase activity is found in close association with acid phosphatase. The discrepancy between the biochemical and electron microscopic results has not been satisfactorily explained. To propose that the enzyme is associated with lysosomal membrane and washed off to resediment with the microsomes during tissue fractionation procedures is convenient but not convincing. A more sensible interpretation is that esterases are present in both lysosomes and endoplasmic reticulum, and that histochemical methods reveal only the lysosomal enzyme, while the biochemical methods provide a more adequate representation of the intracellular distribution of the enzyme to reveal its presence in both the endoplasmic reticulum and the lysosome.

Although acid phosphatase activity was most conspicuous in dense bodies, acid phosphatase could also be demonstrated in endoplasmic reticulum (Novikoff *et al.*, 1963, 1964; Brandes *et al.*, 1965), at least in neurons and ciliates (Brandes *et al.*, 1965) as well as in the prostatic epithelium and sebaceous gland of the *Euglena gracilis* (Brandes, 1965), and in the seminal epithelium and the sebaceous gland of the rat (Brandes *et al.*, 1965).

In addition to having been found in various kinds of dense bodies and the endoplasmic reticulum, acid phosphatase was also found in Golgi vesicles of spermatogonia, hepatocytes, neurons, myelocytes, kidney cells, and sebaceous gland of adult male rats. The relevance of the histochemical distribution of acid phosphatase to the origin and function of lysosomes will be discussed in other sections of this chapter.

Cleaner separation of liver lysosomes from mitochondria were obtained by density equilibration of the particles through sucrose gradients. The picture of the lysosomal pellets obtained by this procedure is unfortunately not representative of biochemical preparation, because only a carefully selected portion of the preparation used for biochemical studies was used to make the electron micrograph.

In conclusion, the combination of painstaking and elegant tissue fractionation and electron microscope studies demonstrated that the peribiliary dense bodies were granules rich in acid hydrolases and that a large proportion of the acid hydrolases found in the granules existed in a latent form. Latent acid hydrolases, as well as acid phosphatase-containing granules resembling dense bodies found in liver, were found in most tissues investigated. The findings made with the electron microscope and the ultracentrifuge, however, did not exclude the presence of acid hydrolases in other cell fractions or conclusively demonstrate that the structural association of the different hydrolases with the dense bodies was identical in all cases. Finally, these studies did not permit any conclusion as to the origin of the hydrolases found in the dense peribiliary bodies. It will soon become obvious that a grasp of the lysosomal concept and its limitations is a prerequisite for adequate understanding of the role of the acid hydrolases in cellular death.

III. Hydrolase Activity and Cell Death

A. HYDROLASES IN AUTOLYSIS

Pathologists have known for a long time that at least some forms of necrosis result in the lysis of cellular components such as proteins and nucleic acid, and that hydrolases, especially nucleases and proteases, catalyze the process. In 1938 Bradley reviewed what was known at the time of the role of enzymes in autolysis and atrophy. It is remarkable that the conclusions reached then are still relevant to modern understanding of the role of hydrolases in necrosis, necrobiosis, and necrocytosis. Bradley proposed that the following sequence of events led to autolysis or atrophy: first, a decrease in oxidizing metabolism resulting from subnormal oxygen tension in the tissues; second, an increase in the concentration of hydrogen ions resulting from the accumulation of acids including lactic acid, CO_2, and other metabolic products; third, an increase in proteolytic activity. The mechanism of activation of proteases was not obvious, although it was already clear in 1932 through the work of Willstatter and Rohdewald (1964) that all cathepsins in tissues do not exist in the free state but that some remain permanently bound to insoluble proteins after attempts at extraction. The bound enzyme was called desmo-cathepsin, the free enzyme lyo-cathepsin. Liberation of the desmo form was achieved by either weak acid or moderate autolysis. A theory popular at the time proposed that in absence of oxygen and in presence of high concentrations of hydrogen ion, S—S groups would be reduced to SH groups and thereby an activator of proteases would be formed. Reduced glutathione was assumed to be such an activator (Bradley, 1938).

Before the lysosomal concept was developed, Stowell and his associates at the University of Kansas examined the fate of a number of biochemicals in mouse livers that underwent autolysis *in vitro* in test tubes, or *in vivo* after transplantation in the peritoneal cavity (Berenbaum *et al.*, 1955 a,b). The advantage of *in vitro* autolysis is that the process can be followed under standard conditions of temperature, humidity, and sterility. As for *in vivo* autolysis, it provides the opportunity for observation of autolysis under conditions close to those existing in the intact animal.

The study included measurements in the activities of peptidases, esterase, alkaline, and acid phosphatase after various periods of autolysis. While the activities of some electron transport enzymes such as cytochrome oxidase and succinic dehydrogenase had dropped to low values, within a few hours after the onset of autolysis minimal changes in acid hydrolase activities took place. Thus, even after 72 hours of autolysis, 20% of the original levels of acid phosphatase activity were measurable. In addition to providing an easy method for the study of autolysis, Stowell's group's investigation yielded a basis for comparison between uncomplicated autolysis and necrosis associated with other experimental conditions, for example, feeding of carcinogens, and starvation. The study also clearly established differential susceptibility of enzymes to autolysis and in particular demonstrated the resistance of hydrolases to prolonged periods of autolysis.

The techniques of *in vitro* autolysis were used by a number of investigators to study the participation of lysosomes in cellular death. Van Lancker and Holtzer (1959a) studied the release of bound acid phosphatase and β-glucuronidase during the first 2 hours after liver autolysis. A drop in bound enzyme activity was observed 30 minutes after onset of autolysis, and after 2 hours bound enzyme activity was approximately one-third of normal. Inasmuch as the acid-soluble phosphorus increased after the rise in acid phosphatase activity had occurred, it appeared that acid phosphatase was in fact acting on its natural substrate during autolysis. Further studies, however, revealed that under similar conditions of autolysis, DNA resisted the hydrolase attack for hours after excision of the liver. Yet 62% of the DNase was released from its granular support with 15 minutes and 92% within 30 minutes after the onset of autolysis (Van Lancker and Holtzer, 1963). No acid phosphatase was released to the supernatant of mouse pancreas even 60 minutes after the onset of autolysis. In contrast, supernatant amylase activity had tripled after that period of autolysis (Holtzer and Van Lancker, 1962). As a result of these studies the authors cautioned readers against hasty conclusions with respect to the role of acid hydrolases in autolysis (Van Lancker and Holtzer, 1963). Using similar methods of autolysis, Goldblatt *et al.* (1963) showed that there was no definite change in the localization of the enzymes until 4 hours after the onset of autolysis.

Thus although it appeared that the acid hydrolases were released in various forms of autolysis, it proved impossible to distill from these experiments a consistent pattern of release or of digestion of substates. Moreover (and of more consequence to the interpretation of the results), the release of the acid hydrolases seemed to take place relatively late after the onset of autolysis.

B. Hydrolases and Necrobiosis

Metamorphosis, organogenesis, and even adult development are associated with extensive histological reorganizations which are often accompanied by the death of large groups of cells or even by resorption of entire organs (Saunders and Fallon, 1966). In some cases the role of hydrolases in the digestive process has been investigated.

An interesting form of necrobiosis is that which takes place when a larva undergoes metamorphosis to the adult form of the species. Some larvae, such as caterpillars, have a well-developed alimentary system and the purpose of the larval stage is primarily nutritional. Other larvae have an especially well-developed locomotion system and as a result travel long distances from the site of hatching. Geographical propagation of the species is believed to be the purpose of that type of larval stage. Tadpoles belong to this second class of larva and have a strong muscular tail. During the metamorphosis of tadpoles to froglets the tail disappears. The resorption of tadpoles tails has been the subject of studies in several laboratories (Weber, 1957a,b; Zollinger, 1948; Weber and Niehus, 1961).

A number of fractions were prepared from tadpole tails by differential centrifugation (Weber, 1963). One fraction contained the myofibrils and the nuclei, another the microsome fractions and a fourth the supernatant. The acid hydrolases were found mainly in the soluble fraction and only relatively small portions (22% in the case of acid phosphatase and 40% in the case of cathepsin) were particulate-bound. The tadpole tail contains several different tissues, including muscle, connective tissue, and cartilage, and therefore it is difficult to homogenize. Consequently it is not surprising that only a small fraction of the lysosomal enzymes were found to be particulate-bound. Osmotic and other forms of structural latencies were demonstrated for acid phosphatase and cathepsin.

The demonstration of the existence of latent hydrolases in the tadpole tail raised the question of their participation in the resorption of the tail during metamorphosis. Interpretation of results *in vivo* are obviously complicated by the participation of lymphocytes and macrophages in the resorption process. Histochemical studies by Weber (1963) and his associates clearly demonstrate that the major increase in enzyme activity resulted from macrophage invasion of the regressing tadpole tail. The tremendous increase in the activities of ribonuclease, deoxyribonuclease, and β-glucuronidase (20- to 40-fold) reported by de Duve and Eeckoudt (see discussion of Weber's paper, 1963) during tadpole tail regression is, therefore, indicative of macrophage invasion.

It was hoped that such interference would not take place when tadpole tails were cultured *in vitro*. Resorption of such tails can be brought about by the addition of thyroxine to the culture medium. Weber measured catheptic activities in control tails and in thyroxine-treated tails incubated *in vitro*. Strangely enough, the catheptic activity of the control tails doubled within 4 days after amputation and reached values three times and sometimes four times as high 12 days after amputation. In thyroxine-treated animals, cathepsin activity rises to values at least three times those measured in the control. The cathepsin activity, however, does not rise right away but only after 1 or 2 days' lag period. Therefore, Weber concluded that the effect of thyroxine could not be on lysosomes but rather on the enzyme itself.

These fascinating results lend themselves to multiple interpretations. The fact is that a considerable amount of catheptic activity not measurable at the time of amputation can be measured days after amputation; and thyroxine enhances this increase in activity. Such findings could result either from a net increase in the number of molecules of cathepsin present at the time of incubation by *de novo* synthesis, or from activation of preexisting molecules. Whether tadpole tails in culture are capable of biosynthesis of specific proteins remains to be seen. Activation of preexisting molecules could result from removal of inhibitors, activation of structurally latent enzyme (possibly bound to the endoplasmic reticulum) or modification of the catalytic properties of the cathepsin molecules under the direct influence of thyroxine. Weber has provided evidence indicating that the catalytic property of the enzyme is modified under the influence of thyroxine. Michaelis-Menten constants (K_m) were determined on homogenates prepared from isolated tail tips and tail tips obtained from intact tadpoles. Although determinations of K_m on crude enzyme preparations are difficult to interpret, it appears that the K_m of the cathepsin measured on tails atrophying *in vivo* or *in vitro* under the influence of thyroxine is about half that measured for tails of nonmetamorphosing larva obtained either at the time of amputation or even after 9 days of culture. In conclusion, it would seem that affinity of cathepsin for the substrate (casein-urea) used in the experiment increases during metamorphosis (Weber, 1963).

One cannot exclude the possibility that the new cathepsin with greater affinity for its substrate results from proliferation of a new type of cell, because macrophages appear even in the cultured tail. Whether these macrophages are derived through proliferation of preexisting macrophages or through transformation of other mesenchymal cells into macrophages has not been established. What seems certain is that during metamorphosis of the larva of *Xenopus*, no granules corresponding to lysosomes appear in the tail. We shall see later that Saunders and Fallon (1966) made similar observations in their studies of necrobiosis of vertebrate limbs.

Two embryonic structures are involved in the development of the genitourinary tract of mammals: the primitive excretory mesonephric or Wolffian, and the paramesonephric or Müllerian ducts. Mesonephric and paramesonephric ducts are of mesenchymal origin. Embryos of both sexes contain both types of ducts at the early stages of development. In the female the excretory-urinary tract develops from the mesonephric duct and the Müllerian duct becomes highly modified, leading to the formation of the uterus, the uterine tubes, and the vagina. In contrast, in the male, the mesonephric ducts connect in due time with the testis and both the genital and the urinary systems use a common pathway for excretion of their products and as a result the Müllerian ducts regress. Regression of Müllerian ducts is a good example of necrobiosis during embryogenesis. Brachet (Brachet et al., 1958) was the first to demonstrate an increase in lysosomal enzyme activity and a conversion of bound to free acid phosphatase during regression of the Müllerian ducts. Scheib (1963) measured the total and free acid phosphatase, β-glucuronidase, cathepsin, and acid ribonuclease activity in homogenates of 8-, 9-, and 10-day-old chick embryo. The total activities of these enzymes (expressed in micro units per organ) during these 3 days doubles for acid phosphatase, more than triples for β-glucuronidase, and doubles for cathepsin, but it remains unchanged for acid ribonuclease in the female. In contrast, small changes in the total activities of these enzymes are observed in the male during that period. However, when the activity of the hydrolases is expressed per milligram of nitrogen found in the organ, the specific activity for all hydrolases is unchanged in the female while it doubles in the male during that period.

Thus while the Müllerian canal grows and differentiates in the female, new hydrolases are synthesized and the ratio of hydrolases to total protein remains constant. In contrast, in the male there is no synthesis of hydrolases but the ratio of hydrolases to other proteins increases as a result of differential protein degradation. Furthermore, there is in the male during that period a shift in the relative ratios of free and bound hydrolases. While at the eighth day, between 10 and 20% of the hydrolases are found in the supernatant, the soluble activities rise to 40 and 60% on the ninth and tenth day of the chick embryo development. Inasmuch as the regression of the Müllerian duct is obviously under hormonal control, Scheib also tested the effect of steroid hormones on the release of hydrolases from their granule support in chick embryo homogenates.

Five micrograms of an alcohol solution of the steroid was added to a millimeter of homogenate. All steroids used released a portion of the enzyme from their granule support. The most effective were testosterone propionate and estradiol benzoate (91% and 74% release after 30 minutes' incubation, respectively). In the face of these results, Scheib was careful to conclude that the activation could not be considered to be a specific effect of the male hormones. Scheib (1963) further extended her studies to explants of Müllerian ducts. Explanted undifferentiated ducts can be maintained in a standard medium for several days. In presence of estrogen, some degree of differentiation of the Müllerian duct is observed. In contrast, the addition of androgen to the cultured medium (embryo juice and phosphate) leads to rapid necrobiosis. However, even 6 hours of exposure to androgen has a surprisingly small effect on the release of acid phosphatase in a free form. Although it seems fair on the basis of these results to conclude that the acid hydrolases participate in the necrobiotic process, these findings shed little information on the role played by hormones in the conversion of bound to free hydrolases or on the event that triggers the cytonecrosis in the Müllerian duct. Results similar to those of Scheib have been reported by Hamilton and Teng (1965).

On the bases of histological examination of the explants, Scheib has claimed that her *in vitro* studies are not vulnerable to the criticism of macrophage invasion. Saunders and Fallon (1966), however, have drawn investigators' attention to the possibility of extensive development of macrophages in explants either as a result of cellular proliferation or because of transformation of preexisting mesenchymal cells. These authors have observed numerous macrophages in the splanchnopleural and connective tissue layers of Müllerian duct explants.

Necrobiosis is not the appanage of the embryo; it occurs during bone development. Osteoclasts have traditionally been incriminated in bone resorption and they are believed to contain lysosomelike granules (see below). Resorption of breast tissue takes place in female mammals after lactation or during menopause. Slater *et al.* (1963b) established, after painstaking efforts to develop adequate methods of homogenization, that the rat breast contains particulate-bound hydrolases with structure-bound latency similar to that of lysosomes. During involutions of the mammary glands, the activity of succinic dehydrogenase, glucose-6-phosphate dehydrogenase, and glutamic aspartic transaminase decreases abruptly. In contrast, acid RNase activity increases slightly, β-glucuronidase activity remains unchanged, and cathepsin activity decreases after a short period of increase. Although these experiments strongly suggest that the acid hydrolases participate in the involution of the breast, they do not prove that the rupture of lysosomes triggers necrocytosis. In fact, as suggested by Slater *et al.* (1963a), the early decrease in the level of respiratory enzyme activity indicates that changes in the bioenergetic pathways may precede the alteration of the activity of the lysosomal enzyme.

C. CHANGES IN THE INTRACELLULAR DISTRIBUTION OF ACID HYDROLASES IN NECROSIS

The activity and intracellular distribution of lysosomal enzymes has been studied in experimental and spontaneous necrosis in animals and humans. Total, free, and bound activities of a number of acid hydrolases have been measured under various conditions of necrosis including ischemia, shock, nutritional imbalances, administration of toxins, and genetic dystrophies. Many of the reports have not included adequate morphological control studies and as such are most difficult to interpret. For example, it is often difficult to decide whether the reported changes in enzyme pattern result from shift in the cellular populations of the organ under investigation. Such changes in liver, muscle, and other organs are the inevitable consequence of necrosis, exudation, and regeneration. The effects of carbon tetrachloride intoxication are too well known to require reviewing here (for review, see Recknagel, 1967). Carbon tetrachloride induces centrolobular necrosis and fatty degeneration of the liver. Ultrastructural and biochemical changes include alteration of the endoplasmic reticulum and of protein synthesis, modification of mitochondria with uncoupling of oxidative phosphorylation, and conversion of bound to free hydrolase activity. Carefully controlled studies by Slater *et al.* (1963a,b), Slater (1961a,b, 1962), Slater and Planterose (1960), and Dianzani (1963) clearly show that the changes in the ratios of free and bound hydrolases occur late in the sequence of events that follow the alterations in the endoplasmic reticulum and protein synthesis.

Because of their potential role in necrosis, the fate of acid hydrolases was investigated in various tissues after the administration of X-radiation. The results of these experiments are often difficult to interpret either because homogenates were prepared in distilled water (Goutier-Pirotte and Thonnard, 1956) or because the initial level of free enzyme in homogenate obtained from nonirradiated animals (Roth, 1956b, 1958b, Goutier, 1963; Feinstein and Balliu, 1953) was unusually high. A release of acid DNase, acid ribonuclease, and cathepsin has been demonstrated in various tissues after irradiation (Goutier, 1963; Brandes *et al.*, 1967; Roth, 1958b; Feinstein and Balliu, 1953). These changes are relatively late events, however, in the sequence of alterations that follow irradiation.

In regenerating liver (Van Lancker, 1969) no significant change in free and bound activity of the enzyme nor of the total activity of β-glucuronidase, deoxyribonuclease, or acid phosphatase could be detected even 24 hours after the administration of 1000 rads whole body dose. Yet, such doses reduced DNA synthesis to 15% of normal within hours after irradiation and changes in the replicating and transcribing properties of the DNA are detectable within 1 hour after irradiation (Ariyama and Van Lancker, 1970). DNase and β-glucuronidase activities are increased in the supernatant of spleen of irradiated mice. Although these changes occur within an hour after irradiation, interference of irradiation with DNA synthesis is detectable prior to the release of acid hydrolases (Fausto *et al.*, 1964).

Electron microscopic studies of irradiated tissues also suggest that lysosomal changes that take place after irradiation are late events in the sequence of steps leading to cytonecrosis. Brandes *et al.* (1967) studied the effect of X-radiation on the lysosomes of mouse mammary gland carcinoma and showed diffusion of acid phosphatase out of the lysosomal granules 3 days after exposure to X-radiation (1000 rads). Careful morphological studies by Novikoff *et al.* (1963) of the lymphocytes of thymus of irradiated animals indicated that cytonecrosis of these cells was not associated with marked activity of the lysosomes.

Tappel *et al.* (1963) has claimed that particulate-bound hydrolases are released when lysosomes are irradiated *in vitro*. The relevance of this *in vitro* effect produced with kilodoses to *in vivo* radiation damage is not clear.

In conclusion, it appears that whatever the role of lysosomes may be in the cytolysis of irradiated cells, their participation in cytonecrosis is always preceded by other important metabolic alterations.

The fate of cytochrome oxidase, glucose-6-phosphatase, and acid phosphatase in the ligated lobes of livers was compared to that of normal liver (de Duve, 1958, 1963). Although these findings provided further evidence for the relative stability of hydrolases compared to other enzymes in necrosis, different degrees of stability were reported for various hydrolases. Thus, while acid phosphatase was stable for days, cathepsin and deoxyribonuclease fell to half their normal value within 6 hours, and β-glucuronidase was found to increase after 6 hours of ligation. In addition, it was shown that bound hydrolases became free up to 40% within 2 or 3 hours after ligation, and 80% after 6 hours of ligation (de Duve, 1958).

The interpretation of measurements of free or bound enzyme activities in animals subjected to prolonged fasting and then injected with carbon tetrachloride (de Duve, 1958) is obscured by the histological changes that are bound to take place, including cytonecrosis, regeneration, and cellular exudation. Therefore it is difficult to decide whether or not the changes in the enzyme patterns that have been reported result from a shift in cellular population or from modification of the intracellular activities.

The pattern of distribution of deoxyribonucleic acid, deoxyribonuclease, ribonuclease, acid phosphatase, and β-glucuronidase is altered in cell fractions of livers of animals fed carcinogenic diets (de Duve et al., 1955). Such results are not surprising because shifts in cellular population of the liver after the administration of hepatic carcinogen have been repeatedly described (E. C. Miller and Miller, 1955). Whatever the significance in the change of the hydrolase activities after administration of chemical carcinogens may be, these findings cannot be construed to mean that hydrolases play a role in carcinogenesis.

Tappel and his associates studied the fate of lysosomal enzyme in two types of muscular dystrophy (Tappel et al., 1962, 1963; Tappel, 1962): genetic muscular dystrophy in mice and chickens, and nutritionally induced dystrophy. Large increases in the lysosomal enzymes in the muscle of the genetically dystrophic mouse and chick, as well as in muscles of vitamin-deficient rabbits were reported. The increase in lysosomal enzyme activities correlated well with the increase in urinary breakdown products such as creatine and amino acids. On the basis of these findings it was proposed that vitamin E deficiency results in peroxidation of cellular and subcellular constituents; peroxidation of the membrane of the lysosome results in its rupture and release of the enzyme. The liberation of the hydrolases leads to necrocytosis which is followed by inflammatory exudation. Although it is possible to describe a superficial correlation between the increase in lysosomal enzymes and degradation of the muscle components, the experiments do not establish whether the destruction of the cellular components results from intracellular release of lysosomal enzyme or from the migration of macrophages and polymorphonuclears in the damaged area.

Van Vleet et al. (1968) did a sequential electron microscopic study of skeletal muscle degeneration in weanling rabbit. Lysosomal formation could not be observed in skeletal muscle during the early stage of degeneration, and the appearance of autodigestive vacuoles coincided with macrophage invasion of degenerated muscle fibers.

Furthermore, Richterich and Schafroth studied muscular dystrophy in mice and showed that the first event is leakage of creatine kinase, aldolase, and lactic dehydrogenase from muscle cells. The increase in lysosomal enzyme activity (β-glucuronidase, acid phosphatase, and sulfatase) is believed to be a secondary reaction to the release of the cytoplasmic enzyme.

Moreover, it seems unlikely that the primary injury in muscular dystrophy in animals or humans is of lysosomal origin. Scanty as it may be, the weight of the evidence suggests a primary alteration of cellular permeability (Richterich and Schafroth, 1963).

Although streptolysin (Hirsch et al., 1963), ultraviolet light (Allison and Paton, 1965; Allison et al., 1966), and viruses (Mallucci and Allison, 1965; Flanagan, 1966; Millson, 1965) may release acid hydrolases from their structural support in vitro, these agents are also known to act on other cell structures (membranes, nucleic acids, etc.). Therefore, to conclude that these agents produce cytonecrosis by direct effects on lysosomes is premature.

D. HYDROLASES IN EXTRACELLULAR DIGESTION

Acid hydrolases are believed to be involved in the digestion of the fundamental substance of cartilage and of necrotic tissues by macrophages. When oxidized papain is injected intravenously in rabbits, the enzyme is reduced and activated specifically in those organs that contain cartilage (L. Thomas et al., 1960). The injected enzyme digests the cartilage matrix and as a result the ears which normally stand up collapse. Histologically there is marked loss of metachromatic material of the cartilage matrix and chondroitin sulfate is released in the blood. The effect of papain is not restricted to ear cartilage but it also affects epiphyseal plates and other cartilagenous tissues. Repeated injections of large doses of vitamin A in rabbits produced similar gross and histological effects (L. Thomas et al., 1960; D. P. Thomas, 1967; McCluskey and Thomas, 1958).

Cartilage was demonstrated to contain particulate-bound cathepsin which is released during incubation of samples of embryonic cartilage in hypotonic media (Dingle et al., 1961; Lucy et al., 1961; Fell and Mellanby, 1952). The similarity between the effect of papain and that of large doses of vitamin A led to the conclusion that vitamin A releases cathepsin from chondrocytes and thereby leads to digestion of the matrix. Moreover because administration of cortisone (L. Thomas *et al.*, 1963) prevents the chondrolytic effect of vitamin A in rabbits, it was proposed that cortisone stabilizes lysosomes; hydrocortisone, however, enhances the chondrolytic effect of vitamin A in tadpoles (Weissman, 1961). No satisfactory explanation for these paradoxical results is yet available.

Dingle (1961) expanded these investigations to direct studies of the effect of vitamin A on lysosomes *in vitro*. In these experiments liver lysosomes were incubated at 37°C, pH 5 for various lengths of time (up to 2 hours). After 2 hours of incubation only 25% of the enzyme activity associated with the lysosomes was released into the supernatant. This is in sharp contrast with the findings of de Duve *et al.* (1961) who demonstrated that close to 90% of the enzyme is in free form after 90 minutes. The differences between the two laboratories probably reside in the choice of methods used to measure free activity. de Duve expresses the free activity as the percentage of the total activity and the total activity is determined after treatment of homogenate with Triton X or distilled water procedures that release all latent enzyme activities. The free activity is determined by incubating the homogenate under conditions where the enzyme present in the intact granules is not accessible to the substrate.

Dingle measured total activity under similar conditions but free activity was determined on the supernatant obtained after sedimentation of the intact organelles. In spite of the discrepancies of the results, both Dingle (1961) and de Duve claim that the addition of increasing doses of vitamin A to the incubation mixture accelerates the release of enzyme from its particulate support. de Duve's experimental design does not reveal whether the vitamin activates the hydrolases by modifying the permeability of the granule to either the enzyme or the substrate or whether it ruptures the granule and releases the enzyme into the incubation mixture. Since Dingle's studies did not include measurement of the activity that remains associated with the pellets, their results are difficult to interpret. Moreover, the results do not permit a conclusion whether vitamin A added to the incubation mixture ruptures cellular structure or facilitates solubilization of enzymes that would otherwise be absorbed on these pellets.

One must also point out that the effect of vitamin A on release of lysosomal enzymes is not likely to be specific. de Duve *et al.* (1961) extended their studies to a number of liposoluble vitamins and steroid compounds. Most of the compounds investigated, except cholesterol, released acid hydrolases from their granular support to a certain degree.

Vitamin A and vitamin E were among the most active. The steroid hormones, in contrast, had little or no effect except for minor effects attributed to progesterone and a deoxycorticosteroid. No correlation between molecular structure and the ability to release hydrolases could be established.

On the basis of such evidence one would hardly dare suggest that hormones or vitamins act *in vivo* by activating lysosomal enzymes, and if one did, as we have pointed out previously, considerable objections would have to be overcome (Van Lancker, 1964).

The effect of vitamin A on cartilage *in vivo* and *in vitro* is most intriguing; it strongly suggests that proteases, including (but not exclusively) cathepsin, digest cartilage. Although *in vitro* experiments suggest that the vitamin may act on lysosomal membranes, the doses used *in vivo* are so considerable that it is impossible to distinguish a primary effect on lysosomal membrane from a lysosomal degradation secondary to cytonecrosis. Neither do these experiments provide conclusive evidence that cortisone acts on the lysosomal membrane. Even if the contradictory findings in rabbit and tadpole are ignored, it still would be necessary to establish that cortisone does not protect the cells against cytonecrosis by mechanisms other than stabilizing lysosomal membranes.

It is difficult to imagine the skeleton, which not only supports but often shields visceral organs, as a system in constant change. Yet, during bone development in the child, the infrastructure of the bone is constantly modified, and new bone is formed at the same time as portions of the old bone are destroyed. Bone resorption is a constant event even in the adult and is, of course, under the influence of parathormone and calcitonin.

Vaes (1965a,b; Vaes and Nichols, 1961, 1962) has shown that acid phosphatase, β-glucuronidase, acetylaminodeoxyglucosidase, acid-free ribonuclease, acid deoxyribonuclease, acid phenylphosphatase, and cathepsin exist in a latent form in bone homogenates and can be partially separated from mitochondria and microsomes by differential centrifugation.

Vaes (1965b) has studied bone resorption *in vitro* in calvaria incubated in presence or absence of parathyroid hormones. Bone resorption under the influence of parathyroid extract *in vitro* is associated with release of lysosomal enzymes in the medium. It takes 2 or 3 days before the enzymes are released. Whether this release of enzyme results from cytonecrosis or from excretion of secretory granules is not certain. Neither do the experiments exclude the possibility of proliferation of macrophages or of osteoclasts. Osteoclasts have for a long time been suspected to play a role in bone resorption and are known to contain acid phosphatase-rich granules.

The great pathologists, Virchow, Koch, and Klebs had seen bacteria inside of cells, but they were blind to the significance of their findings and it was left to Metchnikoff to demonstrate that cellular exudation and phagocytosis were important. Virchow and Metchnikoff were at opposite poles with respect to their interpretation of the presence of bacteria inside cells. The famous pathologist believed that bacteria entered cells and that infected cells served to spread the infection. The great zoologist believed and proved that bacterial infection elicited cellular migration and active engulfment of bacteria inside of the cells. Metchnikoff named this intake of foreign material by living cells "phagocytosis" and described two types of phagocytes: microphages (polymorphonuclears) and macrophages. If it is no longer appropriate to classify phagocytosic cells according to the size of their prey, it is still necessary to distinguish between macrophages and polymorphonuclears on other accounts.

Three features are peculiar to the polymorphonuclear cells: the shape of their nucleus, their poverty in traditional cytoplasmic organelles, and the presence of special cytoplasmic granules. Polymorphonuclears have multilobular nuclei formed by two, three, or more sausage-shaped masses of chromatin linked by thin threads of chromatin. The molecular interactions that are responsible for this most special structure are not known, nor is the purpose of such a special shape understood. An imaginative but unconvincing interpretation was given by Metchnikoff. He believed that the appearance of multilobular nuclei was a feat of evolution that took place when animals with circulatory systems developed. The odd-shaped nucleus of the polymorphonuclear is more readily bent than that of special nuclei of other cells and as a result migration of polymorphonuclears in capillaries is made easier.

A small Golgi, a few mitochondria, and a sparse endoplasmic reticulum form the meager cytoplasmic contingent of the polymorphonuclear. Unique to the polymorphonuclear are special granules found in its cytoplasm. Under the light microscope the granules are large, they pick up neutral, eosinophilic, or basophilic stains. Under the electron microscope, the neutrophil granules appear as round dense masses finely granular but otherwise structureless and bound by a single membrane.

The metabolism of the polymorphonuclear cells is special in several ways. Glucose derived from stored glycogen is their main source of energy and when functioning, the leukocyte derives most of its ATP from glycolysis. Conversion of metabolite coupled to oxygen uptake and CO_2 formation is small and it is not certain that such conversions play any role at all in normal leukocyte function. There is no DNA synthesis in polymorphonuclears and rather puzzling reports have suggested that formate-^{14}C is not incorporated in RNA of polymorphonuclears, while ^{32}P is. If such observations can be confirmed, it would suggest that in polymorphonuclears *de novo* synthesis of purines is impossible.

Incorporation of ^{32}P must thus occur either through the lengthening or turnover of existing RNA chains or through synthesis via the salvage pathway. Rapid turnover of phosphatide during phagocytosis has been linked to the changes in membranes. The high concentrations of free amino acids that have been found in polymorphonuclears remain unexplained. The most significant biochemical feature of the polymorphonuclear cell is the composition of its granules (see above).

This is not the place to describe all steps of the participation of polymorphonuclears in cellular death and inflammation (Hirsch, 1965). The dead tissue sends a chemotactic signal to the polymorphonuclears which then stick to vessel walls, crawl along the endothelial cells, and finally find their way from the bloodstream into the tissues. In tissues bacteria or foreign material are engulfed in invaginations of the membrane which are soon pinched off and thus the bacteria that are found inside the polymorphonuclears are surrounded by an envelope made of the cell membrane. Ultimately the engulfed bacteria must be killed and digested at least in part by the granules that contain the hydrolases. For that purpose the bacteria must be exposed to the content of the polymorphonuclear granules. This can be achieved in at least two ways: either by having the granules release their content (into the cytoplasm indiscriminately or around the phagocytic vacuole) or by having the granules release their enzymic content inside of the phagocytic vacuole after fusion of the two membranes, that of the granule and that of the phagocytic vacuole. Light microscopy studies (Hirsch and Cohn, 1960; Robineaux and Frederiz, 1955) had shown that engulfment of bacteria is followed by disappearance of the granules. Breathtaking phase-contrast cinematography (Hirsch, 1960) pictures reveal the details of the process of degranulation. It is explosive and occurs in the proximity of the ingested granule.

The role that eosinophilic polymorphonuclears play in inflammation is not clear. Yet it would seem that the special features of the eosinophilic granules must be linked to unique functional properties. The size of the eosinophilic granules varies depending upon the type of animal: large (1 µ) in the horse, small (0.2 µ) in the rat. In humans, the size of the eosinophilic granule is intermediate between that of the rat and the horse.

To prepare pellets of clean eosinophilic granules, it is necessary to separate the eosinophilic from the neutrophilic polymorphonuclears. Archer and Hirsch (1963), have successfully separated the two types of cells and have prepared eosinophilic granules by differential centrifugation. Like neutrophilic granules, eosinophilic granules contain latent hydrolases. They differ, however, from neutrophilic granules in at least three ways. After solubilization of most of the content of the granule, the acid phosphatase and a peroxidase remain tightly bound to the structural remains of the granule. As already mentioned, the granule is rich in peroxidase and contains a crystalline structure whose composition and role remain unknown. Under the electron microscope this crystal appears like a rectangle occupying the center or a band spanning the entire width of the otherwise structureless granule.

Macrophages, like polymorphonuclears, play an important role in the inflammatory reaction. They were described by several histologists during the last three decades of the nineteenth century. Metchnikoff, however, was the first to recognize their role in the defense mechanism of the organism. Macrophages readily segregate vital dyes and therefore can easily be stained *in vivo*. One must distinguish three kinds of macrophages: the circulating, the wandering, and the fixed macrophages. Circulating macrophages are found in blood and are referred to as monocytes; wandering types are found in tissues and are usually referred to as macrophages. Fixed macrophages or reticuloendothelial cells are found in liver (Kupffer cells), in spleen and lymph nodes (litoral cells which line venous sinusoid), in lung (alveolar epithelium), and in numerous other tissues.

Monocytes, macrophages, and reticuloendothelial cells are believed to be derived from undifferentiated mesenchymal cells. Polymorphonuclears accumulate during the early stages of inflammation, in contrast to found in liver (Kupffer cells), in spleen and lymph nodes (litoral cells monocytes which accumulate later. Cohn *et al.* (1963) have succeeded in separating blood monocytes from other white cells which have much in common with polymorphonuclears. They are, however, larger monocytes with a bean-shaped nucleus and more abundant cytoplasm. Monocyte cytoplasm is well endowed with the usual cytoplasmic accoutrements: Golgi, mitochondria, endoplasmic reticulum, etc. The endoplasmic reticulum seems to be primarily of the smooth type, although free ribosomes are found in the macrophage cytoplasm. The origin of macrophages in tissues has not been established. Two hypotheses have been proposed—transformation of the blood monocytes or of blood lymphocytes. Like polymorphonuclears, macrophages are capable of phagocytosis and the engulfed material is digested in part.

Special granules rich in acid hydrolases have been prepared from macrophages and it is believed that the acid hydrolases contained in these granules participate in the digestion of the engulfed material.

Fusion of the membranes of hydrolytic and phagocytic vacuoles in kidney has been suggested by Strauss (1962a), who injected peroxidase into rats. The enzyme is engulfed in a phagocytic vacuole stainable in blue by histochemical methods. Other tubular granules containing acid phosphatase can be stained red. Light microscopy examination suggests that fusion of the two types of granules takes place after a certain period of time. In such histochemical studies it is, however, difficult to decide whether the appearance of the enzyme results from fusion of granules or whether it is the result of formation of new foci of autodigestion in the area surrounding the phagocytic vacuole. It can further be objected that the images reported by Strauss could result from overlapping of the two types of granules within the section rather than from genuine coalescence.

Except in cases where short periods of autolysis were studied, biochemical investigations alone did not permit researchers to decide whether the distribution of acid hydrolase in cell fractions was the consequence of shift in cellular population of organelles, or was linked to cytonecrosis. If electron microscopists and histochemists have not filled the gap between the dynamic *in vivo* events and cursory biochemical analysis, they have thrown a new beam of light on the events. As limited as they are in space and time, thousands of electron micrographs from many laboratories have indeed provided a small fragment of a puzzle that is now falling into place.

IV. Autodigestion

A. Morphology of Autodigestion

Several electron microscopists had described vacuoles limited by single and sometimes double membranes which contained cytoplasmic organelles (mitochondria, Golgi, etc.) at various stages of disintegration (F. Miller and Palade, 1964). The first observers of these structures were divided on interpretation. Some believed them to contain phagocytized material (Malet *et al.*, 1964).

Ashford and Porter (1962) observed that numerous vesicles containing mitochondria and surrounded by a single, occasionally a double membrane, appeared in rat livers perfused with glucagon. Novikoff found similar bodies in the liver of rats injected intravenously with Triton X-100 and it was further established that these bodies contained acid phosphatase (Holt, 1959). This finding was soon confirmed in several laboratories (Holt, 1959; Novikoff, 1961; Bitensky, 1962).

Possibly because these structures are morphologically heterogeneous, it seemed that each individual observer felt the need to coin new names, which often reflected individual prejudices. These large vacuoles containing cytoplasmic debris at various stages of degradation were called "lysosomes" (Ashford and Porter, 1962), cytolysomes (Novikoff, 1961), "areas of focal cytoplasmic degradation" (Swift and Hruban, 1964; Biava, 1965; Hruban et al., 1961), "cytosegresome" (Trump and Ericsson, 1965a), "cytolysosomes (Novikoff, 1961; Ghadially and Parry, 1966), "autophagic vacuoles" (de Duve, 1963, 1964; de Duve and Wattiaux, 1966), "glycogenosomes" (Phillips et al., 1967), "autolysosomes," "autolytic vacuoles" (F. Miller and Palade, 1964), and "system macrovacuolaire heterogen geants" (Malet et al., 1964). To distinguish this acid phosphatase-containing granule from the more traditional dense body (lysosome), new and sometimes inspiring qualifications were given to lysosomes, such as "virgin," "true," and "primary" lysosome (de Duve and Wattiaux, 1966) or, more soberly, cytosomes (Trump and Ericsson, 1965a).

At the risk of appearing unimaginative, to avoid prejudicing the reader by giving teleological appellations we shall refer to the original acid phosphatase-containing dense bodies as dense bodies and to acid phosphatase-containing vacuoles, in which cytoplasmic debris is digested, as autolytic vacuoles.

An important contribution was made by Pipa et al. (1962) when they demonstrated that autolytic vacuoles appear in cells of apparently healthy functioning liver, kidney, and nerve cells. Autolytic vacuoles have since been found in the cells of a number of healthy organs (breast, prostate, adipose tissue, etc.). Since cytoplasmic autolysis seems to occur in absence of cellular proliferation or cell loss, it would appear that during their life cycle, cells are capable of digestion and replacement of portions of their cytoplasm. It is, however, after experimental or spontaneous injury that autolytic vacuoles are most conspicuous. The list of pathological conditions in which autolytic vacuoles have been described has become so large that a comprehensive inventory would be pointless (for review, see Hruban and Rechcigl, 1970). Suffice it to say that autolysis takes place in varied tissues after various forms of injury or in most forms of necrobiosis or necrosis.

After a careful study of the morphology of autolytic vacuoles in a number of tissues after administration of varied types of injuries, Hruban *et al.* (1961) proposed an ingenious generalization: the concept of "focal cytoplasmic degradation." The concept proposes that the cell segregates pieces of cytoplasm for autodigestion by surrounding them with membranes. The cytoplasmic components are progressively digested within the confines of the membrane and dense bodies are formed as a result. Thus Hruban and Rechcigl (1970) distinguish three stages in the process of focal cytoplasmic degradation: formation of the autodigestive vacuole, conversion of the autodigestive vacuole into heterogeneous dense bodies (cytolysosomes), and conversion of the double into a single membrane, with the progressive digestion of intravacuolar components. The advantages of the concept are that it has in part domesticated seemingly disparate fauna of intracellular organelles without departing from established facts. Moreover, it raises meaningful questions relating to the origin of the segregating membrane, the mechanism of conversion of a double into a single membrane, the origin of the hydrolases, and the mechanism of elimination of the dense body, without prejudging the facts. More far-reaching problems are the nature of the factors which determine that autolysis shall take place in the restricted area, and a type of signal which tells the surrounding healthy cytoplasm that a portion of cytoplasm is to be digested.

Before we discuss attempts to answer these problems we shall briefly consider the relationship between autodigestion and some forms of cytoplasmic degeneration.

B. Relationship between Autodigestive Vacuole and Nondigestible Accumulating Material

Inborn errors of metabolism which result in the deletion of an enzyme lead to the accumulation of substrate, normally metabolized by the lacking enzyme. If the substrate is a small diffusible molecule, accumulation is detectable not with the microscope but by chemical analysis of the blood. If in contrast the substrate is a macromolecule, it accumulates primarily in those cells specialized in its metabolism. Macromolecular storage diseases are not always the result of inborn errors of metabolism and some are acquired. Cholesterol accumulates in Hans Schuller Christian disease or histocytosis; lipofuscin accumulates in the aged and, in fact, one could consider keratinization to be a form of storage disease. In many cases, material that accumulates in cells, whether it be copper in Wilson's disease, iron in hemochromatoses, glycogen in von Gierke's disease, or mucopolysaccharide in Hurler's disease, becomes part of areas of focal cytoplasmic degradation that are positive for acid phosphatase.

Hers (1965) has proposed that accumulation of macromolecular substances, in at least some cases of inborn errors of metabolism, results from the absence of lysosomal enzymes and that consequently the substrate accumulates inside of the granules in which it was to be digested.

Such a conclusion ignores the fact that in many inborn errors of metabolism which do not result from the absence of acid hydrolase, undigestible material is associated with foci of cytoplasmic degradation. Moreover, even in those diseases in which the absence of lysosomal enzymes is suspected, all the undigestible material is not found inside of membrane-bound structure (Garancis, 1968). Glycogen storage disease II (Pompe's disease) is the one in which lysosomal enzymes have been most extensively implicated (Hers, 1965). This hereditary disease is characterized by muscular weakness with moderate hepatomegaly and great enlargement of the heart. It is assumed to result from a lack of acid mannase. Although a portion of the liver glycogen is segregated into vacuoles surrounded by a single membrane, a large proportion of the glycogen is diffuse in the cytoplasm of liver and kidney and heart (Garancis, 1968).

The findings of Fain-Maurel (1967) on the ultrastructural distribution of particulate glycogen in the hypobranchial gland of planctonic molluscum might well be relevant to this discussion. In these organisms, glycogen-containing vacuoles appear in the endoplasmic reticulum as a result of the invagination of the endoplasmic membranes which seem to lose their ribosomes to become smoother at those sites where glycogen accumulates.

The tissue fractionation studies in which attempts were made to localize some of the hydrolases that are suspected to be missing in some inborn errors, are not always conclusive, such as the studies of Bowen and Radin (1968) on cerebroside galactosidase. This is an enzyme assumed to be missing in cases of Gaucher's disease. As a result of the absence of the enzyme, cerebrosides accumulate in spleen, liver, and nervous tissue, leading to hepatomegaly and splenomegaly associated with mental retardation. Bowen and Radin (1968) found that hydrolase was active at pH 5 and the comparison of intracellular distribution of the galactosidase with that of sulfatase and glucosaminidase demonstrated that the specific activity of the enzyme was twice as high in lysosomal as in mitochondrial fraction. However, since the percentage of proteins in the lysosomal fraction is 7% of total, while it is 24% of total in mitochondria, the total amount of enzyme associated with mitochondria is greater than in lysosomes.

Even if one were to find Hers' logic compelling, caution in extending theory into therapeutic reality is warranted. de Duve has proposed that injection of the missing hydrolase could cure the disease because the enzyme would be phagocytized and be included (by the process of fusion of a primary lysosome with the phagosome) into a secondary lysosome, and thereby digest the accumulating macromolecules. This proposal is based on the assumption (for which there is no evidence) that macromolecules such as glycogen enter the lysosome for digestion. Furthermore, it would appear that the therapy would be self-defeating because the injected enzyme would be included into vacuoles which supposedly should digest foreign material. Whatever the reasoning behind the therapy, before it is undertaken on other humans in view of previous discouraging results it seems imperative that more experimentation on animals be attempted.

In conclusion, it would seem that weight of present evidence is in favor of Hruban and Rechcigl's views (1970) which suggest that the accumulation of undigestible material foreign or genuine to the cell results in segregated material. These areas undergo focal cytoplasmic degradation.

C.   Origin of the Membrane and Enzymes in Autodigestion

Focal cytoplasmic degradation appears to be a process common to all forms of cytoplasmic autodigestion, be it the result of either cytoplasmic turnover or cytonecrosis. Only tentative answers to a few of the questions raised by the concept are presently available. We shall successively review the evidence that has been asembled in support of the origin of the membrane in the areas of focal cytoplasmic degradation and the origin of the enzyme.

The possible origins of the membrane that segregates areas of focal cytoplasmic degradation are numerous. The cell could elaborate a new and special membrane for each area of degradation or it could utilize preexisting membranes to form the sack that surrounds the area of focal cytoplasmic degradation. A new membrane could be made at the site of the injury or elaborated in cell structures and transported to the site of injury. Golgi vesicles or Golgi lamellae would constitute plausible precursors for such a newly formed membrane. The membrane could also be formed by distention of preexisting membranes of the endoplasmic reticulum, or by fusion of smaller vacuoles (for example, primary lysosomes) or simply by engulfment of the condemned area into a primary lysosome.

The problem of the origin of the membrane in focal cytoplasmic degradation is further complicated by the fact that these areas are surrounded by a double membrane at the beginning of the process and by a single membrane later. Because the administration of antimetabolites (Van Lancker, 1969; Schimke *et al.*, 1968), including those that interfere with protein synthesis, does not interfere with the formation of the membrane, it is unlikely that the membranes of foci of cellular autolysis are synthesized *de novo* (Gottlieb *et al.*, 1964; Longnecker and Farber, 1967). Studying chromatolysis in perikaria of rat ganglion nodosum, Holtzman *et al.* (1967) demonstrated that the smooth endoplasmic reticulum provides the membrane of the autodigestive vacuoles, which include the Nissl substance during its chromatolysis. Novikoff and Shin (1963) and previously proposed that the membranes surrounding these vacuoles were formed by flattening and close apposition of the membranes of the cisternae of the endoplasmic reticulum. Using a battery of histochemical tests, Arstila and Trump (1968a) were able to show that the membrane possesses the typical enzyme markers of the endoplasmic reticulum, but not those of the Golgi apparatus or the plasma membrane. Measurement of the thickness of the membrane at the early stages of formation of areas of focal cytoplasmic degradation appears to be compatible with this finding. Even if the endoplasmic reticulum is the origin of the membranes that surround the area of focal cytoplasmic degradation, it is not clear how the double membrane is converted to a single one. Two different mechanisms have been proposed: compaction (Glinsmann and Ericsson, 1966) of inner and outer membrane, and digestion of inner membrane by hydrolases with simultaneous thickening of the outer membrane.

D. ORIGIN OF HYDROLASES IN FOCAL CYTOPLASMIC DEGRADATION

Two principal approaches have been used to study the origin of the acid hydrolases found in autolytic vacuoles: incorporation of labeled amino acid into enzymes purified from both microsomes and lysosomal fractions, and combined histochemistry and electron microscopy. In studying the intracellular distribution of enzymes in regenerating liver, we were struck by the fact that the amounts of acid phosphatase and of β-glucuronidase associated with the microsomal fraction increased at 18 hours after partial hepatectomy, a time when new proteins are synthesized (Walkinshaw and Van Lancker, 1964). Moreover, the increase in microsomal β-glucuronidase activity preceded the increase in total homogenate activity; and studies of the binding of β-glucuronidase demonstrated a different mode of attachment of the enzymes in lysosomal and microsomal fractions (Walkinshaw *et al.*, 1964). These findings suggest that β-glucuronidase is a constituent protein of the endoplasmic reticulum, and that the increased microsomal β-glucuronidase activity after partial hepatectomy reflects *de novo* synthesis of that enzyme. Cytochemical demonstration of β-glucuronidase activity in the membranes of the endoplasmic reticulum supported this hypothesis (Fishman *et al.*, 1967). Perhaps the most convincing proof that some β-glucuronidase is genuine to the microsomes was provided by the procedures used to purify the rat liver enzyme (Van Lancker and Lentz, 1970). While lysosomal enzyme is readily solubilized, the solubilization of rat microsomal and cytosol enzyme requires incubation with ribonuclease. The nature of the RNA complex is not known.

Since several laboratories have shown that β-glucuronidase may exist in various molecular forms (Mills *et al.*, 1953; Smith and Mills, 1953; Plapp and Cole, 1967; Paigen, 1961) it was appropriate to consider whether different forms of the enzyme were present in microsomal and lysosomal fractions. β-Glucuronidases with different pH optima and kinetic characteristics have been isolated from ox spleen (Mills *et al.*, 1953; Smith and Mills, 1953). Moreover, five or more different forms of β-glucuronidase have been separated from bovine liver by ion-exchange chromatography (Plapp and Cole, 1967). These multiple forms have similar protein structure but differ in carbohydrates and possibly other constituents. In contrast, Paigen (1961) could not demonstrate any significant difference on properties of β-glucuronidase derived from microsomal and lysosomal fractions of mouse liver. Starch gel electrophoresis of β-glucuronidase partially purified from mitochondria, microsomes, and cytosol of regenerating rat liver yielded several protein bands associated with β-glucuronidase activity (Van Lancker, unpublished results). However, only one band was obtained when purified lysosomal and nuclear enzymes were used. Mitochondrial cytosol and microsomal enzyme were purified until the preparation yielded single bands on polyacrylamide gel. The single band had a mobility similar to that of the purified lysosomal enzyme. β-Glucuronidase purified from lysosomal and microsomal pellets obtained from 24-hour regenerating liver were found to have similar catalytic and electrophoretic properties. The effects of substrate concentration, pH, thermal activation, and inactivation are identical. The occurrence of a similar form of enzyme in the microsomal and lysosomal fractions of regenerating liver provided an opportunity to determine the site of biosynthesis of lysosomal β-glucuronidase by comparing the specific radioactivity of β-glucuronidase purified from lysosomal and microsomal fractions after various periods of *in vivo* incorporation of amino acid-^{14}C.

The results of these experiments were most unexpected. Although after injection of labeled amino acids the enzyme purified from the endoplasmic reticulum was rapidly labeled, no label appeared in enzyme purified from the lysosomal or other cell fractions, suggesting that in liver regeneration the transfer of enzyme from endoplasmic reticulum to lysosomes was undetectable in the absence of special stimuli. Label suddenly appeared in the lysosomal fraction when animals were also subjected to prolonged periods of hypoxia (2 hours), an event known to be associated with the formation of focal cytoplasmic degradation. These results suggest that in nonhypoxic animals either the regenerating liver enzyme is not transferred at all from microsomes to lysosomes, or the transfer is so slow that it cannot be detected in lysosomes, or finally substantial transfer takes place from endoplasmic reticulum to lysosomes but the newly made lysosomes are rapidly eliminated from the liver cell. If the enzyme is not transferred at all from endoplasmic reticulum to lysosomes, it is difficult to understand how any enzyme ever reaches the lysosomes. Moreover, transfer does take place in liver after hypoxia and in kidney after injection of gonadotropins. In absence of marked transfer of enzyme from ER to lysosomes, however, it is difficult to conceive of a secretion process (similar to that involved in pancreas in the formation of zymogen granules) that would involve the elaboration of lysosomes as the active secretory granules.

It would appear that at the present the most logical interpretation of the results is that slow transfer takes place from endoplasmic reticulum to lysosomes in the regenerating liver of nonhypoxic rats. Rapid incorporation into the endoplasmic reticulum in absence of elimination of the newly made enzyme through secretion or excretion also suggests that a large proportion of the newly made enzyme must remain associated with the endoplasmic reticulum. Association of β-glucuronidase with endoplasmic reticulum has also been observed in kidney. The significance of this observation will be discussed below.

Studies of Fishman et al. (1969), Ide and Fishman (1969), and Kato et al. (1970) in mouse kidney clearly established lysosomal and extralysosomal (mainly endoplasmic reticulum) localization of the enzymes. Histochemical studies of kidney of mice injected with gonadotropins further revealed that the activity of β-glucuronidase expressed as a function of the time of injection increased in the endoplasmic reticulum before its increase in the lysosomes. Purification of lysosomal and microsomal enzyme established that all catalytic properties investigated (Michaelis constant, pH optimum, heat of activation) of the enzyme obtained from the two different sources were identical. Studies of the incorporation of ^{14}C-labeled amino acids into purified lysosomal and microsomal enzyme clearly established that kidney microsomal β-glucuronidase became more rapidly radioactive than the lysosomal enzyme. Thus the results of studies on the origin of β-glucuronidase in kidney of mice injected with gonadotropins and in regenerating liver of hypoxic or nonhypoxic rats are in excellent agreement.

Because Fishman's group used a hormone to stimulate the biosynthesis of the enzyme, their study has raised new and intriguing questions with respect to the origin and the intracellular sites of action of the hydrolases. The investigators measured the activity of β-glucuronidase and acid phosphatase in kidney and liver of male mice injected with gonadotropin. While the activity of β-glucuronidase increases, that of acid phosphatase remains unchanged. Furthermore, it was observed that β-glucuronidase activity uniformly increased (eight times) in all subcellular fractions, including the lysosomes. Only lysosomal enzyme could be released by freezing and thawing.

These studies on the origin of β-glucuronidase in kidney emphasize at least three points. (1) They provide further evidence that newly made enzyme is not all transferred from endoplasmic reticulum to lysosomes. Although the results of Fishman's group show that transfer of β-glucuronidase from lysosomes to microsomes is more significant in kidney of mice injected with gonadotropins than in rat regenerating liver, the findings also indicate that a considerable proportion of the newly elaborated enzyme remains associated with the endoplasmic reticulum. (2) The fact that microsomal β-glucuronidase but not acid phosphatase can readily be released with hyaluronidase clearly proves that the mode of attachment of the enzyme to the membrane varies depending upon the hydrolase under consideration. (3) Because not only the elaboration but also the association of β-glucuronidase in the endoplasmic reticulum is under hormonal influence, the results raise a very important question that has not yet been discussed: do the acid hydrolases function primarily in lysosomes or in endoplasmic reticulum? There seems to be no doubt that the acid hydrolases found in the autolytic vacuoles participate in the autolytic process, nor is there any doubt that acid hydrolases released after cellular death participate in the process of digestion of the necrotic tissues. Such observations, however, do not exclude the possibility that in the intact cell the primary role of acid hydrolases, in particular, β-glucuronidase, is not in the lysosomes but in the endoplasmic reticulum. There is no doubt that the fragments of endoplasmic reticulum or mitochondria captive in foci of cytoplasmic degradation are no longer functional in bioenergetic or biosynthetic pathways. Neither are there *a priori* reasons to assume that the lysosomal enzymes trapped in foci of cytoplasmic degradation are exhibiting their primary metabolic role there. Therefore, the possibility cannot at present be excluded that enzymes found in autodigestive vacuoles are not residual enzymes accumulating passively as a result of cytoplasmic degradation.

Arstilla and Trump (1968a) investigated the origin of the acid hydrolases in autodigestive vacuoles in a correlated morphological and biochemical study of a focal cytoplasmic degradation of liver after the administration of glucagon. The investigators studied the intracellular distribution of acid phosphatase and cytochrome oxidase in the nuclear, mitochondrial, lysosomal, and supernatant of liver homogenates of animals injected with glucagon. The distribution of the enzyme was the same in both cases. The mitochondrial lysosomal pellet was then placed on a discontinuous sucrose density gradient. In control animals most of the acid phosphatase is recovered in a sharp peak, while in glucagon-treated animals this peak is split into two smaller ones. The interpretation of these results is unfortunately complicated for a number of reasons. (1) The sharp peak for acid phosphatase and the rather broad peak for cytochrome oxidase distribution reported by the authors contrast with previous results in which attempts were made to separate these two enzymes by sucrose gradients. In previous studies a rather broad distribution for acid phosphatase and a sharp peak for cytochrome oxidase were found (de Duve, 1963). (2) There is an appreciable increase in the protein content of each fraction in glucagon-treated animals, except in those fractions in which the new acid phosphatase peak appears, suggesting that the absolute activity in each fraction obtained by density equilibration may not be altered significantly by glucagon. (3) Finally, the conclusion made by the authors that the acid phosphatase is transferred from one type of lysosome to another is at present not justified, since there is no way to be certain that the acid phosphatase found in the sharp control peak and the smaller peak from glucagon-injected animals results from the catalytic effect of the same molecular form of the enzyme.

Consequently, it cannot be excluded that the emergence of the two smaller peaks simply results from selective concentration of preexisting acid phosphatase.

E. Transfer of Enzyme from Endoplasmic Reticulum to Foci of Cytoplasmic Degradation

Although there seems to be no doubt on the basis of the results of the biochemical studies that the acid hydrolases must be synthesized in the endoplasmic reticulum, the mechanism of transfer of enzyme from endoplasmic reticulum to autolytic vacuoles is not clear. Complex path-ways of transfer have been proposed involving the endoplasmic reticulum (for review, see Novikoff, 1967) the Golgi apparatus, etc. Most of these mechanisms are based on experiments which assume "guilt" by associa-tion.

Two major arguments invoked in favor of a mechanism involving the Golgi apparatus are: (1) presence of acid hydrolases in the Golgi; (2) the frequent proximity of autolytic vacuoles and Golgi apparatus. The "Gerl" hypothesis (Golgi, endoplasmic reticulum, lysosomes) proposes that the acid hydrolases that are synthesized in the endoplasmic reticulum are in some way transferred to the lamellae and saccules of the Golgi apparatus, where they are packaged into small vacuoles (primary lysosomes) which leave the Golgi complex to move freely in the hyaloplasm until they meet either phagocytic vacuoles or areas of cytoplasm segregated for digestion. The membrane of the Golgi vacuoles and that of the phagocytic vacuoles or of the segregated area of cyto-plasm fuse to form autolytic vacuoles. To be sure, acid phosphatase has been found histochemically in the Golgi region. Furthermore, acid phosphatase was present in rat epididymis Golgi apparatus separated by equilibration in sucrose gradient. However, it is not certain that the acid phosphatase activity of Golgi and lysosome are due to the same mol-ecular form of the enzyme. Moreover, contamination of Golgi membrane prepared biochemically by smooth membrane of the endoplasmic reticulum, lysosome, or lysosomal debris has not been excluded.

Direct transfer from ER to lysosomes cannot be excluded. In such a case the acid hydrolases synthesized in the endoplasmic reticulum would remain in the process of formation of autolytic vacuoles attached to the membranes that delimit areas of focal cytoplasmic degradation. Once anabolic processes have ceased in the segregated area and only catabolic reactions continue, the latent enzymes could be released from their structural support and concentrate in the autolytic vacuoles by default, as a result of digestion of all other cellular components. Obviously, such a mechanism would require that hydrolases be more resistant to hy-drolysis than other cellular components. This was shown to be the case by Berenbaum et al. (1955a). Moreover, studies of effect of puromycin on the activity of acid hydrolases in regenerating liver suggested that the turnover of these enzymes was slow (Gottlieb et al., 1964). Recent electron microscopic studies have provided evidence in support of direct transfer of enzyme from endoplasmic reticulum to lysosomes (Novikoff, 1967).

Although direct transfer could explain the formation of autolytic vacuoles in most cells, whether it also takes place in polymorphonuclears and macrophages is debatable, but not impossible. In the case of the polymorphonuclear, it cannot be excluded that this cell carefully programs the degradation of its cytoplasm, thereby providing it with autolytic vacuoles which explode when presented with the engulfed material. In the case of the macrophage, however, it seems that fusion between phagocytic vacuoles and vacuoles containing the acid hydrolases cannot be put in question. The macrophage may be a very special cell with a special mechanism for forming vacuoles, but the suggestion of Novikoff (1967) that in the macrophage the vacuoles containing the acid hydrolases (which coalesce with phagocytic vacuoles) are in fact autolytic vacuoles (secondary lysosomes) should not be ignored. If such an assumption could be substantiated, a unified mechanism for the formation of autolytic vacuoles in all types of cells, including the macrophage, would be available on the basis of direct transfer of enzymes from endoplasmic reticulum to lysosomes.

V. The Role of Hydrolases in Cell Death and Turnover

A. The Role of Hydrolases in the Point of No Return in Cell Death

The primary causes of cellular death after various types of injury cannot be reviewed here. It is, however, of considerable importance to be able to decide whether the release of acid hydrolases is part of a reversible or an irreversible sequence. Unfortunately, there are few studies in which quantitation of the release of acid hydrolases was coupled to studies on other cell functions which might be responsible for cellular death.

Cell death is often the consequence of injury whether it be infection, administration of toxin, or ischemia. Cell death is not usually an instantaneous event but rather the result of a sequence of molecular and ultrastructural changes in the injured cell. The earliest changes are likely to be reversible, the latest irreversible. It is important to know how long the cell can be exposed to the injurious agent before it enters the inevitable course to death, and what event separates the reversible from the irreversible steps in the sequence of changes brought about by injury. If exposure to the injurious agent is short and diagnosis prompt, it should theoretically be possible to reverse all damage that developed before that critical time, sometimes referred to as the point of no return. We should make clear here that "the point of no return" refers to cellular and not bodily death. The body is capable of surviving extensive areas of cellular death. However, relatively limited cellular death of vital organs (brain and heart) taxes the future of the individual because instead of being restored, the dead cells are replaced by scar tissue of the injured cells. Because these visible changes are reversible, much of the early work on the point of no return was carried out in tissues in which cloudy swelling was produced experimentally (Fonnesu, 1960). The experiments on cloudy swelling will not be reviewed and we shall restrict the discussion on the point of no return to more recent investigation in the field of cellular ischemia.

Because ischemia is such a frequent cause of cellular death in human disease, Vogt and Farber (1968) studied the effect of temporary occlusion of the renal artery on kidney cells. They first established that all morphological changes developing during the first 20 minutes of ischemia are reversible. In contrast, 30 minutes of ischemia results in extensive necrosis. The authors concentrated their studies on two parameters, the levels of ATP and lactic acid. ATP levels dropped quickly and drastically after the establishment of ischemia. Lactic acid levels rose rapidly during the first 20 minutes of ischemia, (23% of normal levels at 10 minutes, 600% of normal levels at 20 minutes) and continued to rise steadily up to at least 2 hours of ischemia (1200% of normal). When the occlusion of the renal artery was interrupted, for example, after 10 minutes, ATP levels rose to 80% of normal and lactic acids dropped to normal or slightly subnormal levels within 10 minutes of recovery. Recovery takes a little longer after 20 minutes of arterial occlusion.

A critical observation is that although extensive necrosis takes place after 30 minutes of occlusion, partial biochemical recovery of ATP levels (50% of normal) and restoration of normal levels of lactic acid occur within 10 minutes after lifting the occlusion. In fact, partial biochemical recovery is possible even after 2 hours of ischemia. In the absence of oxygen, the levels of ATP drop and most of the ATP made is derived from glycolysis. Most cells are capable of surviving under those conditions for at least 20 minutes. What happens at 20 minutes to set these cells on the irreversible path to death is not known. We have seen that the levels of ATP are low but remain unchanged, even after 2 hours of ischemia. In contrast, the levels of lactic acid continue to increase and therefore it would seem logical to implicate lactic acid accumulation as the trigger that changes a reversible to an irreversible course of events. Yet when the levels of lactic acid were reduced by treatment of the animals with 2-deoxyglucose or iodoacetate, the kidney still did not recover after 30 minutes of ischemia. Of course the interpretation of these results is somewhat complicated by the fact that the inhibitors of glycolysis further reduce the ATP levels in the ischemic kidney (almost 2% of normal with iodoacetate and approximately 6% with deoxy-glucose).

As pointed out by the authors, comparison of the results of recovery after 20 and 30 minutes of occlusions show a clear-cut difference in the ability to restore ATP levels. Therefore, these studies on the pathogenesis of ischemia were extended to that of mitochondria isolated from ischemic kidney before and after recovery. Although mitochondria after 20 or 30 minutes of ischemia presented a drop in oxygen uptake or in their ability to couple oxidation and phosphorylation, they were perfectly capable of recovering normal function if the occlusion was lifted and the kidney allowed to recover for 20 or 30 minutes. In fact, mitochondria could recover normal functions even when they were obtained from kidney already doomed to necrosis (30 minutes ischemia plus 30 minutes recovery). Two hours of ischemia induced a marked drop in dinitrophenol-stimulated ATPase. ATPase activity returned to normal when the kidney was allowed to recover before isolation of the mitochondria. Mitochondria obtained from ischemic kidney had not lost their ability to actively swell in the presence of calcium or to contract in presence of ATP.

The results of the study of Vogt and Farber (1968) are intriguing, but difficult to interpret. Indeed, it would appear that what distinguishes the recoverable kidney cell from unrecoverable kidney is that the latter cannot integrally restore its ability to make ATP, even though its mitochondria recover readily. Consequently, the reduction in ATP level must result from either: (1) the existence of mitochondrial inhibitors in the living ischemic kidney; (2) interference with the biosynthesis of ATP precursor; (3) accelerated ATP breakdown; but thus it would appear that recovery of the ability to synthesize ATP is essential for survival. It is unlikely, however, that low levels of ATP are responsible for irreversible cellular death. Hepatic cells are known to survive up to 48 hours in presence of even lower levels of ATP.

Earlier studies of Gottlieb and Van Lancker are relevant to the findings of Vogt and Faber. Because occluding of the kidney artery is simple and readily interrupted and because another kidney of the same animal is available as a control, the ratio of free and bound acid hydrolases as studied in kidney after occlusion of the renal artery showed that no significant amounts of acid phosphatase or β-glucuronidase were released from their granular support even after 1 hour of clamping of the renal artery. Judah et al. (1964) proposed that cellular death might be triggered by changes in the membrane leading to excessive sodium entry and potassium loss; thus the cell would accelerate sodium pumping and consequently ATP reserves would soon be exhausted. When the intracellular ATP drops to excessively low values, mitochondrial changes take place, death follows, and alteration of the plasma membrane permeability leads to sodium and calcium accumulation inside the damaged cell.

The studies of Trump and Bulger on isolated flounder kidney tubules treated with cyanide provide experimental support of the theory expounded by Judah and his associates (1964). Time-lapse cinematography indicated that cells in which the electron transport chain had been blocked went into two different stages of swelling. The first stage is a moderate form of swelling associated with water and sodium intake, which is likely to result from ineffective performance of the sodium pump, because of lack of ATP due to mitochondrial damage. This stage is reversible but if not reversed, it explodes into massive swelling with cellular intake of sodium and calcium. The second phase of the swelling resembles changes produced by direct damage to the plasma membrane. Electron microscopic studies of the system revealed early swelling and densification of the mitochondrial matrix associated with reduction of histochemically detectable adenosine triphosphatase in the membrane. Other changes included reversible clumping of the nuclear chromatin similar to that observed in ischemia or hypoxia. Of interest is the fact that the dense bodies do not burst even after irreversible swelling of the cell, indicating that release of hydrolases cannot be responsible for triggering cellular death.

The findings of Vogt and Farber (1968) are similar to previous studies of E. W. Busch et al. (1964) in liver except that liver tissue is capable of overcoming ischemia of 1, 2, or even 3 hours' duration. After these periods, the ability to recover normal levels of ATP is maintained. However, if the period of ischemia is prolonged for 4 hours, the ability to restore ATP falls to 40% of normal; and if ischemia is maintained for 5 hours, all oxidative phosphorylation disappears. Infusion of inosine prior to restoration of normal blood circulation renders the liver capable of surviving 4 or 5 hours of ischemia.

B. THE ROLE OF HYDROLASES IN CELLULAR TURNOVER AND PROTEIN DEPLETION

There is a great deal of evidence indicating that tissue proteins are in a dynamic state. Most of the studies on protein turnover have been done in liver (Schimke et al., 1968). Because the life span of liver cells is 160 to 400 days, while the average replacement time for liver proteins is approximately 20 days, the conclusion of intracellular replacement of proteins is inescapable. In a remarkable paper on the subject, Schimke and his associates have summarized results obtained in their laboratories and other laboratories: all proteins turn over, the rate of turnover differs from one protein to another, and turnover can be modified by genetic or environmental factors.

Intracellular turnover of individual proteins is a discrete, specific, and well-timed event. It is not likely that such events could be the result of massive degradation of relatively large segments of cytoplasm as they occur in the formation of autolytic vacuoles. Selective and specific uptake of proteins by primary lysosomes could explain the specificity and heterogeneity of turnover, but it requires the existence of primary lysosomes in all cells of all tissues. Moreover, it would also be necessary to establish the occurrence of selective intake by lysosomes. Schimke *et al.* (1968) have proposed a more plausible mechanism for degradation of protein, involving neutral peptidases and proteases, and possibly highly specific proteolytic enzymes, such as those believed to be attacking insulin or glucagon molecules.

Although even less is known of the turnover of other macromolecular constituents, conclusions reached in the case of protein turnover are likely to apply also to RNA phospholipid and glycoprotein (Van Lancker, 1964).

Even if it is unlikely that acid hydrolases are involved in discrete turnover, it still needs to be explained why autolytic vacuoles are found in apparently normal cells. Although when restricted the formation of autolytic vacuoles is compatible with life, it would appear that such development is rather a drastic one. What are the cellular events that trigger this massive deterioration of cytoplasm? Are they not likely to be associated with structural and chemical reorganization of the cytoplasm? Two such cellular events come to mind, mitosis and changes in pattern of differentiation. Electron microscopists have reported that the incidence of autolytic vacuoles is increased in dividing cells. Exact quantitative data is difficult to obtain and even if real, the significance of the changes needs to be evaluated. Moreover, there is no change in the relative distribution of free and bound hydrolases in 24-hour regenerating liver (Van Lancker and Sempoux, 1958). But since in regenerating liver only a small number of cells divide at a given time, it is possible that changes in free enzymes are not readily detectable by biochemical methods.

Although the changes in macromolecular constituents that take place after prolonged starvation seem to fall into a category different from discrete turnover, it cannot be excluded that the earliest events in starvation, while quantitatively more extensive, are in fact qualitatively identical to those events that break down individual proteins in the course of their turnover.

Some of the critical events that regulate protein synthesis and breakdown during protein deprivation have been reviewed by Munro (1968). In spite of the fact that the total levels of RNA in liver decrease after prolonged periods of starvation, ^{32}P uptake in RNA is increased during the first 24 hours of starvation, suggesting equal or increased rate of synthesis in starved and normal rats. Studies of glycine-^{14}C uptake revealed that the pool of precursor was increased in the starved animals. Increased synthesis in the presence of an enlarged pool clearly suggested that the decrease in absolute RNA levels was the result of accelerated breakdown. More detailed information on the mechanism controlling RNA breakdown was obtained when a protein-free diet, supplemented with essential amino acid, was administered. Such studies established that the quality of the amino acid diet regulated the rate of RNA breakdown.

A striking observation was made with diets free of tryptophan. When that essential amino acid was withheld from the diet, incorporation of either orotic-^{14}C or glycine-^{14}C in liver RNA was reduced. A systematic investigation of the effect of tryptophan deletion on the machinery for protein synthesis revealed changes in the polysome profiles of animals fed such a diet. Omission of tryptophan from the diet was responsible for the appearance of a ratio of monosomes to polysomes comparable to that found in livers of normal rats. It was further established that this shift from polyribosomes to monosomes was less likely to result from a reduction in messenger production than from an accelerated breakdown of ribosomes, and it was suggested that the breakdown of polysomes to monosomes was followed by disruption of the ribosomes into subunits with the concomitant activation of the ribonuclease that is associated with ribosomes.

In vitro studies by Hori *et al.* (1967) with reticulocytes by Elliasson *et al.* (1967) in tissue culture, and by Munro (1968) in cell-free systems substantiate the role of essential amino acid in the conversion of ribosomes to polysomes. With the cell-free protein synthesizing system, it was shown that the lack of free amino acid induced cessation of protein synthesis, which could be restored to normal after addition of free amino acids to the incubation mixture. These observations suggest that ribosomal and messenger RNA remain intact in spite of cessation of protein synthesis. Yet, when the profiles of polyribosome distribution were studied after cessation of protein synthesis, a shift from polyribosomes to monosomes had taken place. The shift would occur with the withdrawal of any amino acid from the incubation mixture. On the basis of these findings, Munro concluded that polysome aggregation and desegregation was closely linked to the availability of amino acids for protein synthesis.

The following mechanism was proposed. As a result of increased protein synthesis, ribosomes are strung on the messenger to form polyribosomes and therefore their rate of degradation is reduced. Such interpretations of the results, fascinating as they are, leave much unexplained. Do amino acids really protect activating and other enzymes against breakdown? Are monosomes less stable than polysomes? What explains the very special effect of tryptophan deficiency on polysome degradation *in vivo*?

Two explanations have been offered for the unique *in vivo* effect of tryptophan: (1) the pools of tryptophan are smaller than that of any other amino acid and as a result the intact animal is more sensitive to deprivation of tryptophan than of any other amino acids; (2) incorporation of tryptophan in protein is rare compared to that of other amino acids because the average protein contains only small amounts of tryptophan. Consequently, in the average protein, large structures of polypeptide chains will be synthesized before a tryptophan residue is reached, but when it is reached in the absence of tryptophan, then synthesis of the rest of the chain is slowed down and free ribosomes remain trapped. Thus, one may anticipate that more ribosomes will be immobilized in the case of tryptophan deficiency than of any other amino acid deficiency.

C. The Role of Nonlysosomal Hydrolases in Cellular Death and Turnover

As mentioned previously, acid hydrolases constitute only a small contingent of the large population of hydrolases found in the cell. Nonlysosomal hydrolases may be found in the microsomes, supernatant, nuclei, mitochondria, Golgi apparatus, plasma membrane, or other ultrastructures separated by differential centrifugation. These enzymes catalyze the hydrolytic splitting of ester, phosphoric diesters, or sulfuric ester bonds. They attack terminal peptide bonds (amino or carboxy terminal amino acid) and peptide in dipeptides or polypeptides. They catalyze the breakdown of carbon–nitrogen bonds, and hydrolyze acid anhydrides. They are capable of separating glycogen moieties from the peptide chain. A partial list of these enzymes and their intracellular distribution is presented in Table I. The list is not intended to be com-

TABLE I

STITUATIONS IN WHICH AUTOLYTIC VACUOLES HAVE BEEN OBSERVED[a]

Organ	Type of event	Animal	References
Hepatic cell	β-3-Furylamine administration	Rats	Hruban et al., 1965b
	Hypoxia	Rats	Abraham et al., 1968
	Glucagon perfusion	Rats	Ashford and Porter, 1962; Arstila and Trump, 1968b
	Starvation	Rats	Herdson et al., 1964
	Kwashiorkor	Humans	Blackburn and Vinijchaikal, 1969
	Triparanol	Rats	Hruban et al., 1965a
	Diethanolamine	Rats	Hruban et al., 1965a
	Triton X-100	Rats	Holt, 1959
	Ligation of bile duct	Rats	Novikoff and Essner, 1960
	Heliotrine	Rats	Kerr, 1967
	Wilson's disease	Human	Goldfischer and Sternlieb, 1968
	Riboflavin deficiency	Mouse	Tandler et al., 1968
	Endotoxin shock	Dogs	Goldenberg et al., 1967
Pancreas	β-3-Thienylamine	Rat	Hruban et al., 1962a
	β-3-Furylamine	Rat	Hruban et al., 1965b
	D,L-Ethionine	Rat	Herman and Fitzgerald, 1962; Ekholm et al., 1962
	Triparanol	Rats	Hruban et al., 1965a
	Puromycin	Rats	Longnecker et al., 1968
	Diethanolamine	Rats	Hruban et al., 1965a
	Actinomycin D	Mouse	Rodriguez, 1967
	Amino acid deficiencies	Rats	Scott, 1966
	Choloraguine	Rats	Fedorko, 1968
Kidney	Ureter ligation	Rat	Novikoff, 1959
	Chronic potassium deficiency	Human	Muehrcke and Rosen, 1964
Nervous system	Anoxemia	Rats	Sulkin et al., 1968
Brain	Colchicine	Rabbits	Wisniewski and Terry, 1967
	Cholesterol inhibitors	Rats	Schutta and Neville, 1968
	Autonomic—ganglia scurvy	Guinea pig	Sulkin et al., 1968

Organ/Tissue	Condition	Species	Reference
	Peripheral nerve	Rat	Holtzman *et al.*, 1967
	Schwann cell—crushed nerve	Rats	Casley-Smith and Reade, 1965
	Reticuloendothelial cells		
	Foreign particles	Rats	Kluge and Hovig, 1969
	Thorotrast	Rat	Brandes, 1963
Prostate	Aging	Rat	Brandes *et al.*, 1962; Brandes and Portela, 1960
	Castration		
Lung	Pulmonary adenomatosis	Human	Schulz, 1963
	Fetal and newborn	Rabbits	Kikkawa *et al.*, 1968
	Vagotomized	Rat	Goldenberg *et al.*, 1967
Intestine Duodenum	X-radiation	Mice	Muehrcke and Rosen, 1964; Hugon and Borgers, 1966a,b
	Starvation	*Euglena gracilis*	Brandes and Bertini, 1964
Jejunum	Development	Rat	Behnke, 1963
	Fatty acid deficiency	Rats	Snipes, 1968
Miscellaneous Erythrocyte	Maturation	*Amphiuma tridactylum*	Tooze and Davies, 1965
Lymphocyte	Phytohemagglutin	Human	Hirschhorn *et al.*, 1965
	Antigenic sensitization	Human	Weissmann, 1967
Sertoli cells	Degenerating Seminiferous Epithelium	Rats	Hugon and Borgers, 1966a,b
	Whole-body irradiation	Mice	Reddy and Svoboda, 1967
Meristem		Young leaves of *Triticum vulgare*	Poux, 1963
Mammary gland tumors	X-radiation	Mice	Anton and Brandes, 1968
Uterus	Cyclophosphamate	Mice	Brandes and Anton, 1966, 1969
	Involution	Rats	

[a] The purpose of this list is to be illustrative rather than comprehensive.

plete, but mainly to illustrate the widespread distribution of these enzymes. The role that these enzymes play in metabolism, let alone in the sequence of steps that lead to cellular death, is not clear. But if the primary event in cellular death is one that interferes with bioenergetic or anabolic pathways, interference with those pathways will make more substrate available to the neutral hydrolases. Moreover, if we can assume with Schimke that enzyme turnover is regulated by the relative concentrations of substrate and catabolic enzyme, then it is likely that any injury to bioenergetic or anabolic pathways would quickly be followed by an attack on the unutilized substrate by catabolic enzyme acting at neutral pH. In fact, it is not unlikely that the more specific enzymes which act at neutral pH might play critical roles in determining irreversibility of the process of necrosis, leaving the scavenging of the remains of the murderous attack to the lysosomal enzymes.

There is no evidence that the cell sets apart a group of enzymes for the purpose of scavenging. In contrast, there is a good deal of evidence indicating that enzymes with obvious specialized physiological functions may act as scavengers under very unique circumstances. That the secretion of the gastrointestinal tract and its annexes may play a role under special circumstances in necrosis is clearly indicated by the lipid necrosis induced by pancreatic lipase and the role that trypsin plays in acute pancreatitis, as well as the role of pepsin in the autolysis of the stomach mucosae after death.

VI. Conclusion

Among cellular hydrolases, the acid hydrolases are unique. They exhibit structural latency and are found in association with cytoplasmic components that can be separated by differential centrifugation. Electron microscopy indicates that at least an ample portion of the enzyme activity can be readily demonstrated in foci of cytoplasmic autodigestion, whether these result from necrosis, necrobiosis, phagocytosis, or cytoplasmic resorption. Biochemically, appreciable amounts of acid hydrolases are found in the endoplasmic reticulum where the enzyme is synthesized. The transfer of enzyme to autolytic vacuole is not understood for sure. Two mechanisms have been proposed: (1) transfer of newly synthesized enzymes to Golgi with formation of hydrolytic vacuoles whose membranes fuse with phagocytic vacuoles or with areas of focal cytoplasmic degradation to form the autolytic vacuoles; and (2) direct transfer from endoplasmic reticulum as a consequence of segregation of some portion of the cytoplasm for physiological or pathological reasons. The major difference between these two mechanisms is that the first requires formation of primary lysosomes and the second does not, thus excluding regulation of death at the level of the lysosome.

Acid hydrolases are released from their structural support in almost all forms of necrosis and necrobiosis. Yet it is not likely that this event is part of the triggering mechanism, or of the reversible sequence of cellular death. Acid hydrolases are not involved in the sentence or execution of the victim, but they act as scavenger of the remains. Therefore, the two major problems of cellular death are still unsolved: the mechanism triggering cellular death, and the molecular type and position in the sequence of events leading to the point of no return. The solution to these problems is a matter of speculation, but the question can be raised whether the same or similar mechanisms for triggering the events obtain in all forms of necrosis and necrobiosis.

The formation of autolytic vacuoles is clearly a fundamental cellular function compatible with life, accelerated in cellular death. The uniformity of the response raises the possibility of a unique triggering mechanism which obtains in all forms of focal cytoplasmic degradation.

There is a constant association between the formation of autolytic vacuoles and cellular death. Autolytic vacuoles appear, however, under other circumstances, namely, cytoplasmic resorption associated with drastic cellular change and phagocytosis, principally in polymorphonuclears and macrophages, followed or not by digestion.

Let us assume that there are no qualitative differences between cell death and focal cytoplasmic degradation compatible with life and that these are only quantitative aspects of the same phenomenon. In cytonecrosis, because of programming or injury, large segments of the cytoplasm are converted into autolytic vacuoles, while in focal cytoplasmic degradation compatible with life, the process is limited. If we can now further assume that the vacuoles in polymorphonuclears and macrophages are in fact autolytic vacuoles (see above), we can proceed to ask ourselves what molecular event might be responsible for the formation of autolytic vacuoles. But before we attack the problem, we will make one further assumption that the concentrations of macromolecular substrate or other substrates are not uniform throughout the cell. Thus, some areas of cytoplasm will be well supplied, while other areas will be at the threshold of deprivation of substrate. Most of the cytoplasm is likely to be in between abundance and famine, with respect to a given substrate. Obviously whenever spontaneous or induced deprivation of a specific substrate occurs, the threshold zones for that substrate are the first to suffer.

If a uniform mechanism for the formation of cytoplasmic autolytic vacuoles exists, physiological necrosis would seem to be the most adequate model for such a mechanism.

A superficial examination of the ways by which cells die, as part of the conditions for maintenance of a steady state in the organism, reveals two major mechanisms. One is hyperdifferentiation, followed by shedding (keratin in the skin), or phagocytosis (hemoglobin in the red cells). The other is partial self-digestion followed by phagocytosis (tail of the tadpole, wing formation in insects and chickens). Both types of death are rigidly programmed and in fact cannot readily be dissociated, because, as mentioned previously in the case of the erythrocytes, the process of maturation is associated with the formation of autolytic vacuoles.

Rigid programming in time and space suggests that information for triggering and implementing death must be stored within cells, although it is possible that the condemned cells are unaware of the massacre that awaits them. If cellular death is programmed like differentiation and mitosis, is it not possible that factors that are involved in differentiation might also play a role in cellular death?

A plausible interpretation of the mechanism triggering the event in differentiation proposes that the event is preceded by a reassortment in the patterning of templates that are transcribed. Such a triggering mechanism may well function also in cellular death associated with necrobiosis. Thus, in necrobiosis, cytonecrosis could be the result (as in the case of the slow death of the erythrocytes) of template deprivation of the cytoplasm. In the absence of templates ribosomes are more susceptible to catabolic destruction by ribonuclease, and the permeability of organelles, or plasma membranes could be modified because of a lack of replacement of essential building blocks. Consequently, withdrawal of templates could easily cascade into a situation in which the dynamic steady state is switched from anabolic to catabolic pathways.

The extent to which the cell is deprived of templates would determine the extent of autodigestion. Moreover, some degree of specificity as to the type of molecules that will be broken down would be indirectly imparted by the coding properties and the stability of the templates that are withdrawn. For example, if as a result of preprogramming the genome were to stop the elaboration or the transcription of templates A, B, C, and D, those areas of the cytoplasm that have reached threshold levels of A, B, C, and D would be the first to be deprived. Furthermore, if because of a special molecular configuration template A is less stable than template B, then deprivation in template A in threshold zones would also be more rapid than deprivation in template B. The genome would thus have the power to select areas of cytoplasm that need to be autolyzed simply by rearranging the pattern of template synthesis. In necrobiosis, template deprivation is likely to be extensive and as a result large segments of cytoplasm undergo focal cytoplasmic degradation. In contrast, the template deprivation is restricted in focal cytoplasmic degradation compatible with life. Focal cytoplasmic degradation compatible with life would thus be regulated by mechanisms similar to those that regulate cellular death but very different from those that regulate discrete turnover. Discrete turnover is not likely to be controlled by the genome. The kinetic data of Schimke et al. (1968) suggest that enzyme breakdown is regulated by the availability of enzyme and substrate. It is, however, not unlikely that an acceleration of discrete turnover could be one of the first consequences of template deprivation (see above).

The portion of cytoplasm that is on the catabolic trend in some way becomes segregated from the surrounding healthy cytoplasm as a result of the appearance of a membrane of smooth endoplasmic reticulum that is either made *de novo* or built from preexisting membranes. A newly synthesized membrane would have to be made by the surrounding healthy tissue. The signal for such a synthesis could be a feedback loop from dead cytoplasm to genome and from genome to healthy cytoplasm. This is unlikely because such membranes are formed in the presence of actinomycin. A direct stimulus from dying to healthy cytoplasm triggering a preexisting machinery for membrane formation could exist.

A more plausible explanation for the formation of the membrane would be that of a molecular reorganization of preexisting structures as a consequence of their partial hydrolysis. The same process would be responsible for release and concentration of acid hydrolase in the segregated area.

It is difficult to imagine how any of the steps following the formation of the membrane could be reversible. The reversible sequence is more likely to precede the formation of the membrane. The only available clue to the molecular event associated with irreversibility comes from the studies of ischemic kidney and liver and studies of the effect of X-radiation on cells, but we shall return to this matter later. First let us examine whether a mechanism similar to that one we have postulated to explain cellular death in necrobiosis could also operate in other forms of cellular death.

The type of injury most likely to lead to cell death by a sequence of events similar to that just described is that produced by X-radiation. In mammalian tissue, as in bacteria, it seems now well established that DNA is the primary target for X-radiation administered at lethal doses.

In regenerating liver and spleen, the damage to DNA is followed by interference of incorporation of precursor into rapidly labeled nuclear RNA and in 18 S cytoplasmic RNA. Later events are the breakdown of polyribosomes, which has been demonstrated both biochemically and ultrastructurally. The formation of autolytic vacuoles is a late step, far removed from the primary injury (Van Lancker, 1969). In conclusion, it would appear that X-radiation, by damaging DNA, distorts the normal programming of production of templates, thus depriving threshold zones in nucleus and cytoplasm.

Obviously all forms of cellular damage cannot result from primary injuries of the genome. Some injurious agents do interfere with translation, others with the operation of one or more biosynthetic or bioenergetic pathways or with permeability of the membrane. Let us consider the results of a primary injury to the cell membrane. If the insult is extensive, it is likely to modify cellular permeability with sodium loss and potassium intake. Hyperactivity of the sodium pump and exhaustion of ATP reserves will follow. Intracellular concentrations of ATP are not likely to be the same in all parts of the cell. Consequently at any given moment in the cell life, there are likely to be areas of cytoplasm which are at the threshold of their ATP reserve; those will naturally be the first victim of ATP losses. In the presence of a constant derivation of ATP for sodium pumping, all biosynthetic pathways are likely to cease function in those threshold areas and the irreversible trend to focal cytoplasmic degradation is set.

If the injury to the membrane is limited, and sodium pumping functions adequately to prevent further damage, repair may be possible. Repair is likely to involve enzyme processes and possible elaboration of polypeptide chains. If the life of the enzyme is finite, increased activity could well be associated with increased breakdown. Consequently the area of cytoplasm involved in repair becomes a threshold zone for the enzyme involved in molecular repair of the membrane. New enzyme molecules must be made or brought in in time to maintain the steady dynamic state. Any insult which results in a primary injury of a bioenergetic pathway would result in reduced ATP levels with inadequate levels appearing first in the threshold zone. This would be followed by the shift from anabolic to catabolic pathways: first, an acceleration of discrete molecular breakdown, later the formation of focal cytoplasmic degradation. Reversibility of the damage would depend not so much on restoration of ATP levels as upon the stability of existing templates and the ability to bring new templates to the threshold area.

It also cannot be excluded that the appearance of phagocytic or pinocytic vacuoles of a critical size would present a mechanical barrier to the flow of templates, thereby generating a threshold area and depriving some cytoplasm of the needed templates and again the trend from anabolic to catabolic pathways would automatically set in.

In conclusion, the simplest way to explain focal cytoplasmic degradation is to assume that the cytoplasm, like a beehive, is divided into innumerable unique honeycombs. Each comb is unique because of its special enzyme mosaic and permeabilities of the membrane.

The concentration of any substrate varies constantly within limits. Thus, at any given time while some honeycombs are rich in substrate, others are low. If injury interferes with availability of the substrate, the enzymes involved in the metabolism of the substrates are no longer protected and disintegrate by acceleration of discrete turnover.

Restoration of anabolic activities can only be brought about by activation of existing templates or introduction of new templates. If existing templates are unstable, if new templates cannot be synthesized or brought into the area, or if neither existing nor new templates can be activated, the catabolic trend persists and the area becomes segregated, possibly by compaction of the membranes of the endoplasmic reticulum. Finally, enzymes resistant to hydrolysis concentrate in the segregated area by default. Limited segregation and autodigestion is compatible with life, extensive autodigestion leads to cell elimination; thus, the availability of templates would make the difference between life and death.

ACKNOWLEDGMENTS

This paper is dedicated to Dr. Harry S. N. Greene, former Professor and Chairman of the Department of Pathology at Yale, who died while this manuscript was in preparation.

The author is grateful for the help and suggestions of Dr. Babette Stewart who kindly read the manuscript before its publication.

REFERENCES

Abraham, R., Dawson, W., Grasso, P., and Goldberg, L. (1968). *Exp. Mol. Pathol.* 8, 370.

Allison, A. C., and Paton, G. R. (1965). *Nature* 207, 370.

Allison, A. C., Magnus, I. A., and Young, M. R. (1966). *Nature* 209, 874.

Anton, E., and Brandes, D. (1968). *Cancer* 21, No. 3, 483.

Appelmans, F., and de Duve, C. (1955). *Biochem. J.* 59, 426.

Appelmans, F., Wattiaux, R., and de Duve, C. (1955). *Biochem. J.* 59, 438.

Archer, G. T., and Hirsch, J. C. (1963). *J. Exptl. Med.* 118, 277.

Ariyama, K., and Van Lancker, J. L. (1970). In preparation.

Arstila, A., and Trump, B. F. (1968a). *Am. J. Pathol.* 53, 687.

Arstila, A. U., and Trump, B. F. (1968b). *Abstr. 65th Ann. Meeting Am. Assoc. Pathologists Bacteriologists.*

Ashford, T. P., and Porter, K. R. (1962). *J. Cell Biol.* **12**, 198.

Bainton, D. F., and Farquhar, M. G. (1968a). *J. Cell Biol.* **39**, 286.

Bainton, D. F., and Farquhar, M. G. (1968b). *J. Cell Biol.* **39**, 299.

Baudhuin, P., Beaufay, H., Rahman-Li, Y., Sellinger, O. Z., Wattiaux, R., Jacques, P., and de Duve, C. (1964). *Biochem. J.* **92**, 179.

Beaufay, H. (1957). *Arch. Intern. Physiol. Biochim.* **65**, 155.

Beaufay, H., Baudhuin, P., and Bendall, D. S. (1958). *Proc. 4th Intern. Congr. Biochem., Vienna, 1958*, p. 71. Pergamon Press, Oxford.

Behnke, O. (1963). *J. Cell Biol.* **18**, 251.

Berenbaum, M., Chang, P. I., and Stowell, R. E. (1955a). *Lab. Invest.* **4**, 315.

Berenbaum, M., Chang, P. I., Betz, H. E., and Stowell, R. E. (1955b). *Cancer Res.* **15**, 1.

Berthet, J., and de Duve, C. (1951). *Biochem. J.* **50**, 174.

Berthet, J., Berthet, F., Appelmans, F., and de Duve, C. (1951). *Biochem. J.* **50**, 182.

Biava, C. (1965). *Am. J. Pathol.* **46**, 435.

Bitensky, L. (1962). *Quart. J. Microscop. Sci.* **103**, 205.

Blackburn, W. R., and Vinijchaikul, K. (1969). *Lab. Invest.* **20**, 305.

Bowen, D. M., and Radin, N. S. (1968). *Biochim. Biophys. Acta* **152**, 599.

Bowers, W., and de Duve, C. (1967). *J. Cell. Biol.* **32**, 339.

Brachet, J., Decroly-Briers, M., and Hoyez, J. (1958). *Bull. Soc. Chim. Biol.* **40**, 2039.

Bradley, H. C. (1938). *Physiol. Rev.* **18**, 173.

Brandes, D. (1963). *Lab. Invest.* **12**, 290.

Brandes, D. (1965). *J. Ultrastruct. Res.* **12**, 63.

Brandes, D., and Anton, E. (1966). *Lab. Invest.* **15**, 987.

Brandes, D., and Anton, E. (1969). *J. Cell Biol.* **41**, 450.

Brandes, D., and Bertini, F. (1964). *Exptl. Cell Res.* **35**, 194.

Brandes, D., and Portela, A. (1960). *Anat. Record* **136**, 169.

Brandes, D., Groth, D. P., and Gyorkey, F. (1962). *Exptl. Cell Res.* **28**, 61.

Brandes, D., Bartini, F., and Smith, E. W. (1965). *Exptl. Mol. Pathol.* **4**, 245.

Brandes, D., Sloan, D. W., Anton, E., and Bloedorn, F. (1967). *Cancer Res.* **27**, 731.

Busch, E. W., von Habel, G., and Wichert, P. V. (1964). *Biochem. Z.* **341**, 85.

Busch, H. (1959). "Chemistry of Pancreatic Disease." Thomas, Springfield, Illinois.

Casley-Smith, J. R., and Reade, P. C. (1965). *Brit. J. Exptl. Pathol.* **46**, 475.

Chauveau, J., Moule, Y., and Rouiller, C. (1957). *Bull. Soc. Chim. Biol.* **39**, 1521.

Cohn, Z. A., and Hirsch, J. G. (1960). *J. Exptl. Med.* **112**, 983.

Cohn, Z. A., Hirsch, J. G., and Wiener, E. (1963). *Ciba Found. Symp., Lysosomes* p. 126.

Cooper, E. J., Trautmann, J. L., and Laskowski, M. (1950). *Proc. Soc. Exptl. Biol. Med.* **73**, 219.

Cruz-Coke, E., Plaza de Los Reyes, M., Martens, J., Del Nide, J., and Araya, J. (1951). *Soc. Biol. Santiago de Chili* **9**, 38.

Dabrowska, W., Cooper, E. J., and Laskowski, M. (1949). *J. Biol. Chem.* **177**, 991.

de Duve, C. (1958). *In* "Subcellular Particles" (T. Hayashi, ed.), p. 128. Ronald Press, New York.

de Duve, C. (1959). *In* "Subcellular Particles" (T. Hayashi, ed.), p. 128. Ronald Press, New York.

de Duve, C. (1963). *Ciba Found. Symp., Lysosomes* p. 1.

de Duve, C. (1964). *Federation Proc.* **23**, 1045.

de Duve, C. (1965). *Harvey Lectures* **59**, 49.

de Duve, C., and Wattiaux, R. (1966). *Ann. Rev. Physiol.* **28**, 435.

de Duve, C., Giannetto, R., Appelmans, F., and Wattiaux, R. (1953). *Nature* **172**, 1143.

de Duve, C., Passau, L., and Maisin, J. (1955). *Acta. unio. internat. contra Cancrum* **11**, 638.

de Duve, C., Wattiaux, R., and Wiho, M. (1961). *Biochem. Pharmacol.* **8**, 30.

de Reuck, A. V. S., and Cameron, M. P. (1963). *Ciba Found. Symp., Lysosomes* p. 446.

Desnuelle, P. (1960). *Enzymes* **4**, 93.

Dianzani, M. U. (1963). *Cibra Found. Symp., Lysosomes* p. 335.

Dianzani, M. U., Baccino, F. M., and Comporti, M. (1966). *In* "Biochemical Pathology" (E. Farber and P. N. Magee, eds.), p. 149. Williams & Wilkins, Baltimore, Maryland.

Dingle, J. T. (1961). *Biochem. J.* **79**, 509.

Dingle, J. T., Lucy, J. A., and Fell, H. B. (1961). *Biochem. J.* **79**, 479.

Dorfman, A. (1960). *In* "The Metabolic Basis of Inherited Disease" (J. B. Stanbury, J. B. Wyngaarden, and D. S. Frederickson, eds.), p. 963. McGraw-Hill, New York.

Eisenberg, S., Stein, Y., and Stein, O. (1968). *Biochim. Biophys. Acta* **164**, 205.

Ekholm, R., Edlund, Y., and Zelander, T. (1962). *J. Ultrastruct. Res.* **7**, 102.

Elliasson, E., Bauer, G. E., and Hultin, T. (1967). *J. Cell Biol.* **32**, 287.

Errera, M. (1968). *Advan. Biol. Med. Phys.* 333.

Evans, W. H., and Karnovsky, M. L. (1961). *J. Biol. Chem.* **236**, 30.

Fain-Maurel, A. (1967). *Compt. Rend.* **265**, 126.

Farber, E., Shull, K. H., Villa-Trevino, S., Lombardi, B., and Thomas, M. (1964). *Nature* **203**, 34.

Farber, E., Molly, T., and Vogt, E. (1968). *Am. J. Pathol.* **53**, 1.

Fausto, N., Smoot, A. O., and Van Lancker, J. L. (1964). *Radiation Res.* **22**, 288.

Fedorko, M. E. (1968). *Lab. Invest.* **18**, 27.

Feinstein, R. N., and Balliu, J. C. (1953). *Proc. Soc. Exptl. Biol. Med.* **56**, 6.

Fell, H. B., and Mellanby, E. (1952). *J. Physiol. (London)* **116**, 320.

Field, R. A. (1966). *In* "The Metabolic Basis of Inherited Diseases" (J. B. Stanbury, J. B. Wyngaarden, and D. S. Fredrickson, eds.), p. 141. McGraw-Hill, New York.

Fishman, W. H., DeLellis, R. A., and Goldman, S. S. (1967). *Nature* **213**, 457.

Fishman, W. H., Ide, H., and Rufo, R. (1969). *Histochemie* **20**, 287.

Flanagan, J. F. (1966). *J. Bacteriol.* **91**, 789.

Fonnesu, A. (1960). *In* "The Biochemical Response to Injury" (H. B. Stoner and C. J. Threlfall, eds.), p. 85. Thomas, Springfield, Illinois.

Fredrickson, D. S. (1966). *In* "The Metabolic Basis of Inherited Disease" (J. B. Stanbury, J. B. Wyngaarden, and D. S. Fredrickson, eds.), p. 565. McGraw-Hill, New York.

Fredrickson, D. S., and Trams, E. G. (1966). *In* "The Metabolic Basis of Inherited

Disease" (J. B. Stanbury, J. B. Wyngaarden, and D. S. Fredrickson, eds.),
p. 523. McGraw-Hill, New York.
Frieden, K. H., Harper, A. A., Chin, F., and Fishman, W. H. (1964). *Steroids* **4**, 786.
Gahan, P. B. (1967). *Intern. Rev. Cytol.* **21**, 1.
Garancis, J. C. (1968). *Am. J. Med.* **44**, 289.
Ghadially, F. N., and Parry, E. W. (1966). *Cancer* **19**, 1989.
Gianetto, B., and de Duve, C. (1955). *Biochem. J.* **59**, 433.
Glinsmann, W. H., and Ericsson, J. L. (1966). *Lab. Invest.* **15**, 762.
Goldblatt, P. J., Trump, B. F., and Stowell, R. E. (1963). *Lab. Invest.* **12**, 861.
Goldenberg, V. E., Buckingham, S., and Sommers, S. C. (1967). *Lab. Invest.* **16**, 693.
Goldfischer, S., and Sternlieb, I. (1968). *Am. J. Pathol.* **53**, 883.
Goldfischer, S., Caresso, M., and Favard, P. (1963). *J. Microscopie* **2**, 621.
Gomori, G. (1952). "Microscopic Histochemistry: Principles and Practice." Univ. of Chicago Press, Chicago, Illinois.
Gottlieb, L. I., and Van Lancker, J. L. Unpublished results.
Gottlieb, L. I., Fausto, N., and Van Lancker, J. L. (1964). *J. Biol. Chem.* **239**, 555.
Goutier, R. (1963). *Progr. Biophys. Biophys. Chem.* **11**, 53.
Goutier, R., and Bacq, Z. M. (1963). *In* "Metabolic Inhibitors" (R. M. Hochster and J. H. Quastel, eds.), Vol. 2, p. 631. Academic Press, New York.
Goutier-Pirotte, M., and Thonnard, A. (1956). *Biochim. Biophys. Acta* **22**, 396.
Gross, P. R. (1968). *Ann. Rev. Biochem.* **37**, p. 631.
Hamilton, T. H., and Teng, C. S. (1965). *In* "Organogenesis" (R. De Haan and H. Ursprung, eds.), p. 681. Holt, New York.
Harding, B. W., and Samuels, L. T. (1961). *Biochim. Biophys. Acta* **54**, 42.
Henstell, H. H., Freedman, R. L., and Ginsbury, B. (1952). *Cancer Res.* **12**, 346.
Herdson, P. B., Garvin, P. J., and Jennings, R. B. (1964). *Am. J. Pathol.* **45**, 157.
Herman, L., and Fitzgerald, P. J. (1962). *J. Cell Biol.* **12**, 227.
Hers, H. G. (1964). *Advan. Metab. Disorders* **1**, 2.
Hers, H. G. (1965). *Gastroenterology* **48**, 625.
Hirsch, J. G. (1960). *J. Exptl. Med.* **3**, 323.
Hirsch, J. G. (1965). *In* "The Inflammatory Process" (B. W. Zweifach, L. Grant, and R. T. McCluskey, eds.), p. 245. Academic Press, New York.
Hirsch, J. G., and Cohn, Z. A. (1960). *J. Exptl. Med.* **112**, 1005.
Hirsch, J. G., Bernheimer, A. W., and Weissmann, G. (1963). *J. Exptl. Med.* **118**, 223.
Hirschhorn, R., Kaplan, J. M., and Weissmann, G. (1965). *Science* **147**, 55.
Holt, S. J. (1959). *Exptl. Cell Res.* **7**, 1.
Holtzer, R. L., and Van Lancker, J. L. (1962). *Am. J. Pathol.* **40**, 331.
Holtzer, R. L., Van Lancker, J. L., and Swift, H. (1963). *Arch. Biochem. Biophys.* **101**, 434.
Holtzman, E., Novikoff, A. B., and Villaverde, H. (1967). *J. Cell Biol.* **33**, 419.
Hori, M., Fisher, J. M., and Rabinowitz, M. (1967). *Science* **155**, 83.
Hruban, Z., and Rechcigl, M. (1970). *Intern. Rev. Cytol.*
Hruban, Z., Swift, H., and Wissler, R. W. (1961). *Am. J. Pathol.* **38**, 761.
Hruban, Z., Swift, H., and Wissler, R. W. (1962). *J. Ultrastruct. Res.* **7**, 359.
Hruban, Z., Swift, H., Dunn, F. W., and Lewis, D. E. (1965a). *Lab. Invest.* **14**, 70.
Hruban, Z., Swift, H., and Slesers, A. (1965b). *Lab. Invest.* **14**, 1652.

Hruban, Z., Villa-Trevino, S., and Farber, E. (1965c). *Lab. Invest.* **14**, 468.

Hugon, J., and Borgers, M. (1966a). *Anat. Record* **155**, 15.

Hugon, J., and Borgers, M. (1966b). *J. Microscopie* **5**, 649.

Ide, H, and Fishman, W. H. (1969). *Histochemie* **20**, 300.

Judah, J. D., Ahmed, K., and McLean, A. E. M. (1964). *Ciba Found. Symp., Cellular Injury* p. 187.

Judah, J. D., Ahmed, K., and McLean, A. E. M. (1965). *Federation Proc.* **25**, 1217.

Judah, J. D., Ahmed, K., McLean, A. E. M., and Christie, G. S. (1966). *In* "Ion Transport in Ethionine Intoxication" (E. Farber and P. N. Magee, eds.), p. 167. Williams & Wilkins, Baltimore, Maryland.

Kato, K, Ide, H., Shirahama, T., and Fishman, W. H. (1970). *Biochem. J.* (in press).

Kerr, J. F. R. (1967). *J. Pathol. Bacteriol.* **93**, 167.

Kikkawa, Y., Motoyama, E. K., and Gluck, L. (1968). *Am. J. Pathol.* **52**, 177.

King, D. W., Paulson, S. R., Puckett, N. L., and Krebs, A. T. (1959). *Am. J. Pathol.* **35**, 835.

Kirk, J. M. (1960). *Biochim. Biophys. Acta* **42**, 167.

Kluge, T., and Hovig, T. (1969). *Am. J. Pathol.* **54**, 355.

Koenig, H. (1962). *Nature* **195**, 782.

Koenig, H., Gaines, D., McDonald, T., Gray, R., and Scott, J. (1964). *J. Neurochem.* **11**, 729.

Kuff, E. L., Hogeboom, G. H., and Dalton, A. J. (1956). *J. Biophys. Biochem. Cytol.* **2**, 33.

Kurnick, N. B., Schwartz, L. I., Pariser, S., and Lee, S. L. (1953). *J. Clin. Invest.* **32**, 193.

La Farge, C., Frayssinet, C., and Simard, R. (1966). *Compt. Rend.* **263**, 1011.

Lombardi, B. (1966). *In* "Biochemical Pathology" (E. Farber and P. N. Magee, eds.), p. 1. Williams & Wilkens, Baltimore, Maryland.

Longnecker, D. S., and Farber, E. (1967). *Lab. Invest.* **16**, 321.

Love, R., Studzinski, G. P., and Ellem, K. A. (1965). *Federation Proc.* **24**, 1206.

Lucy, J. A., Dingle, J. T., and Fell, H. B. (1961). *Biochem. J.* **79**, 500.

McCluskey, R. T., and Thomas, L. (1958). *J. Exptl. Med.* **108**, 371.

MacFarlane, M. B., and Datta, N. (1954). *Brit. J. Exptl. Pathol.* **35**, 191.

Magee, P. N. (1966). *In* "Biochemical Pathology" (E. Farber and P. N. Magee, eds.), p. 111. Williams & Wilkins, Baltimore, Maryland.

Malamed, S. (1966). *Z. Zellforsch. Mikroskop. Anat.* **75**, 272.

Malet, P., Joyon, L., and Turchini, J. P. (1964). *Compt. Rend.* **258**, 5067.

Mallucci, L., and Allison, A. C. (1965). *J. Exptl. Med.* **121**, 477.

Miller, E. C., and Miller, J. A. (1955). *J. Natl. Cancer Inst.* **15**, 1571.

Miller, F., and Palade, G. E. (1964). *J. Cell Biol.* **23**, 519.

Mills, G., Paul, J., and Smith, E. (1953). *Biochem. J.* **53**, 232.

Millson, G. C. (1965). *J. Neurochem.* **12**, 461.

Muehrcke, R. C., and Rosen, S. (1964). *Lab. Invest.* **13**, 1359.

Munro, H. N. (1968). *Federation Proc.* **27**, 1231.

Nosman, M., Hamdy, M. K., and Caster, W. O. (1966). *Bone Marrow Radiation Protect. Conf., Ann. Arbor, Michigan.*

Novikoff, A. B. (1958). *In* "Subcellular Particles" (T. Hayashi, ed.), p. 1. Ronald Press, New York.

Novikoff, A. B. (1959). *J. Biophys. Biochem. Cytol.* **6**, 136.

Novikoff, A. B. (1961). *In* "The Cell" (J. Brachet and A. E. Mirsky, eds.), Vol. 2, p. 423. Academic Press, New York.
Novikoff, A. B. (1963). *Ciba Found. Symp., Lysosomes* p. 36.
Novikoff, A. B. (1967). *In* "The Neuron" (H. Hyden, ed.), p. 215. Elsevier, Amsterdam.
Novikoff, A. B., and Essner, E. (1960). *Am. J. Med.* **29**, 102.
Novikoff, A. B., and Shin, W. Y. (1963). *J. Microscopie* **3**, 187.
Novikoff, A. B., Podber, J., and Ryan, E. (1953). *J. Histochem. Cytochem.* **1**, 27.
Novikoff, A. B., Beaufay, H., and de Duve, C. (1956). *J. Biophys. Biochem. Cytol.* **2**, 179.
Novikoff, A. B., Essner, E., and Quintana, M. (1963). *J. Microscopie* **2**, 3.
Novikoff, A. B., Quintana, M., Villaverde, H., and Forscheirm, R. (1964). *J. Cell Biol.* **23**, 68.
Novikoff, A. B., Roheim, P. S., and Quintana, M. (1966). *In* "Biochemical Pathology" (E. Farber and P. N. Magee, eds.), p. 27. Williams & Wilkins, Baltimore, Maryland.
Ottesen, M. (1967). *Ann. Rev. Biochem.* **36**, Part 1.
Paigen, K. (1961). *Exptl. Cell. Res.* **25**, 286.
Pettengill, O. S., and Fishman, W. H. (1962). *Exp. Cell Res.* **28**, 248.
Phillips, M. J., Unakar, N. J., Doornewaard, G., and Steiner, J. W. (1967). *J. Ultrastruct. Res.* **18**, 142.
Pipa, R. L., Nishioka, R. S., and Bern, H. A. (1962). *J. Ultrastruct. Res.* **6**, 164.
Plapp, B., and Cole, R. (1967). *Biochemistry* **6**, 3676.
Poux, N. (1963). *Compt. Rend.* **257**, 736.
Recknagel, R. O. (1967). *Pharmacol. Rev.* **19**, 145.
Reddy, K. J., and Svoboda, D. J. (1967). *Am. J. Pathol.* **51**, 1.
Reid, E., and Stevens, B. M. (1958). *Biochem. J.* **68**, 376.
Richterich, R., and Schafroth, H. (1963). *Enzymol. Biol. Clin.* **3**, 165.
Robineaux, J., and Frederiz, J. (1955). *Compt. Rend. Soc. Biol.* **149**, 486.
Rodriguez, T. G. (1967). *J. Ultrastruct. Res.* **19**, 116.
Rosen, S., Coughlan, M., and Barry, K. G. (1967). *Proc. Soc. Exptl. Biol. Med.* **124**, 909.
Rosenberg, M., and Janoff, A. (1968). *Biochem. J.* **108**, 889.
Roth, J. S. (1956a). *Biochim. Biophys. Acta* **21**, 34.
Roth, J. S. (1956b). *Arch. Biochem. Biophys.* **60**, 7.
Roth, J. S. (1958a). *J. Biol. Chem.* **231**, 1085.
Roth, J. S. (1958b). *Radiation Res.* **9**, 173.
Roth, J. S., and Bachmurski, D. (1957). *Biol. Bull.* **113**, 332.
Rozenszajn, L., Epstein, Y., Shoham, D., and Arber, I. (1968). *J. Lab. Clin. Med.* **72**, 786.
Saunders, J. W., Jr., and Fallon, J. P. (1966). "Cell Death in Morphogenesis. Major Problems in Developmental Biology," p. 289.
Scarpelli, D. G., and Kanczak, N. M. (1965). *Intern. Rev. Exptl. Pathol.* **4**, 55.
Scheib, D. (1963). *Ciba Found. Symp., Lysosomes* p. 264.
Schimke, R. T., Ganschow, R., Dowyle, D., and Arias, I. M. (1968). *Federation Proc.*, **27**, No. 5, 1223.
Schulz, H. (1963). *Lab. Invest.* **12**, 616.
Schutta, H. S., and Neville, H. E. (1968). *Lab. Invest.* **19**, 487.
Scott, E. F. (1966). *Arch. Pathol.* **82**, 119.

Sheldon, H., Zetterqvist, H., and Brandes, D. (1955). *Exptl. Cell Res.* **9**, 592.

Sidbury, J. B. (1961). *In* "Molecular Genetics and Human Disease" (T. Gardner, ed.), Chapter V, p. 61.

Slater, T. F. (1961a). *Biochem. J.* **78**, 500.

Slater, T. F. (1961b). *Biochim. Biophys. Acta* **51**, 193.

Slater, T. F. (1962). *Arch. Intern. Physiol. Biochim.* **70**, 167.

Slater, T. F., and Planterose, D. N. (1960). *Biochem. J.* **74**, 584.

Slater, T. F., Greenbaum, A. L., and Wang, D. Y. (1963a). *Ciba Found. Symp., Lysosomes* p. 311.

Slater, T. F., Sawyer, B. C., and Strauli, U. (1963b). *Biochem. J.* **86**, 303.

Smith, E. W., and Mills, G. (1953). *Biochem. J.* **54**, 164.

Smuckler, E. A. (1966). *In* "Biochemical Pathology" (E. Farber and P. N. Magee, eds.), p. 157. Williams & Wilkins, Baltimore, Maryland.

Snipes, R. L. (1968). *Lab. Invest.* **18**, 179.

Stoner, H. B., and Threlfall, C. J. (1960). *In* "The Biochemical Response to Injury, A Symposium," (H. B. Stoner and C. J. Threlfall, eds.), p. 105. Thomas, Springfield, Illinois.

Strauss, W. (1954). *J. Biol. Chem.* **207**, 745.

Strauss, W. (1956). *J. Biophys. Biochem. Cytol.* **2**, 513.

Strauss, W. (1957). *J. Biophys. Biochem. Cytol.* **3**, 1037.

Strauss, W. (1959). *J. Biophys. Biochem. Cytol.* **5**, 193.

Strauss, W. (1962a). *J. Cell Biol.* **12**, 231.

Strauss, W. (1962b). *Exptl. Cell Res.* **27**, 80.

Strauss, W. (1963). *Ciba Found. Symp., Lysosomes* p. 151.

Strauss, W. (1964a). *J. Histochem. Cytochem.* **12**, 470.

Strauss, W. (1964b). *J. Cell Biol.* **20**, 497.

Strauss, W. (1964c). *J. Cell Biol.* **21**, 295.

Sulkin, D. F., Sulkin, N. M., and Rothrock, M. L. (1968). *Lab. Invest.* **19**, 55.

Swift, H., and Hruban, Z. (1964). *Federation Proc.* **23**, 1026.

Tandler, B., Erlandson, R. A., and Synder, E. L. (1968). *Am. J. Pathol.* **52**, 69.

Tappel, A. L. (1962). *In* "Symposium on Foods: Lipids and Their Oxidation" (H. W. Schultz, ed.), p. 367. Avi Publ. Co., Westport, Connecticut.

Tappel, A. L., Zalkin, H., Caldwell, K. A., Desai, I. D., and Shibko, S. (1962). *Arch. Biochem.* **96**, 340.

Tappel, A. L., Sawant, P. L., and Shibko, S. (1963). *Ciba Found. Symp., Lysosomes* p. 78.

Thomas, D. P. (1967). *Rheumatol. (Basel)* **1**, 29.

Thomas, L., McCluskey, R. T., Potter, S. L., and Weissmann, G. (1960). *J. Exptl. Med.* **11**, 705.

Thomas, L., McCluskey, R. T., Li, J., and Weissmann, G. (1963). *Am. J. Pathol.* **42**, 271.

Tooze, J., and Davies, H. G. (1965). *J. Cell Biol.* **24**, 146.

Trump, B. F., and Bulger, R. E. (1967). *Lab. Invest.* **16**, 453.

Trump, B. F., and Bulger, R. E. (1968). *Lab. Invest.* **18**, 721.

Trump, B. F., and Ericsson, J. L. (1965a). *In* "The Inflammatory Process" (B. Zweifach, L. Grant, and R. T. McCluskey, eds.). Academic Press, New York.

Trump, B. F., and Ericsson, J. L. (1965b). *Lab. Invest.* **14**, 200.

Ugazio, G., and Pani, P. (1963). *Exptl. Cell Res.* **31**, 424.

Vaes, G. (1965a). *Biochem. J.* **97**, 393.

Vaes, G. (1965b). *Exptl. Cell Res.* **39**, 470.

Vaes, G., and Jacques, P. (1965). *Biochem. J.* **97**, 389.

Vaes, G., and Nichols, G., Jr. (1961). *J. Biol. Chem.* **236**, 3323.

Vaes, G., and Nichols, G., Jr. (1962). *Endocrinology* **70**, 546.

Van Lancker, J. L. (1959). *Biochim. Biophys. Acta* **45**, 63.

Van Lancker, J. L. (1964). "Concluding remarks." *Fed. Proc.* **23**, 1050.

Van Lancker, J. L. (1966). *In* "Biochemical Pathology" (E. Farber and P. Magee, eds.), p. 192. William & Wilkins, Baltimore, Maryland.

Van Lancker, J. L. (1969). *In* "Symposium on Biochemistry of Cell Division." Thomas, Springfield, Illinois.

Van Lancker, J. L. (1970). *Fed. Proc.* (in press).

Van Lancker, J. L. Unpublished results.

Van Lancker, J. L., and Holtzer, R. L. (1959a). *Am. J. Pathol.* **35**, 563.

Van Lancker, J. L., and Holtzer, R. L. (1959b). *J. Biol. Chem.* **234**, 2359.

Van Lancker, J. L., and Holtzer, R. L. (1963). *Lab. Invest.* **12**, 102.

Van Lancker, J. L., and Holtzer, R. L. (1964). Unpublished results.

Van Lancker, J. L., and Lentz, P. (1970). *J. Hist. Cytochem.* (in press).

Van Lancker, J. L. and Sempoux, D. G. (1958). *Arch. Biochem. Biophys.* **77**, 129.

Van Vleet, J. F., Hall, B. V., and Simon, J. (1968). *Am. J. Pathol.* **52**, 1067.

Viala, R., and Giannetto, R. (1955). *Can. J. Biochem. Phys.* **33**, 839.

Vogt, M. T. and Farber, E. (1968). *Am. J. Pathol.* **53**, 1.

Wacker, A. (1963). *Progr. Nucleic Acid Res.* **1**, 369.

Walkinshaw, C. H., and Van Lancker, J. L. (1964). *Lab. Invest.* **13**, 513.

Walkinshaw, C. H., McClue, H. M., and Van Lancker, J. L. (1964). *Lab. Invest.* **13**, 524.

Wattiaux, R., and de Duve, C. (1956). *Biochem. J.* **63**, 606.

Weber, R. (1957a). *Rev. Suisse Zool.* **64**, 326.

Weber, R. (1957b). *Experientia* **13**, 153.

Weber, R. (1963). *Ciba Found. Symp., Lysosomes* p. 282.

Weber, R., and Niehus, B. (1961). *Helv. Physiol. Pharmacol. Acta* **19**, 103.

Weissmann, G. (1961). *J. Exptl. Med.* **114**, 581.

Weissmann, G. (1967). *Ann. Rev. Med.* **18**, 97.

Willstatter, R., and Rohdewald, M. (1964). *Z. Physiol. Chem.* **208**, 258.

Wisniewski, H., and Terry, R. D. (1967). *Lab. Invest.* **17**, 577.

Woodin, A. M. (1961). *Biochem. J.* **82**, 9.

Zeya, H. I., and Spitznagel, J. K. (1966). *Science* **154**, 1049.

Zeya, H. I., and Spitznagel, J. K. (1969). *Science* **163**, 1069.

Zollinger, H. U. (1948). *Am. J. Pathol.* **24**, 569.

AUTHOR INDEX

Numbers in italics refer to the pages on which the complete references are listed.

A

Aarts, E. M., 187, 188, 197, 199, 217, *225*
Abdel-Aziz, M. T., 24, 26, 31, *49*
Abe, H., 246, *319*
Abraham, A., 246, 248, *313*
Abraham, R., 404, *411*
Abrahams, H. E., 297, 322, 343, *353*
Acheson, R. M., 251, *313*
Ackerman, W. W., 132, *153*
Ackermann, D., 11, *16*
Ackers, J., 285, *319*
Acocella, G., 217, 218, *225*
Adachi, K., 225, *226*
Adams, G. A., 249, 300, *319*
Adams, J. B., 135, *153*, 246, 250, 255, 259, 260, 281, 294, 310, *313*
Adamson, L., 248, *313*
Agranoff, B. V., 175, 176, *227*
Ahmed, K., 361, 399, 400, *415*
Airaksinen, M. M., 168, *225*
Aitio, A., 168, *229*
Albrecht, G. J., 160, *225*
Ali, S. Y., 297, *313*
Allen, E., 265, *313*
Allfrey, V. G., 139, *153*
Allison, A. C., 380, *411*, *415*
Alonso, C., 259, *316*
Alsalt, E., 282, *322*
Ammon, R., 268, *321*
Anast, C., 248, *313*
Anastisi, A., 252, 275, *313*
Andersen, R. A., 80, 97, *114*
Anderson, B., 289, 290, 291, *313*, *321*, *323*

Anderson, I. G., 280, *313*
Anderson, L., 175, *225*, *230*
Anderson, S. O., 263, *313*
Andrew, R. H., 46, *49*
Ansanelli, V., 215, *228*
Anton, A. H., 137, *153*
Anton, E., 378, 405, *411*, *412*
Antoniades, H. N., 139, *155*
Antonopolous, C. A., 292, *313*
Anzai, M., 181, 184, 224, 225, *232*, *237*
Aoki, K., 147, 148, *154*
Appel, W., 58, 61, *114*
Appelmans, F., 343, 352, 363, 364, *411*, *412*, *413*
Appolonio, T., 212, *227*
Aqvist, S., 248, *315*
Araki, M., 107, *116*
Araya, J., 358, 366, *412*
Arber, I., *416*
Archer, G. T., 385, *411*
Arcos, J. C., 84, 112, *114*
Arcos, M., 23, 26, 44, *49*, 279, *313*
Argus, M. F., 84, 112, *114*
Arias, I. M., 163, 164, 168, 171, 199, 204, 216, 217, 218, *225*, *226*, *228*, 391, 400, 401, 409, *416*
Ariyama, K., 378, *411*
Armand, G., 291, *323*
Armas,-Merino, R., 217, 218, *225*
Armstrong, D., 268, 305, *314*
Armstrong, M. D., 11, *16*
Arnoldt, R. I., 13, *17*, 244, *316*
Aronson, N. N., 297, 298, 299, *313*, *314*, 341, 343, 344, 345, *352*

419

SUBJECT INDEX